transfer operations

R.A. GREENKORN
Professor of Chemical Engineering
Purdue University

D.P. KESSLER
Associate Professor of Chemical Engineering
Purdue University

transfer operations

McGRAW-HILL BOOK COMPANY

New York St. Louis San Francisco Düsseldorf Johannesburg Kuala Lumpur London
Mexico Montreal New Delhi Panama Rio de Janeiro Singapore Sydney Toronto

CHEMISTRY

*This book was set in Times Roman by Textbook Services, Inc. and
printed and bound by Kingsport Press, Inc. The drawings were done
by John Cordes, J. & R. Technical Services, Inc. The editors were B. J.
Clark and M. E. Margolies. Matt Martino supervised production.*

transfer operations

Library of Congress Catalog Card Number 70-168450

07-024351-4

1234567890KPKP798765432

to
Rosemary
and
Carolyn

contents

9736

preface

This book is intended as a textbook, not a reference book or handbook. Our objective is to present momentum, heat, and mass transfer from a macroscopic point of view. Specifically, we present material (1) to provide the student with a basic knowledge of the mechanisms by which momentum, heat, and mass are transferred and (2) to introduce the student to the design equations for momentum, heat, and mass and to the use of these equations. The course for which this textbook is used in the School of Chemical Engineering at Purdue University is the first of a three-course transport sequence. The second course in the sequence is one in transport phenomena—a microscopic point of view—and the third a laboratory where both transfer coefficients and transport coefficients are determined. We believe this sequence—presenting the macroscopic approach before the microscopic—to be the soundest from a pedagogical point of view. To use with understanding the transport approach, with its more sophisticated mathematics, requires the physical insight and the motivation of recognized applications which this text provides.

Although the course taught from this book is in a sequence in the School of Chemical Engineering, it does stand alone in that the techniques presented are immediately applicable. The course is required for juniors in chemical engineering, industrial engineering, and materials science and metallurgical engineering. Only the chemical engineers take the three-course sequence. Students from other departments, such as agricultural, nuclear, and mechanical engineering, and food science and home economics students who have sufficient background and who wish a course in momentum, heat, and mass transfer with practical application, also take this course.

The text is built around a large number of examples which are worked in detail. Many of the examples are, of course, idealized because their purpose is to illustrate elementary principles, but we have kept them as realistic as possible. The text is arranged in a matrix fashion as shown in Table 1.2-1 on page 5. Within reason, either the columns or rows of this matrix stand alone. In other words, one could go through the introductory material of column 1 as one "subject," the momentum transfer of column 2 as one "subject," etc., or one could look at mechanism in row 1 as one "subject," the design equation in row 2 as one "subject," etc.

At Purdue the course is taught by successive columns (which is the sequence for chapter numbers). Each column represents approximately one-fourth of a four-semester-credit course. The chapters on applications (Chap. 4, 7, 10, and 13) are not taught in their entirety since they do not contain principles; rather, they are assigned as reading and specific parts may be employed as illustrations by the instructor. These four chapters are designed to show how overall balances are used, how transfer coefficients are measured, how the design equations are used in a real system, and how these calculations are integrated into a real system. (The industrial and material science and metallurgical engineers at Purdue have the option of ending the course for three semester credits by omitting the fourth part on mass transfer.)

We have taken the position in writing this book that it is *part* of the instructional material available to the student—the instructor teaching the course is the other part. The procedures presented are common ones which provide the base for various, but limited, solution procedures on which the instructor may choose to build further, or not as he sees fit. We have accepted equipment as a basic but secondary part of transfer operations; we concentrate on the principles and we use practical examples and system applications. The equipment is included as a matter of fact. A particular instructor may wish to emphasize or deemphasize this point depending on his own approach to the topics.

We owe an original debt to Professors Jack Myers and Alexander Sesonske in whose association both the authors were introduced to this general area of undergraduate teaching. We also have been helped in construction of examples and review of the notes by Professors Ron Barile, Neal Houze, and Theo Theofanous of the School of Chemical Engineering at Purdue, by Larry Hochreiter, a graduate student in Nuclear Engineering, and by Dr. Tom Sifferman, a former graduate student in Mechanical Engineering. We wish to thank Susie O'Dore and Linda Rhodes for their patient efforts in typing and retyping the manuscript. In addition, the Chemical Engineering Department of the University of California, Berkeley, deserves our thanks for permitting one of the authors to teach a course from the notes as a visiting professor at a critical time in the text development. The Pan American Petroleum Corporation was most cooperative. It provided the flow sheet for the Natural Gasoline Plant and reviewed the application examples in Chaps. 7, 10, and 13. A last but most significant debt goes to Snooker Table Number 4 in the Purdue Billiard Room—where the emotions and frustrations associated with writing the text found release.

R. A. GREENKORN
D. P. KESSLER

list of symbols

A	= area, L^2
a	= interfacial area per unit volume of bed, L^{-1}
C	= discharge coefficient, dimensionless
$\hat{C}_p$	= heat capacity at constant pressure, per unit mass, $L^2/t^2 T$
$\hat{C}_v$	= heat capacity at constant volume, per unit mass, $L^2/t^2 T$
C_p	= pitot tube coefficient, dimensionless
c	= total molar concentration, moles/L^3
C_V	= venturi coefficient, dimensionless
c_i	= molar concentration of species i, moles/L^3
D	= characteristic length in dimensional analysis or diameter of sphere or cylinder, L
D_p	= particle diameter, L
$\mathscr{D}_{AB}$	= binary diffusivity for system A-B, L^2/t
$\mathscr{D}_{ij}$	= binary diffusivity for system i-j, L^2/t
E	= $U + K + \phi$ = total fluid energy, ML^2/t^2
e	= 2.71828...
e	= emissivity, dimensionless
e	= total energy flux relative to stationary coordinates, M/t^3
F	= force of a fluid on an adjacent solid, ML/t^2
F_{12}	= direct view factor, dimensionless
$\bar{F}_{12}$	= indirect view factor, dimensionless
$\mathscr{F}_{12}$	= combined emissivity and view factor, dimensionless
f	= friction factor or drag coefficient, dimensionless
G	= molar velocity, moles/tL^2
G	= $H - TS$ = Gibbs free energy, or "free enthalpy," ML^2/t^2
g	= gravitational acceleration, L/t^2
g_c	= force-mass conversion factor
H	= $U + pV$ = enthalpy, ML^2/t^2
h	= heat transfer coefficient, $M/t^3 T$
h	= Planck's constant, ML^2/t

h = elevation, L

i = $\sqrt{-1}$

J_i = molar flux of i relative to mass average velocity, moles/tL^2

J_i^* = molar flux of species i relative to the molar average velocity, moles/tL^2

j_i = mass flux of i relative to mass average velocity, M/tL^2

j_i^* = mass flux of species i relative to the molar average velocity, M/tL^2

j_D = Chilton-Colburn j factor for mass transfer, dimensionless

j_H = Chilton-Colburn j factor for heat transfer, dimensionless

K = kinetic energy, ML^2/t^2

k = roughness, L

k = shape factor, dimensionless

k = thermal conductivity, $ML/t^3 T$

k_n = homogeneous chemical reaction rate constant, moles$^{1-n}/L^{3-3n} t$

k_x = mass transfer coefficient in a binary system, moles/tL^2

k_{xi} = mass transfer coefficient of species i in a multicomponent mixture, moles/tL^2

L = length of tube or other characteristic length, L

L = molar velocity, moles/tL^2

M = molar mean molecular weight, M/mole

M_A = molecular weight of A, M/mole

$\mathscr{M}$ = moles of material

$\mathscr{M}_i$ = moles of component i

m = mass of a molecule, M

m = mass of flow system, M

m_i = mass of component i in flow system, M

$\tilde{N}$ = Avogadro's number, (g mole)$^{-1}$

N = rate of rotation of a shaft, t^{-1}

N_i = molar flux with respect to stationary coordinates, moles/$L^2 t$

n_i = mass flux with respect to stationary coordinates, $M/L^2 t$

$\mathbf{n}$ = molecular concentration or number density, L^{-3}

$\mathbf{n}$ = outward normal

P = momentum, ML/t

$\mathscr{P}$ = $p + \rho g h$ (for constant ρ and g), M/LT^2

p $\qquad$ = fluid pressure, M/Lt^2

p' $\qquad$ = vapor pressure, M/Lt^2

$\bar{p}$ $\qquad$ = partial pressure, M/Lt^2

Q' $\qquad$ = amount of heat transfer, ML^2/t^3

Q $\qquad$ = volumetric flow rate, L^3/t

Q $\qquad$ = rate of energy flow across a surface, ML^2/t^3

Q_{12} $\qquad$ = radiant energy flow from surface 1 to surface 2, ML^2/t^3

Q_{12} $\qquad$ = net radiant energy interchange between surface 1 and surface 2, ML^2/t^3

q $\qquad$ = energy flux relative to mass average velocity, M/t^3

R $\qquad$ = gas constant, $ML^2/t^2 T$ mole

R $\qquad$ = radius of sphere or cylinder, L

R_h $\qquad$ = hydraulic radius, L

R_A $\qquad$ = molar rate of production of species A, moles/tL^3

r $\qquad$ = radial distance in both cylindrical and spherical coordinates, L

r_A $\qquad$ = mass rate of production of species A, M/tL^3

S $\qquad$ = cross-sectional area, L^2

$\mathbf{S}$ $\qquad$ = vector giving cross-sectional area and its orientation, L^2

s $\qquad$ = $R - r$ = distance into fluid from solid boundary in cylindrical coordinates, L

T $\qquad$ = absolute temperature, T

t $\qquad$ = time, t

U $\qquad$ = internal energy, ML^2/t^2

U $\qquad$ = overall heat transfer coefficient, $M/t^3 T$

V $\qquad$ = characteristic speed in dimensional analysis, L/t

V $\qquad$ = volume, L^3

v $\qquad$ = mass average velocity, L/t

v_i $\qquad$ = velocity of species i, L/t

v_∞ $\qquad$ = approach velocity, L/t

v^* $\qquad$ = molar average velocity, L/t

W $\qquad$ = rate of doing work on surroundings, ML^2/t^3

W' $\qquad$ = amount of work done

$\mathscr{W}$ $\qquad$ = molar flow rate, moles/t

$\mathscr{W}_A$ $\qquad$ = molar flow of species A through a surface, moles/t

$\mathbf{w}$ $\qquad$ = vector giving mass flow rate and its direction, M/t

w $\qquad$ = mass flow rate, M/t

w_A	= mass flow of species A through a surface, M/t
x	= rectangular coordinate, L
x_i	= mole fraction of species i, dimensionless
y	= rectangular coordinate, L
y_i	= mole fraction of species i, dimensionless
z	= rectangular coordinate, L
α	= $k/\rho\hat{C}_p$ = thermal diffusivity, angle, absorptivity
β	= thermal coefficient of volumetric expansion, T^{-1}
γ	= $\hat{C}_p/\hat{C}_v$, reflectivity, dimensionless
Δa	= $a_2 - a_1$, in which 1 and 2 refer to two control surfaces
δ	= film thickness, L
ϵ	= fractional void space, emissivity, dimensionless
η	= non-Newtonian viscosity, M/Lt
θ	= angle in cylindrical or spherical coordinates, radians
λ	= wavelength of electromagnetic radiation, L
μ	= viscosity, M/Lt
μ_p	= parameter in Bingham model, M/Lt
ν	= frequency of electromagnetic radiation, t^{-1}
ν	= μ/ρ = kinematic viscosity, L^2/t
π	= $3.14159\ldots$
ρ	= nm = fluid density, M/L^3
ρ_i	= mass concentration of species i, M/L^3
σ	= Stefan-Boltzmann constant, $M/t^3 T^4$
τ	= transmissivity, dimensionless
τ_0	= parameter in Bingham model, $M/t^2 L$
τ_0	= magnitude of shear stress at fluid-solid interface, $M/t^2 L$
Φ	= potential energy, ML^2/t^2
ϕ	= angle in spherical coordinates, radians
ψ	= stream function; dimensions depend on coordinate system
Ω	= potential function
ω_i	= mass fraction of i, dimensionless
Overlines	
$\sim$	= per mole
$\wedge$	= per unit mass
——	= partial molal
——	= time smoothed

Brackets

$<a>$ = weighted value of a over a flow cross section

Superscripts

* = reduced with respect to some characteristic dimension

$'$ = deviation from time-smoothed value

t = turbulent

l = laminar

Subscripts

A, B = species in binary systems

av = arithmetic mean driving force or associated transfer coefficient

b = bulk or "mixing-cup" value for enclosed stream

G = gas

L = liquid

i, j, k = species in multicomponent systems

lm = logarithmic mean driving force or associated transfer coefficient

loc = local transfer coefficient

m = mean transfer coefficient for a submerged object

tot = total quantity in a macroscopic system

0 = quantity evaluated at a surface

1,2 = quantity evaluated at cross sections "1" and "2"

transfer operations

1
Introduction

The ultimate goal of the engineer is, as it always has been, *change*—that is, a tangible alteration in the course of the "real world." In order to plan and effect such change, and, for that matter, in order to detect whether changes have occurred, it is necessary to *describe* the world.

Originally, this description could be performed in terms of intuitively appealing quantities—size, shape, number, etc. For example, it was readily apparent to the designer of a battering ram that if he made it twice as heavy and had twice as many men swing it, the gates of the city might yield somewhat sooner. As the years have passed, however, this happy but unsophisticated state of affairs has evolved to one aesthetically more pleasing and intuitively less satisfying. (It is a far cry from deciding the effect of one more man on the battering ram team to describing the effect of a change in crude oil composition on the product distribution of a refinery.) Nature, alas, though simple in detail has proved to be complex in toto. This state of affairs leaves us (in principle) with a choice—we may either represent the processes of nature by a multitude of manipulations of simple relationships or we may retain simplicity and generality in the manipulations by making the relations we manipulate more abstract.

For example, in design of a pipeline to carry a specified flow rate of a liquid or gas we might wish to know the pressure drop required to force the fluid through at the required rate. One approach to this problem would be to take (in the laboratory) a series of pipes of varying diameters and lengths, apply various pressure drops, and measure the resulting flow rates using various fluids. This information could be bound into an encyclopedic set of volumes and used for design.[1] The work (manipulation) involved would be staggering and the information would be hard to store and retrieve, but one would need only such intuitively satisfying concepts as diameter, length, and flow rate, and this brute-force empirical method would work, albeit without regard to economy. However, if we are willing to give up these elementary variables in favor of two abstractions called Reynolds number and friction factor we can do the whole business with perhaps 20 or 30 experiments and summarize everything on a single page. (We will learn how this is done later in our course of study.) This is an obvious use of abstraction to reduce the number of manipulations.

1.1 MATHEMATICAL MODELS AND THE REAL WORLD

As the quantities which we manipulate grow more abstract, our rules for manipulating these quantities grow more formal, that is, more and more prescribed by that system of logic we call mathematics and less by that accumulation of experience and application of analogy called physical intuition. In its purest form, this approach leads us to replace our real world process by a mathematical model, to operate on that model without use of physical intuition, and at the end to translate the answer back to the real world.

Such a procedure is exemplified by the way one balances his checkbook—the manipulation of the symbols on paper does not come from observing the physical flow of money to and from the bank (real world), but, if the manipulations are performed correctly, the checkbook balance (model) will agree with the amount in the bank at the end of the month. In this case the mathematical model represents *one* aspect of the real world *exactly*—that is, account balance. This is not true in general for models (and is seldom true in engineering models). We get useful answers only in the area where the performance of the mathematical model and the real world coincide.

---

[1] If we use only 10 different diameters, 100 different fluids, and 10 different pressure drops, we have to run at least 10,000 experiments—and we have still neglected the effects of temperature and pipe roughness.

For example, a mathematical treatment of fluid flow which neglects fluid viscosity may give excellent results for real problems at locations where velocity *gradients* are small, since viscosity operates via such gradients, but the same model is useless in problems involving sizable gradients in velocity. We frequently use just such a model for flow around airfoils by neglecting the presence of velocity gradients "far from" the airfoil. In the case of a plane flying at, say, 550 mph, "far from" may be only the thickness of a sheet of paper from the wing surface. In this case, the relative velocity between wing and air goes from 0 to almost 550 mph in this short distance, and remains relatively constant outside this "boundary layer." Obviously *within* this layer velocity gradient cannot be neglected; a model which neglected it would give erroneous results.

The idea of abstract manipulation of a mathematical model is fine *so long as the mathematical model reproduces the essential features of the real world.* A bad or incomplete model will, of course, guarantee bad or incomplete predictions regardless of how elegantly it is manipulated—as, for example, if I try to predict *where* (i.e., at what location) my checks will be cashed by looking at my checkbook—this is simply not a feature included in the model. Unfortunately, as models become more and more complex it becomes progressively easier (even among experienced engineers) to lose sight of this elementary fact. *We shall attempt in the following material to remind the reader as often as possible of the distinction between model and real world.*

Practicing engineers apply models in a variety of levels of sophistication. Some work at a purely experimental level, treating variables in their physically most appealing form and determining correlations of the behavior of these variables by exhaustive experimental studies in the laboratory. Other engineers work with mathematical models of relatively low sophistication which are then fitted to experimental data taken in the laboratory by adjustment of various parametric quantities. Still other engineers work almost completely in abstractions, with mathematical models which sometimes describe processes for which data simply cannot be obtained in the laboratory—perhaps because sensing devices to acquire this data are not available or because the environment in which the data must be acquired is too corrosive, too hot, etc. The most effective *engineer* is the one who chooses a single approach listed above or the combination of these approaches that is most economical in obtaining the answer he *needs*, that is, an answer sufficiently accurate but of no higher order of accuracy than is useful in drawing conclusions appropriate to the problem.[1] For example, in soft-boiling an egg it might be useful to know

---

[1] For an excellent discussion of experiment vis-à-vis theory, see S. W. Churchill, *Chem. Eng. Progr.*, **66**(7):86(1970).

the effect of barometric pressure so that one can adjust the heat transfer calculation (for how long to cook the egg) for the change in boiling point of water with temperature. It would be useless, however, to determine the barometric pressure effect to six significant figures because the uncertainty in other factors far exceeds this (egg-to-egg size variation, ambient temperature when the egg is removed from the water, how long it is until the egg is broken after removing it from the water, and so on). It is very embarrassing and not uncommon for engineers to find themselves refining without limit an irrelevant factor, and the reader is cautioned to attempt always to work on the limiting step in the process.

We can think of the real world as a black box as in Fig. 1.1-1. The box has certain inputs which are not at our disposal to change, for example, composition of the crude oil we pump from the earth. It also has a set of knobs which represent the variables (parameters) under our control: cooling water flow rate, etc. (an input for one problem may be a variable in another and vice versa). Finally, the output of the box (our objective) can be measured on some sort of meter—a large dial reading, for example, yield, or dollars per month, or pounds of product. The ultimate question we need to answer as engineers is "For given inputs and given knob settings, what does the dial read?" As mentioned above in the pressure-drop example, one way to answer this question is to set the knobs at various combinations, read the meter, and never mind what is in the box (trial and error experiments). The other extreme approach corresponds to replacing the black box with something else (here, our mathematical model), making the settings on *that*, and then reading the meter. If our model is good, the meter reads about the same as if we had the "real world" black box in place.

In general, of course, our task as engineers is to maximize or minimize the meter reading. This is the province of optimization theory, and usually depends on our first being able to simulate the system. *In this course we treat primarily simulation, not optimization.*

1.2 ORGANIZATION OF TEXT MATERIAL

We have two objectives in our treatment. The first is to give sufficient information that useful model building for design of momentum, heat, and mass transfer systems (through assembly of various design equations) can be performed by one who uses this text as a terminal course. The second is to give the elementary insight into physical mechanism on which subsequent transport phenomena courses may be based, and by acquaint-

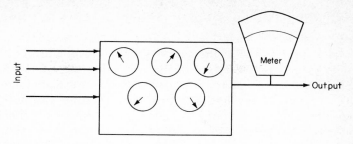

Fig. 1.1-1 Black box model.

ing the reader with a variety of design equations to motivate him to continue study to learn how to construct the more sophisticated models on which these equations are based. The overall organization is shown in Table 1.2-1.

In our experience the direct introduction of transport phenomena to an undergraduate without first showing him (1) the utility of the subject for design and (2) its relation to the general field of engineering leads to a singular lack of motivation. We try to supply this motivation by indicating the molecular transport mechanism, by demonstrating the application of equations based on transport principles (justifying the general composition of these equations by dimensional analysis), and by pointing toward the usefulness of transport phenomena which show in detail how to select the significant variables and how to determine the functional form of their interrelation.

Table 1.2-1   Text organization

	Overall balances	*Momentum*	*Heat*	*Mass*
Mechanism	Mass Energy Momentum	Fluid behavior Laminar flow Turbulence	Conduction Convection Radiation	Molecular diffusion Eddy diffusion
Coefficients and design equations	Dimensional analysis	Drag coefficient friction factor correlations	Heat transfer coefficient correlations	Mass transfer coefficient correlations
System applications	Flow measurement	Pipes and packed beds	Heat exchangers	Continuous contactors

1.3   THE NATURAL GASOLINE PLANT

We continue in Chaps. 4, 7, 10, and 13 with selected examples many of which are referred to the natural gasoline plant shown in Fig. 1.3-1 and Table 1.3-1 in order continually to remind the reader that the models which we construct and treat as independent usually interact strongly.

A natural gasoline plant produces "natural" gasoline, that is, a fraction taken directly from a producing well without chemical alteration by cracking, polymerizing, etc. The basic function of the plant is to separate the fractions that are too light (volatile) for gasoline—methane, ethane, etc.—from the heavier components. Since gasoline is an extremely heterogeneous mixture, the separation is not complete, but partial.

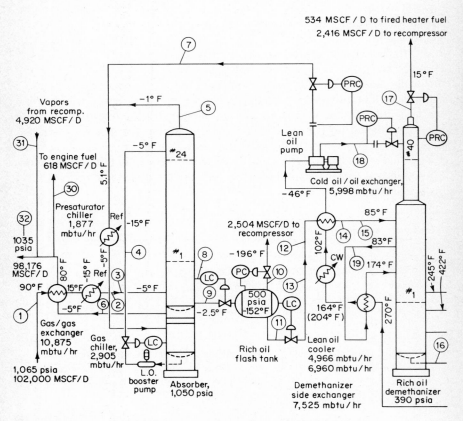

Fig. 1.3-1 Natural gasoline plant. (*Pan American Petroleum Corporation.*)

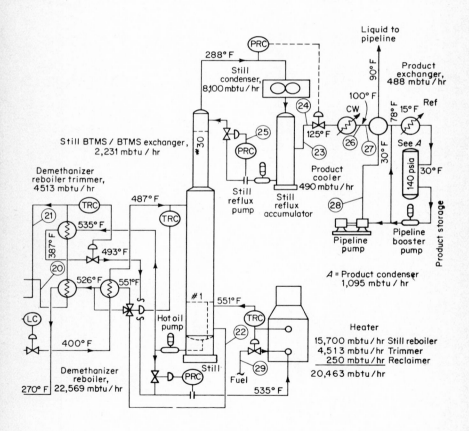

Table 1.3-1 Statistics for Fig. 1.3-1

	1	2	3	4	5	6	7	8	9	10	11	12	13	14	15	16
	Inlet gas stream	Absorb. feed	Chiller cond.	Presat. lean oil	Gross ovhd. vapor	Absorb. resid. bottom feed	Lean oil to absorb. presat.	Absorb. rich oil	Feed 500 flash	500 flash vapor	500 flash liq.	Vapor oil/oil exch.	Liq. oil/oil exch.	Vapor feed dem.	Liq. feed dem.	Dem. btms. liq. frms.
N_2	29.12	29.12			29.12	29.12		13.19	13.48	3.61	9.87	0.89	8.98	5.02	4.85	1.60
CO_2	85.11	84.82	0.29	10.68	82.31	71.63		551.49	567.95	258.54	309.41	61.94	247.47	216.85	92.56	1.26
C_1	10,481.3	10,421.8	16.46	497.15	10,420.5	9,923.35		104.40	106.63	9.68	96.77	2.64	94.13	25.60	71.17	89.23
C_2	366.21	363.98	2.23	78.99	338.57	259.58		100.67	102.91	2.33	100.58	0.55	100.03	10.66	89.92	100.58
C_3	125.43	123.19	2.24	24.81	47.33	22.52		32.95	34.10	0.30	33.80	0.07	33.73	1.77	32.07	33.80
IC_4	34.72	33.57	1.15		0.62	0.62		23.47	24.58	0.15	24.43	0.03	24.40	0.98	23.45	24.43
NC_4	24.64	23.53	1.11		0.06	0.06		7.40	8.96	0.04	8.92	0.01	8.91	0.34	8.58	8.92
IC_5	8.96	7.40	1.56					8.27	8.96	0.01	8.95		8.95	0.10	8.85	8.95
NC_5	8.96	8.27	0.69					4.51	6.72		6.72		6.72	0.05	6.67	6.72
C_6	6.72	4.51	2.21					12.83	17.92		17.92		17.92	0.02	17.90	17.92
C_7	17.92	12.83	5.09					682.5	682.5		682.5		682.5	0.14	682.36	935.5
L.O.				682.5			682.5									
m/hr	11,199.09	11,166.06	33.03	1,294.13	10,918.51	10,306.8	682.5	1,546.68	1,574.71	274.84	9,292.81	66.13	1,233.74	261.53	1,038.34	1,248.91
lb/hr	196,640	195,166	1,474	104,051	183,891	171,976	92138	115.326	116.802	4737	112.063	1143	11.0920	5157	106.906	143.689
gal/hr 60°			333	18,401			14,128	21,003	21,336		19,533		19,099		17911	23086
gal/min 60°			5.6	306.6			235.5	350.0	355.5		325.8		318		295	348
gal/min T			5.1	270.3			230.4	333.6	340.3		308.8		299.8		295.8	502.2
sp.gr. 60°	0.6042	0.6034	0.5285	0.6782	0.5814	0.5758	0.7818	0.6578	0.6558	0.5948	0.6873	0.5938	0.6958	0.6804	0.7231	0.7441
sp.gr. T			0.580	0.718			0.8	0.69	0.685		0.725		0.738		0.721	.569
T°F	90°	-5°	-5°	-5°	-.09°	-5°	3°	-255°	-255°	-14.4°	-14.4°	17.82°	17.82°	82.96°	82.96°	422.09°
mw	17.55	17.48	44.68	80.40	16.84	16.68	135.0	74.81	74.17	17.23	86.21	17.20	69.90	19.71	103.45	114.81
p/psia	1,065	1,055	1,055	1,070	1,045	1,035	1045	1,045	500	500	500	390	390	390	390	390
MSCF/D	102,000	101,701			99,446	93,875				2,504		602		2382		

	17 Dem. ovhd. vapor	18 Lean oil to dem.	19 Liq. to side htr.	20 Liq. to reboil	21 Vapor from reboil	22 Still btms.	23 Still net ovhd. liq.	24 Still net ovhd. vapor	25 Still reflux	26 Still liq. prod., 100°F	27 Still vapor prod., 100°F	28 Plant prod.	29 Fired htr. fuel	30 Eng. fuel	31 Recycle vapor	32 Plant resid. gas
N_2	8.27													0.19		28.93
CO_2	308.15		16.90	7.92	6.32		0.07	1.53	0.34	0.11	1.49	1.60	1.50	0.47	10.39	81.54
C_1	7.54		157.22	9.15	7.89		0.04	1.22	0.19	0.05	1.21	1.26	55.77	65.73	510.95	10,368.5
C_2	0.0		157.64	310.36	221.13		9.36	79.87	42.91	14.24	74.99	89.23	1.37	1.72	16.03	273.59
C_3			125.71	274.34	173.76		25.72	74.86	117.94	37.90	62.68	100.58		0.15	2.33	24.70
IC_4			37.98	73.57	39.77		13.82	19.95	63.39	18.81	14.99	33.80			0.30	0.92
NC_4			26.82	49.06	24.63		11.76	2.67	53.93	15.73	8.70	24.43			0.15	0.21
IC_5			18.78	20.77	11.85		5.93	2.99	27.35	7.09	1.83	8.92			0.04	0.04
NC_5			6.98	13.82	3.87		6.33	2.57	29.31	7.44	1.51	8.95			0.01	0.01
C_6			9.13	10.60	3.88		5.35	0.59	25.60	6.22	0.50	6.72				
C_7			9.04	20.24	2.72		17.33	0.54	39.04	17.68	0.24	17.92				
L.O.		273.6	956.63	1,053.53	0.50	5.50										
m/hr	393.96	273.00	1,522.83	1,842.76	593.85	773	96.39	197.12	440.39	125.27	168.14	293.41	58.64	68.26	539.36	10,778.4
lb/hr	5524	36,855	150,010	176,843	33,454	178	6,173	8.223	29,312	7,606	6,790	14,396	1000	1133	9,267	180,111
gal/hr 60°		5,651	25,194	30,233		29	1,236		5,771	1,562		3,308				
gal/min 60°		94	420	504		1.9	20		90	26		55.1				
gal/min T		91.9	428.5	590		820	21.8		101.5	27.3		52.7				
sp.gr. 60°	0.5893	0.7823	0.7142	0.7001	1.9447	57	0.5983	1.4403	0.6093	0.5834	1.3940	0.5213	0.5893	0.5758	0.5921	0.5787
sp.gr. T		0.8	0.700	0.598		26.7°	0.548		0.567	0.555		0.545				
T°F	15.48°	3°	82.90°	245.38°	422.09°	5.6	125°	125°	125°	100°	100°	30°				
mw	17.07	135.00	98.51	95.97	56.33	5.5	64.04	46.72	66.56	60.72	40.38	49.06	17.07	16.68	17.15	8.75
p/psia	390	390	390	390	390		160	160	160	150	150	140				
MSCF/D	2950				5409			1795			1531		534	618	4,920	9,814

9

A feed stream, stream 1, consisting mostly of gasoline-range hydrocarbons plus a great deal of C_1 to C_3 hydrocarbons, is chilled and absorbed completely under high (1,050 psia) pressure in a higher-boiling oil. This mixture is then flashed (exposed to a lower pressure) and a substantial amount of the methane recovered. The oil is then heated and most of the remaining methane recovered. The natural gasoline is then separated from the oil by distillation and the gasoline sent to the product pipeline while the oil is recycled into the process.

This plant offers the pedagogical advantage of being based on simple physical separations without the complication of chemical reaction, while at the same time offering examples of the interactions of various momentum, heat, and mass transfer processes. It also illustrates a small subsystem which (in a fairly continuous fashion) ties into an entire oil company operation including production, refining, and pipelining.

Before we begin with our formulation of various mathematical models, it is necessary to develop some basic concepts which will be used throughout the remainder of the text. These are treated below.

1.4 MACROSCOPIC VERSUS DIFFERENTIAL APPROACH

In this textbook, we take mainly a macroscopic approach; for example, we do not worry about what the temperature is at many specific points within a piece of equipment, but rather we are interested primarily in the temperature of the stream flowing in and the temperature of the stream flowing out. The determination of the detailed temperature, velocity, and concentration gradients within pieces of equipment involves mathematical models which are usually partial differential equations, and this is the concern of a subsequent course usually called transport phenomena. By limiting ourselves to the macroscopic approach here, we achieve the simplification that most of the differential equations which we encounter are ordinary differential equations and therefore the reader need not confront an excess of mathematical manipulation at the same time we are trying to introduce new physical concepts.

1.5 GENERALIZED CONCEPT OF AVERAGE

In this text we make extensive use of the concept of average, particularly in conjunction with bulk velocity and bulk or mixing-cup temperature, which are average properties of flowing streams. All these types of averages may be subsumed under a generalized concept of average which we will discuss below.

Consider a function $y = f(x)$. By the *average value* of y with respect to x we mean a number such that, if we multiply it by the range on x, we get the same result as by *integrating* y over the same range. Denoting the average by $\bar{y}$, we write this definition as

$$\bar{y}(x_b - x_a) = \int_{x_a}^{x_b} f(x)\ dx \tag{1.5-1}$$

Another way of writing this is

$$\bar{y} = \frac{\displaystyle\int_{x_a}^{x_b} f(x)\ dx}{\displaystyle\int_{x_a}^{x_b} dx} \tag{1.5-2}$$

The same definition can be extended to the case where y is a function of many independent variables: $y = f(x_1, x_2, x_3, x_4, \ldots, x_n, \ldots, x_N)$. By the average of y with respect to x_n we mean

$$\bar{y} = \frac{\displaystyle\int_{x_{na}}^{x_{nb}} f(x_1, x_2, x_3, \ldots, x_n, \ldots, x_N)\ dx_n}{\displaystyle\int_{x_{na}}^{x_{nb}} dx_n} \tag{1.5-3}$$

Sometimes we take the average with respect to more than one independent variable; to do this we simply integrate over the second, third, etc., variables as well. For example, the average of $y = f(x_1, x_2, x_3, \ldots, x_n, \ldots, x_N)$ with respect to x_1, x_6, and x_{47} (N must be greater than 47) is

$$\bar{y} = \frac{\displaystyle\int_{x_{1a}}^{x_{1b}} \int_{x_{6c}}^{x_{6d}} \int_{x_{47e}}^{x_{47h}} f(x_1, x_2, x_3, \ldots, x_n, \ldots, x_N)\ dx_{47}\ dx_6\ dx_1}{\displaystyle\int_{x_{1a}}^{x_{1b}} \int_{x_{6c}}^{x_{6d}} \int_{x_{47e}}^{x_{47h}} dx_{47}\ dx_6\ dx_1} \tag{1.5-4}$$

where (x_{1a}, x_{1b}), (x_{6c}, x_{6d}), and (x_{47e}, x_{47h}) are the ranges of the respective variables. *Notice that notation for averages usually does not distinguish with respect to which variables they are taken. This information must be known in order to apply the average properly.*

For example, suppose I drive from Detroit to Chicago (300 miles) and upon arrival am asked "What was your average speed?" If I answer, "60 mph," I am probably thinking somewhat as follows: An idealized plot of my speed versus time might look like Fig. 1.5-1. The average of 60 mph means to me that 60 mph times the elapsed time of 5 hr equals 300

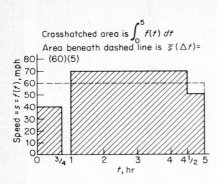

Fig. 1.5-1 Time-averaged distance.

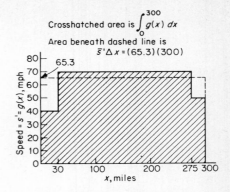

Fig. 1.5-2 Space-averaged distance.

miles. Here my dependent variable is speed s and it is a function of time t:

$$s = f(t)$$

The range on time is 5 hr, so that

$$\bar{s} = \frac{\displaystyle\int_0^5 f(t)\,dt}{\displaystyle\int_0^5 dt} \tag{1.5-5}$$

    In other words, $\bar{s}$ multiplied by the time interval gives me the same area as the integration as shown in Fig. 1.5-1.

    Note that we could, however, equally well regard speed as a function of *distance*, not *time*, and get another function $s' = g(x)$* as shown in Fig. 1.5-2, where this new function tells me how fast I was going as a function of *where* I was rather than what *time* it was. Note that I can define an average based on this function:

$$\bar{s}' = \frac{\displaystyle\int_0^{300} g(x)\,dx}{\displaystyle\int_0^{300} dx} = 65.3 \text{ mph} \tag{1.5-6}$$

This average speed is not the same as $\bar{s}$, since it is intended to be used differently—instead of multiplying it by elapsed *time* we multiply by elapsed *distance*. Thus, $\bar{s}$ is the *time* average, $\bar{s}'$ is the *space* average. By conven-

---

* The prime does *not* denote differentiation.

tion we use the time average in talking about speeds of cars [because $\int_0^{300} g(x)\, dx$ has no intuitive appeal; $\int_0^5 f(t)\, dt$ does since it is merely the total distance].

1.6 LAMINAR VERSUS TURBULENT FLOW

One fascinating aspect of fluid behavior which is little understood but occurs constantly in engineering applications is the difference between *laminar* and *turbulent* flow of fluids. If fluid flows steadily through a pipe, and we introduce a stream of dye into the flowing fluid, there are two possible flow situations that can result. If the fluid flow is slow enough, we will notice that the stream of dye will very gradually disperse, but for all practical purposes will follow along in a straight line as shown in Fig. 1.6-1a. It is almost as if we could paint "particles" of fluid with the dye and notice that they marched along one behind the other much like a string of railroad cars going along a track. This type of flow we call *laminar*. If, however, we perform the same experiment with a higher fluid velocity we will notice that the fluid literally appears to "explode" as it leaves the tip of the hypodermic needle with which we introduce the dye and instantly is mixed across the entire cross section of the pipe with violent eddying motions. This is a type of flow which we will call *turbulent*. If we carry out the experiment in a different manner, first using a low flow rate and gradually increasing the flow, we will see what appears in Fig. 1.6-1b; that is, the flow will remain in straight lines until we reach a certain velocity and then suddenly undergo an abrupt transition and begin to mix across the entire pipe cross section. This transition has puzzled scientists and engineers for generations. We now understand some of the fundamentals of why such a transition should occur, but by and large we are still vastly ignorant on this subject. We do know, however, that we can describe such behavior in terms of a "Reynolds number" which we will discuss in detail later in this text.

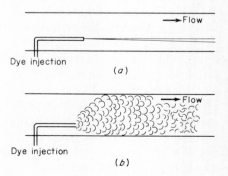

Fig. 1.6-1 Schematic—laminar and turbulent flow.

1.7 REYNOLDS NUMBER

The Reynolds number is a product of a characteristic length (here the pipe diameter), a characteristic velocity (here the bulk velocity in the pipe), and the density of the fluid divided by its viscosity. The result is a dimensionless number which represents the ratio of the inertial to the viscous forces in the fluid. We can determine, for example, that transition from laminar to turbulent flow will normally occur in commercial pipes at about a Reynolds number of 2,100 or slightly above. Treatment of transition from laminar to turbulent flow and treatment of turbulent flow itself in terms of the mathematics required is quite a sophisticated subject, and is reserved for advanced courses in fluid mechanics. For the moment it is only necessary that the reader appreciate that the two types of flow exist and that they may be described in terms of certain dimensionless groups such as the Reynolds number. What is required at this point is a qualitative, rather than a quantitative, insight into the situation.

1.8 VELOCITY PROFILES

It is an experimentally observed fact that fluid usually sticks to solid surfaces. If fluid flows in a pipe, this implies that the velocity rises from zero relative to the wall (at the wall) and as one traverses a diameter rises to a maximum and then declines to zero again at the opposite wall. In order to obtain volumetric flow rate from such information it is necessary to integrate the velocity profile across the cross-sectional area.

Such a procedure is time-consuming, and frequently one does not, in fact, know the velocity profile, or is not really interested in such detail. The variable of prime interest is frequently the volumetric flow rate. Engineers are used to thinking in terms of velocity, however, and velocity frequently is the limiting variable in pipeline design, so that instead of speaking of volumetric flow rate much engineering work is done in terms of the *bulk velocity*. The bulk velocity will be considered in detail later, but for the moment it can be considered simply as a number that, multiplied by the flow cross section, gives volumetric flow rate. It is really a way of specifying volumetric flow. It is a particular type of average velocity.

A little thought should convince the reader that, if the velocity profile were flat at the constant value of bulk velocity (which means one supposes the fluid does *not* stick to the wall), the same volumetric flow is obtained as for the velocity profile from which the bulk velocity is derived. (This is the so-called plug flow assumption; it assumes that the liquid flows like a solid plug down the pipe.) This assumption is sometimes made, but, as we will see later, it represents no real fluid in its momentum behavior—only in its volumetric flow behavior.

1.9 MOLECULAR VERSUS CONVECTIVE TRANSPORT

Momentum, heat, and mass are ultimately transferred via interaction at the molecular level—so-called molecular transport. In fluids, however, many thousands of molecules sometimes will move temporarily as aggregates, for example, in eddies. These aggregates of fluid carry with them momentum, heat, and mass. The gross motion of these large "clumps" of fluid we refer to as "convective" rather than "molecular" transport. This concept will be considered in detail later. For now, we simply need the concept for descriptive purposes.

2
Macroscopic Balances

In our study we will be concerned with the transport of mass, energy, and momentum much in the same way that a businessman is concerned with the transport of goods and services via his manufacturing and sales efforts. Just as the businessman needs an accounting procedure to permit him to describe the inventory and movement of goods, we need an accounting procedure to help us describe the storage and movement of mass, energy, and momentum.

This accounting procedure, coupled with some physical laws—for mass and energy, conservation laws; and for momentum, Newton's second law—permits us to develop *macroscopic balances* which tell us *how much* mass, energy, and momentum has been transferred, but not how the *rate* of this transfer relates to the physical variables in the system (that is, to velocity, temperature, and concentration differences, and to the physical properties of the fluid). The macroscopic balances are analogous to the part of a company's annual report which tells how much the assets have changed and the current rate of sales, but not the effect of the market on the sales volume.

As a transition for the study of transport phenomena we include a brief discussion of how the differential balances and subsequent transport equations follow from each of the integral balances.

2.1 THE ENTITY BALANCE[1]

In our course of study, we shall be faced with the manipulation of many abstractions: "lost work" and "enthalpy," to name two. Many of these entities will be handled by the simple accounting procedure which we will now describe—in particular, those relating to mass, energy, and momentum. This accounting procedure will lead us in a natural way to the mathematical model representing the description of our process in terms of, for example, conservation of mass.

Our accounting procedure (the entity balance) must be applied to a *system* or *control volume*. By system or control volume we mean some unambiguously defined region of space.[2] For example, if we fill a balloon with air and release it, we might take as our system the space interior to the balloon. This would provide a perfectly *acceptable* system even though it moves about and changes size in a peculiar way as it flies about the room. Although *acceptable*, it might or might not be the most *convenient* system and *one of our problems in many situations is the selection of the most convenient system.* One sometimes uses as a system a specific quantity of matter all parts of which remain in proximity; for example, a "clump" of fluid. This is consistent with the concept of a system since this matter unambiguously defines a region in space.

The idea of a system can also be illustrated by considering fluid flowing in a pipe. We can choose as our system the region bounded by two planes normal to the pipe axis and the *inner* wall of the pipe. We could equally well use the region between the two planes and the *outer* wall of the pipe. Or, to take yet a third system, we could select a certain pound of fluid and follow the pound of fluid as it moves in space.

The entity balance used in our accounting system is

$$\text{Input} + \text{generation} = \text{output} + \text{accumulation} \qquad (2.1\text{-}1)$$

By *input* we mean that which crosses the system boundary *from* outside *to* inside in time Δt; by *output* the converse. (There is no real need to

---

[1] An entity is a thing which has reality and distinctness of being, either in fact or for thought ("Webster's Third New International Dictionary of the English Language," G. & C. Merriam Company, Springfield, Mass., 1969).

[2] Mechanical engineers frequently restrict the name "system" to a *closed* system, that is, one where mass does not cross the boundary. See A. G. Hansen, "Fluid Mechanics," p. 96, John Wiley & Sons, Inc., New York, 1967.

define a second term, output—one could instead speak of positive and negative input—it is simply traditional.[1]) By *accumulation* we mean the result of subtracting that which was in the system at some time *t from* that which was there at some *later* time $t + \Delta t$. As opposed to the input-output case, we do *not* define a term called "depletion"; we speak instead of a "negative accumulation." By *generation* we mean that which *appears* within the system without either being present initially or being transferred in across the boundary—it materializes, somewhat as the ghost of Hamlet's father, but in a far more predictable fashion.

You will note that the above paragraph is full of "that which"; we have carefully avoided saying just what it is for which we are accounting. This is deliberate, and is done to stress the generality of the procedure, which, as we shall see below, is applicable to people and money as well as to mass, energy, and momentum.

Example 2.1-1 An entity balance The entity balance can be applied to very general sorts of quantities. When it is applied to discrete entities it is frequently called a *population* balance. One such entity, of course, is people, although automobiles, tornadoes, buildings, trees, bolts, marriages, etc., can also be described using the entity balance approach.

If the entity balance is applied to people with a country as the system, the input terms are calculable from immigration statistics and the output from emigration statistics. The accumulation term is calculable from census figures (the quantity within the system). The accumulation term is made up of births and deaths.

For example, suppose that a political unit (the system) had a population of 1,000,000 people at the beginning of the year. At the end of the year (Δt) the population was 1,010,000. Thus the *accumulation* is 10,000. During this time 10,000 people died and 18,000 were born, giving *generation* of $(18,000 - 10,000) = 8,000$ people. If 4,000 people immigrated (input), how many emigrated (output)?

Application of the entity balance shows that

$$4,000 + 8,000 = \text{output} + 10,000$$

$$\text{Output} = 2,000$$

Therefore 2,000 people emigrated. An interesting exercise is to check statistics on population, birth/death, and immigration/emigration for consistency by using the entity balance. Such a calculation performed

---

[1]As Tevye says in "Fiddler on the Roof" before singing "Tradition," "You may ask, how did this tradition start? I'll tell you—I don't know! But it's a tradition. Because of our traditions, everyone knows who he is and what God expects him to do."

on the world population leads to the large accumulation term which is referred to as the "population explosion."

□ □ □

Conserved quantities

Another interesting fact is that when our equation is applied to some entities, there is never any generation. For example, if I apply the equation to my checking account as the system, I never find money spontaneously appearing. If I didn't put it in, it isn't there to remove. Such a quantity we call a *conserved* quantity, that is, one which is neither created nor destroyed. (We note in passing that an embezzler may be regarded as an output or a negative accumulation term but the total amount of money is still conserved.)

Quantities which are conserved during some processes may not be conserved under other processes. In the course of this book, we will assume that matter and energy are conserved quantities, as they are for all practical purposes in the processes we will consider. We know, however, in processes where nuclear reactions are taking place or where things move at speeds approaching the speed of light that matter and energy are not conserved. They can, instead, be transmuted into one another, and only their totality is conserved.

We now will consider in some detail application of Eq. (2.1-1) to generate the equations describing the *conservation* of total mass and energy. We will also consider its application to momentum, and to mass of an individual species, which are not in general conserved in our processes but which can be described by our accounting procedure. Another difference in applying Eq. (2.1-1) to momentum is that we will generate a *vector* equation rather than a *scalar* equation.

2.2 THE MACROSCOPIC MASS BALANCE

We now apply the entity balance equation (2.1-1) term by term to *total mass*. Since total mass will be *conserved* in processes we consider, we know that generation is zero.

For the sake of generality, let us consider a control volume or system of arbitrarily changing shape moving at some changing velocity in space during a time interval Δt, as in Fig. 2.2-1. We will first examine the *accumulation* term in Eq. (2.1-1).

The mass present in the system

Accumulation is the difference between the mass in the system at time t and that at $t + \Delta t$. Our first step, therefore, is to write down an expression for the mass in the control volume at an arbitrary time t.

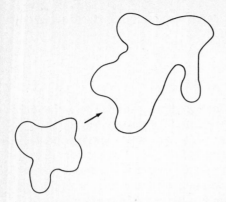

Fig. 2.2-1 System that moves and changes shape.

The obvious way to write down the mass, as with counting the number of people in the country, is to count the number of molecules of each species and multiply by the species mass. This is another example of an *acceptable* procedure but an *inconvenient* one. (Assuming it were actually possible to count molecules in this manner at a rate of one million per second it would take time about equivalent to the present age of the earth to count the molecules in one mole of gas.) Instead, we introduce the very convenient concept of density, or mass per unit volume. Since the control volume is of finite size, density can, in general, vary in position as well as time:

$$\rho = \rho(x, y, z, t) \tag{2.2-1}$$

The idea of density *at a point* involves a very fundamental assumption—that the material behaves as a continuum.[1]

---

[1] We define the density at a point as the mass per unit volume, but since a point has no volume it is not at all clear what we mean. By our definition we *might* mean

$$\rho = \lim_{\substack{\Delta x \to 0 \\ \Delta y \to 0 \\ \Delta z \to 0}} \left[\frac{\Delta m}{\Delta x\, \Delta y\, \Delta z} \right] = \lim_{\Delta V \to 0} \left[\frac{\Delta m}{\Delta V} \right]$$

where Δm is the mass in an elemental rectangular volume Δx by Δy by Δz.

Let us consider what happens in applying this definition to, say, a gas at *uniform* temperature and pressure. Assume for the moment that we can observe $\Delta m / \Delta V$ instantaneously, that is, that we can "freeze" the molecular motion as we make any *one* observation. Suppose that for a range of values for ΔV we make many observations of $\Delta m / \Delta V$ at *various* times but at about the same point.

At large values of ΔV, the density from successive observations would be constant for all practical purposes. As we approach smaller volumes, however, our observed value would no longer be *unique*, because the gas is not a *continuum* but rather is made up of small regions of large density (the molecules) connected by regions of zero density (empty space).

The mass present at t in some arbitrary *small portion* ΔV of a control volume may be written *approximately* as $\rho \, \Delta V$ where the value used for ρ is equal to ρ at some *arbitrary* point in ΔV and at t. The total mass in the control volume is the mass in the sum of all the small elemental volumes that make up the control volume. This approximation becomes exact for a continuum if we let ΔV approach zero, since ρ will approach ρ at a point, which quantity we restricted to a unique value in Eq. (2.2-1):

$$m = \lim_{\Delta V \to 0} \sum_{V} \rho \, \Delta V \qquad (2.2\text{-}2)$$

which from calculus defines the integral[1]

$$m = \int_{V} \rho \, dV \qquad (2.2\text{-}3)$$

where again $\rho = \rho(x,y,z,t)$.

---

[1] We adopt the usual shorthand notation of writing multiple integrals with a single integral sign indicating the range of the integral and the differentials of the variables replaced by a single letter; e.g.,

$$\int_{V} \rho \, dV \equiv \int_{x} \int_{y} \int_{z} \rho \, dz \, dy \, dx$$

For example, if our ΔV were of the same order of volume as an atomic nucleus, the observed density would fluctuate from zero to the tremendous density of the nucleus itself, depending on whether or not a nucleus were present in our ΔV at the time of the observation. This is annoying, and means that the limit in Eq. (2.2-2) has no utility since its value is not *unique*. To avoid this problem we make the arbitrary restriction that we will treat only problems where we may assume material behaves as a *continuum*, that is, at any point

$$\lim_{\Delta V \to 0} \frac{\Delta m}{\Delta V}$$

is a unique value for any time at which we make an observation. This replaces our real process with a *mathematical model* and means that the model will not be valid for regions of space the same order of size as the mean free path of molecules. We gain, however, the important advantage that we may now write

$$\lim_{\Delta V \to 0} \left[\frac{\Delta m}{\Delta V} \right] = \frac{dm}{dV}$$

and the door to the use of calculus is now open.

The accumulation

The change of m in time Δt, or the accumulation, is then

$$\left. \int_v \rho \, dV \right|_{t + \Delta t} - \left. \int_V \rho \, dV \right|_t = \text{accumulation} \qquad (2.2\text{-}4)$$

The input and output

The input and output terms in Eq. (2.1-1) may be evaluated by examining the input or output for an arbitrary small area ΔA on the surface of the control volume and summing over the entire surface. (See Fig. 2.2-2.)

The flow velocity vector is not necessarily normal to the surface of the control volume. We describe the direction of the velocity vector by its angle with the *outward* normal to the surface at the appropriate point in ΔA. The component of the velocity normal to the surface at that point is then

$$\mathbf{v} \cdot \mathbf{n} = |\mathbf{v}| \cos \alpha = v \cos \alpha \qquad (2.2\text{-}5)$$

where $\mathbf{n}$ is the outward *unit* normal and α is the angle between $\mathbf{v}$ and $\mathbf{n}$.

The velocity $\mathbf{v}$ is the relative velocity between the material flowing and the surface of the control volume at that point, that is, the velocity with respect to a coordinate system sitting on the surface of the control volume at the point in question.

The volumetric flow rate Q, through an elemental area ΔA, may be written approximately as the product of the velocity normal to the surface, evaluated at some point in ΔA, and ΔA, which is the approximate

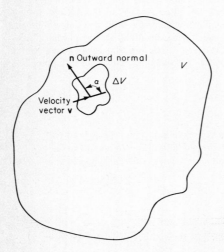

Fig. 2.2-2 Control volume.

area normal to flow. The product of this volumetric flow rate and density gives the approximate mass flow through ΔA. Since both $\mathbf{v}$ and ρ are functions of x, y, z, and t, we must select a definite point in space and time to get a unique value for $\mathbf{v}$ and ρ. This is done later in this development when we take the limit as elements of *volume*, *area*, and *time* go to zero; until then all the equations are approximate.

We can rearrange the entity balance equation (2.1-1) to read if there is no generation:

$$\text{Output} - \text{input} + \text{accumulation} = 0 \tag{2.2-6}$$

We can *approximately* account for the output − input term by writing

$$\text{Output} - \text{input} \cong \sum_A \rho(\mathbf{v} \cdot \mathbf{n}) \, \Delta A \, \Delta t = \sum_A \rho v (\cos \alpha) \, \Delta A \, \Delta t \tag{2.2-7}$$

where $\mathbf{v}$ and ρ are the values somewhere in ΔA and Δt. (Note: Since we are talking about flows of tangible material, at each point we may have *either* output or input but not both, and further, the sign of $\cos \alpha$ takes care of whether we have output or input, since $\cos \alpha$ is positive for output and negative for input.)

If we let ΔA approach zero, Eq. (2.2-7) becomes

$$\text{Output} - \text{input} \cong \lim_{\Delta A \to 0} \sum_A \rho(\mathbf{v} \cdot \mathbf{n}) \, \Delta A \, \Delta t = \Delta t \int_A \rho(\mathbf{v} \cdot \mathbf{n}) \, dA$$

$$\tag{2.2-8}$$

The macroscopic total mass balance

Substituting Eqs. (2.2-8) and (2.2-4) in Eq. (2.2-6) gives

$$\Delta t \int_A \rho(\mathbf{v} \cdot \mathbf{n}) \, dA + \int_V \rho \, dV \bigg|_{t+\Delta t} - \int_V \rho \, dV \bigg|_t \cong 0 \tag{2.2-9}$$

If we divide by Δt and take the limit as Δt approaches zero, the equation is no longer approximate and reads

$$\int_A \rho(\mathbf{v} \cdot \mathbf{n}) \, dA + \frac{d}{dt} \int_V \rho \, dV = 0 \tag{2.2-10}$$

This is the macroscopic total mass balance. Note: It is now written in terms of *rates*, but these rates are not yet related by the equation to the physical variables of the system. This relationship is the topic of later chapters.

Simplified forms of the macroscopic total mass balance

In many applications we are interested only in the total flow into or out of the system rather than the flow or velocity at any one point on the surface. Such problems are usually discussed in terms of bulk velocities. The bulk velocity $\langle v \rangle$ is simply a number that, if multiplied by the area exposed to flow, gives the total volumetric flow rate (which we denote by Q):

$$\langle v \rangle A = \int_A |(\mathbf{v} \cdot \mathbf{n})| \, dA = Q \qquad (2.2\text{-}11)$$

This obviously may be written in the form of the general average for $\cos \alpha$ constant:

$$\langle v \rangle = \frac{\int_A |(\mathbf{v} \cdot \mathbf{n})| \, dA}{\int_A dA} = \frac{|\cos \alpha| \int_A v \, dA}{\int_A dA} \qquad (2.2\text{-}12)$$

showing it is simply an *area* average of the normal component of velocity. The bulk velocity is usually taken by convention to be a *positive* number. This permits us, for example, to speak of the bulk velocity in a pipe without reference to a system. When we use this velocity in equations relating to a particular system we must, of course, associate it with the appropriate sign.

The *mass* rate of flow w is

$$w = \int_A \rho |(\mathbf{v} \cdot \mathbf{n})| \, dA \qquad (2.2\text{-}13)$$

where w is taken by convention to be positive and associated with the appropriate sign when referred to a specific system for similar reasons.

If we have a constant density across dA, and if $\cos \alpha = \pm 1$ the first term in Eq. (2.2-10) simplifies, using Eq. (2.2-11)

$$\int_A \rho(\mathbf{v} \cdot \mathbf{n}) \, dA = \rho \int_A (\mathbf{v} \cdot \mathbf{n}) \, dA = \pm \rho \langle v \rangle A \qquad (2.2\text{-}14)$$

Since ρ, $\langle v \rangle$, and A are all positive numbers, their product is also positive. This quantity is the *magnitude* of the mass inflow or outflow and must be preceded by the appropriate sign. If we always denote outlet conditions by subscript 2 and inlet conditions by subscript 1 we can write

Rate of mass output − rate of mass input = Δw

$$= \Delta(\rho \langle v \rangle A) = \rho_2 \langle v \rangle_2 A_2 - \rho_1 \langle v \rangle_1 A_1 \qquad (2.2\text{-}15)$$

Frequently we can also simplify the second term in Eq. (2.2-10). If the density of the fluid is always uniform *in* the system,

$$\frac{d}{dt}\int_V \rho\, dV = \frac{d}{dt}\left(\rho \int_V dV\right) = \rho\,\frac{d}{dt}\left(\int_V dV\right) = \rho\,\frac{dV}{dt} \qquad (2.2\text{-}16)$$

This term obviously vanishes for a constant-volume system.

The integral in the second term of Eq. (2.2-10) is the mass in the system, and this quantity is denoted by m:

$$\frac{d}{dt}\int_V \rho\, dV = \frac{dm}{dt} \qquad (2.2\text{-}17)$$

The macroscopic mass balance can also be written, therefore, as

$$\Delta w + \frac{dm}{dt} = 0 \qquad (2.2\text{-}18)$$

Note that this equation is valid for systems where density *varies* since the constant-density assumption was not used in defining w and m.

Example 2.2-1 Simple application of mass balance Water is being pumped at a rate of 10 ft³/min into a 10-ft-diameter tank initially empty. Water leaves the tank at a rate that is dependent on the height of liquid in the tank as described by the following relation:

$$Q_{out} = 2h \qquad \text{(Note that 2 has units of ft}^2\text{/min)}$$

where Q_{out} is in ft³/min and h is the height of liquid in feet.

Find the height of liquid in the tank as a function of time.

Solution Since this is equivalent to asking the amount of liquid accumulated in the tank as a function of time, we apply the overall mass balance using Eqs. (2.2-11), (2.2-15), and (2.2-16), and dividing through by the constant density

$$Q_{out} - Q_{in} + \frac{d}{dt}\,V = 0$$

The volume of liquid in the tank in ft³ is

$$V = \frac{\pi D^2 h}{4} = \frac{\pi}{4}\,100\ \text{ft}^2\,h\ \text{ft} = 78.7h\quad \text{ft}^3$$

Substituting for V and Q, we obtain

$$2h - 10 + \frac{d}{dt}(78.7h) = 0$$

or

$$78.7\frac{dh}{dt} = 10 - 2h = 2(5 - h)$$

Since at $t = 0$, $h = 0$

$$39.4\int_0^h \frac{dh}{5 - h} = \int_0^t dt$$

by integrating, we obtain

$$-39.4 \ln\frac{5 - h}{5} = t$$

Then,

$$\frac{5 - h}{5} = e^{-t/39.4}$$

Solving for h,

$$h = 5(1 - e^{-t/39.4})$$

Note that as $t \to \infty$, the level stabilizes at 5 ft.

□ □ □

The macroscopic mass balance for an individual species

We said above that in the course of our study we will assume that *total* mass is *conserved*. We noted that we can, however, also apply the entity balance equation (2.1-1) to an individual species (for example, to propane in a propane-air mixture). An individual species is *not*, in general, *conserved* under processes of interest to us because frequently we are interested in carrying out chemical reactions. A chemical reaction in a system can give positive or negative generation of an individual species. For example, if we feed a propane-air mixture into a combustion furnace (the system) there will be a negative generation of propane; i.e., propane disappears. (As this happens, water and C and/or CO and/or CO_2 also spontaneously *appear*.)

In order to apply Eq. (2.1-1) we must, therefore, have a way of describing the rate of generation of a species by chemical reaction. This is a complicated problem in itself, and is the concern of the discipline of chemical kinetics. We could spend many chapters in discussion of how to relate the generation of an individual species to the variables in a system, but for the sake of brevity we will give only a general discussion.

Call the *rate* of mass generation *per unit volume* of the ith component r_i. Note that r_i can be a function of x, y, z, and t:

$$r_i = r_i(x,y,z,t) \tag{2.2-19}$$

The mass density of the ith component we will denote by $\rho_i(\rho_i =$ mass of i per unit volume—a concentration of sorts). Similarly we will refer to the mass rate of flow of the ith component as w_i, and to the mass of the ith component in the system as m_i.

We now apply Eq. (2.1-1). The argument for the input, output, and accumulation terms goes just as before except that ρ is replaced by ρ_i, w by w_i, and m by m_i.[1] In order to obtain the generation term the same sort of argument is followed as with the other terms, first writing the generation for a small element ΔV in time Δt:

$$\text{Generation in } \Delta V \text{ during } \Delta t \cong r_i\,\Delta V\,\Delta t \tag{2.2-20}$$

where r_i is the value anywhere in ΔV and Δt. Summing over all the ΔV's to get the approximate total generation in the system, and taking the limit as $\Delta V \to 0$,

$$\lim_{\Delta V \to 0} \sum_V r_i\,\Delta V\,\Delta t = \Delta t \int_V r_i\,dV \cong \text{generation} \tag{2.2-21}$$

Rewriting Eq. (2.1-1) as

$$\text{Output} - \text{input} + \text{accumulation} = \text{generation} \tag{2.2-22}$$

and substituting,

$$\Delta t \int_A \rho_i(\mathbf{v}\cdot\mathbf{n})\,dA + \int_V \rho_i\,dV\Big|_{t+\Delta t} - \int_V \rho_i\,dV\Big|_t = \Delta t \int_V r_i\,dV \tag{2.2-23}$$

[1] Strictly speaking, we must also replace v by v_i, since the individual species does not move at v in the presence of concentration gradients. The difference between v and v_i is negligible, however, for most problems involving flowing streams, which are the type considered here. Chapter 11 discusses the differences in v and v_i.

Dividing by Δt and taking the limit as $\Delta t \to 0$ yields

$$\int_A \rho_i (\mathbf{v} \cdot \mathbf{n}) \, dA + \frac{d}{dt} \int_V \rho_i \, dV = \int_V r_i \, dV \qquad (2.2\text{-}24)$$

This is the macroscopic mass balance for an individual species where the species moves at essentially the stream velocity. There is one of these equations for each species; their sum is the macroscopic total mass balance.

Another way to write this equation is

$$\Delta w_i + \frac{d}{dt} \, m_i = \int_V r_i \, dV \qquad (2.2\text{-}25)$$

In order to evaluate the right-hand side of this equation, we must have a way to express r_i as a function of V. This is usually done by expressing r_i as a function of ρ_i and then ρ_i as a function of V. This problem is the usual concern of a course in chemical kinetics. We frequently assume perfect mixing, which implies that ρ_i is independent of position. This simplifies the right-hand side of Eq. (2.2-25). An example follows.

Example 2.2-2 Mass balance with reaction Consider a perfectly mixed tank where the reaction $B \to C$ takes place. (By *perfectly mixed* we mean that ρ_i is not a function of x, y, and z; that is, ρ_i is uniform throughout the tank at any time t. No real vessel is perfectly mixed; however, this is a limiting case which frequently gives a good approximation to real problems.) The *inlet* concentration of B is $\rho_{B1} = 15$ lb$m$/ft^3 and the *initial* concentration in the tank is $\rho_{B0} = 5$ lb$m$/ft^3. The volume of liquid in the tank is 100 ft^3, and remains constant because the flow in and the flow out are each 10 ft^3/min and no volume change is associated with the reaction. Assume we can describe the generation of B in lb$m$/sec ft^3 by the first-order equation (which we would obtain from kinetic data):

$$(a) \qquad r_B = -k\rho_B$$

where $k = 0.1$ min^{-1}. The negative sign indicates that B disappears via the reaction.

How does the concentration in the tank behave with time?

Solution Applying Eq. (2.2-24) and substituting for r_B from equation (*a*)

$$(b) \qquad \int_A \rho_B (\mathbf{v} \cdot \mathbf{n}) \, dA + \frac{d}{dt} \int_V \rho_B \, dV = \int_V -k\rho_B \, dV$$

The first term on the left-hand side of equation (b) may be simplified by rewriting the integral over the total surface as the sum of three integrals:

1. The integral over the inlet cross section A_1.
2. The integral over the outlet cross section A_2.
3. The integral over all remaining surface. This integral will be identically zero because the velocity *normal* to the surface $(\mathbf{v} \cdot \mathbf{n})$ is everywhere zero except at the inlet and outlet.

This gives

$$\int_A \rho_B(\mathbf{v} \cdot \mathbf{n}) \, dA = \int_{A_1} \rho_B(\mathbf{v} \cdot \mathbf{n}) \, dA + \int_{A_2} \rho_B(\mathbf{v} \cdot \mathbf{n}) \, dA$$

Assuming that ρ_B does not vary across the inlet or outlet cross section gives

$$\int_A \rho_B(\mathbf{v} \cdot \mathbf{n}) \, dA = \rho_{B1} \int_{A_1} (\mathbf{v} \cdot \mathbf{n}) \, dA + \rho_{B2} \int_{A_2} (\mathbf{v} \cdot \mathbf{n}) \, dA$$

But $\int_A (\mathbf{v} \cdot \mathbf{n}) \, dA$ is the volumetric flow rate, so that

$$\int_A \rho_B(\mathbf{v} \cdot \mathbf{n}) \, dA = \rho_{B2} Q_2 - \rho_{B1} Q_1$$

Since the outlet concentration will be the same as the concentration in the tank for perfect mixing,

$$\rho_B = \rho_{B2}$$

and

$$\int_A \rho_B(\mathbf{v} \cdot \mathbf{n}) \, dA = \rho_B(10) - (15)(10)$$

This is the simplified form of the first term in equation (b).

Considering the second term on the left in equation (b), we note that ρ_B is independent of *where* we are in the tank by virtue of the perfect mixing, and so

$$(c) \qquad \frac{d}{dt} \int_V \rho_B \, dV = \frac{d}{dt} \left(\rho_B \int_V dV \right) = \frac{d}{dt} (\rho_B V) = \rho_B \frac{dV}{dt} + V \frac{d\rho_B}{dt}$$

However, if we write a macroscopic *total* mass balance

$$\rho_2 Q_2 - \rho_1 Q_1 + \frac{d}{dt}(\rho_2 V) = 0$$

and observing that $\rho_1 = \rho_2 = $ constant and that $Q_2 = Q_1$

$$\frac{dV}{dt} = 0$$

Thus from equation (c)

$$\frac{d}{dt}\int_V \rho_B \, dV = V\frac{d\rho_B}{dt} = 100\frac{d\rho_B}{dt}$$

Next, examination of the right-hand side of equation (b) shows that we can remove the mass concentration from the integral (the tank is perfectly mixed), and since the reaction rate constant also does not depend on where one is in the tank,

$$\int_V - k\rho_B \, dV = -k\rho_B\int_V dV = -k\rho_B V = -0.1\rho_B(100)$$

Substituting the simplified expressions above into equation (b) yields

$$10\rho_B - (10)(15) + 100\frac{d\rho_B}{dt} = -0.1\rho_B(100)$$

Rearranging and integrating:[1]

$$\int_5^{\rho_B} \frac{d\rho_B}{15 - 2\rho_B} = \int_0^t \frac{dt}{10}$$

Multiplying and dividing by -2:

$$\frac{1}{-2}\int_5^{\rho_B} \frac{(-2)d\rho_B}{15 - 2\rho_B} = \int_0^t \frac{dt}{10}$$

or

$$\ln\frac{15 - 2\rho_B}{5} = \frac{-2}{10}t$$

---

[1]We use the common but imprecise practice of using the same symbol for the dummy integration variable as in the limit.

Taking the antilogarithm of both sides

$$3 - 0.4\rho_B = e^{-0.2t}$$

or

$$\rho_B = 2.5(3 - e^{-0.2t})$$

which gives us ρ_B as a function of t. As $t \to \infty$, ρ_B approaches 7.5 $\text{lb}m/\text{ft}^3$.

□ □ □

2.3   THE DIFFERENTIAL MASS BALANCE[1]

In the following we discuss the transition from the *integral* equation for the mass balance (conservation of mass) to the *differential* equation for the mass balance—from the *macroscopic* to the *microscopic* equation. The resulting differential balance is one of the starting equations for the study of transport phenomena. In contrast to the macroscopic balances, which are usually applied to determine input/output relationships, the microscopic balances permit calculation of profiles (velocity, temperature, concentration) within systems.

The macroscopic total mass balance was determined above to be

$$\int_A \rho(\mathbf{v} \cdot \mathbf{n}) \, dA + \frac{d}{dt} \int_V \rho \, dV = 0 \qquad (2.3\text{-}1)$$

The velocity $\mathbf{v}$ was defined as the relative velocity—the velocity relative to the surface of the control volume. In order to avoid a redefinition of velocities we now assume the control volume to be stationary. For a stationary control volume, the limits on the volume integral in Eq. (2.3-1) do not depend on time, and we can write:

$$\frac{d}{dt} \int_V \rho \, dV = \int_V \frac{\partial \rho}{\partial t} \, dV \qquad (2.3\text{-}2)$$

We also have the divergence theorem from vector calculus:

$$\int_V (\boldsymbol{\nabla} \cdot \mathbf{s}) \, dV = \int_A (\mathbf{s} \cdot \mathbf{n}) \, dA \qquad (2.3\text{-}3)$$

---

[1] This section provides a transition for subsequent courses in transport phenomena—the differential approach referred to in Sec. 1.4.

which is valid for an arbitrary vector **s**. We apply this theorem to the vector $\rho\mathbf{v}$ in the first term of Eq. (2.3-1) and apply Eq. (2.3-2) to the volume integral term, giving

$$\int_V [\nabla \cdot (\rho\mathbf{v})] \, dV + \int_V \frac{\partial \rho}{\partial t} \, dV = 0 \qquad (2.3\text{-}4)$$

or

$$\int_V \left[\frac{\partial \rho}{\partial t} + (\nabla \cdot \rho\mathbf{v}) \right] dV = 0 \qquad (2.3\text{-}5)$$

Since the limits of integration are arbitrary, the term in brackets must be zero:

$$\frac{\partial \rho}{\partial t} + (\nabla \cdot \rho\mathbf{v}) = 0 \qquad (2.3\text{-}6)$$

which is the *differential total mass balance* (or continuity equation). In rectangular Cartesian coordinates this equation is

$$\frac{\partial \rho}{\partial t} + \frac{\partial}{\partial x} \rho v_x + \frac{\partial}{\partial y} \rho v_y + \frac{\partial}{\partial z} \rho v_z = 0 \qquad (2.3\text{-}7)$$

We can develop a microscopic mass balance for an individual species in the same general way that we developed the microscopic total mass balance, the only difference being that (1) the individual species may be created or destroyed by reaction, that is, there is a *generation* term, and (2) the individual species may not move at the same velocity as the bulk fluid.

We obtained for the macroscopic mass balance for an individual species:

$$\int_A \rho_i(\mathbf{v} \cdot \mathbf{n}) \, dA + \frac{d}{dt} \int_V \rho_i \, dV = \int_V r_i \, dV \qquad (2.3\text{-}8)$$

This equation is restricted to cases where the velocity of the bulk fluid and the species velocity are virtually identical, as was noted in the derivation. The microscopic balance, however, is frequently applied to cases where this is not true, and so we begin here with the more general equation

$$\int_A \rho_i(\mathbf{v}_i \cdot \mathbf{n}) \, dA + \frac{d}{dt} \int_V \rho_i \, dV = \int_V r_i \, dV \qquad (2.3\text{-}9)$$

By incorporating the species velocity $\mathbf{v}_i$ instead of the velocity of the bulk fluid $\mathbf{v}$, we can handle cases where $\mathbf{v}_i$ and $\mathbf{v}$ differ—for example, as is shown in Chap. 11, where one species may move faster than the bulk fluid because of a concentration gradient.

As before, we restrict our development to stationary volumes so that we can move the differentiation inside the integral in the second term in Eq. (2.3-9). We also apply the divergence theorem to the first term, and write

$$\int_V \mathbf{\nabla} \cdot (\rho_i \mathbf{v}_i) + \int_V \frac{\partial \rho_i}{\partial t} \, dV = \int_V r_i \, dV \qquad (2.3\text{-}10)$$

or

$$\int_V \left[\frac{\partial \rho_i}{\partial t} + \mathbf{\nabla} \cdot (\rho_i \mathbf{v}_i) - r_i \right] dV = 0 \qquad (2.3\text{-}11)$$

Again, since this holds for an arbitrary volume the integrand must be zero:

$$\frac{\partial \rho_i}{\partial t} + \mathbf{\nabla} \cdot (\rho_i \mathbf{v}_i) - r_i = 0 \qquad (2.3\text{-}12)$$

This is the microscopic mass balance for an individual species. (There is one such equation for each species and the sum of these must be consistent with the total mass balance.) The investigation of solutions of this equation for various systems (boundary conditions) where components diffuse and disperse is the topic of mass transport as a transport phenomenon.[1]

Equation (2.3-12) is seldom used in the form given, because what we are usually attempting to obtain is some mass flux in terms of concentration profiles. To do this we combine Fick's law, which is discussed in Chap. 11, with Eq. (2.3-12) and obtain a variety of differential equations. (Fick's law relates the transport properties of the fluid to the component flux produced by a concentration gradient.) For example, the resulting equation of change for component A moving in one dimension in a mixture of A and B is

$$\frac{\partial x_A}{\partial t} + v_z \frac{\partial x_A}{\partial z} = \mathscr{D}_{AB} \frac{\partial^2 x_A}{\partial x^2} \qquad (2.3\text{-}13)$$

where $\mathscr{D}_{AB}$ is the diffusion coefficient, discussed in Chap. 11.

---

[1] R. B. Bird, W. E. Stewart, and E. N. Lightfoot, "Transport Phenomena," John Wiley & Sons, Inc., New York, 1960.

2.4 THE MACROSCOPIC ENERGY BALANCE

We now consider the application of our entity balance to the second of our *conserved* quantities, energy. Just as we did for the mass balance, we consider an arbitrary system moving in some prescribed manner. Since mass carries with it associated energy because of its position, motion, or physical state, we will find that the *energy* balance will, therefore, have an energy term corresponding to each term of the macroscopic total mass balance. In addition, though, we can transport energy across the boundary of a system in a form *not associated with mass*. This fact introduces terms into the overall energy balance that have no counterpart in the overall mass balance. We still, however, deal with scalars.

The entity balance for a *conserved* quantity reads:

$$\text{Output} - \text{input} + \text{accumulation} = 0 \qquad (2.4\text{-}1)$$

First let us examine the accumulation term.

Accumulation of energy within the system

Although energy not associated with any mass may *cross* the boundary of a system, the only way we will consider that energy may be present *within* a system is *associated with mass*. We do not attempt to make our equations valid, for example, *within* systems where large, varying amounts of electromagnetic radiation are present.

Energy associated with mass can be classified in three ways:

1. Energy present because of the *position* of the mass in a field (gravitational, magnetic, electrostatic, etc.). This energy is called potential energy Φ.
2. Energy present because of translational or rotational motion of the mass. This energy is called kinetic energy K.
3. All other energy associated with the mass, e.g., rotational and vibrational energy in chemical bonds, etc. This catchall classification is called internal energy U.

Remember that earlier we restricted our mathematical models to *continua*. If we had not, the same problem would appear now in defining the density of energy that we had in defining the density of mass earlier (Sec. 2.2). The continuum view of energy implies that the energy of the individual molecules is regarded as "smeared out" or distributed locally through the system. This means that the energy of the random molecular translation is lumped in the *internal energy*, *not* the kinetic energy, even though it is kinetic energy. Our continuum assumption implies that we cannot look at matter on the molecular scale, and therefore we do not see the kinetic energy of molecular motions as such but rather we simply see internal energy possessed by a local region of space.

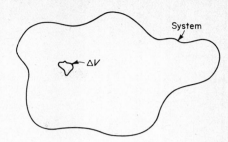

Fig. 2.4-1 System for energy balance.

We shall denote the amount of a quantity associated with a *unit mass* of material by placing a caret above the quantity. Thus $\hat{U}$ is the internal energy per unit mass and $\hat{W}$ is the rate of doing work per unit mass flowing.

Since the system is of arbitrary extent and is moving in a prescribed manner in space, we know that U, Φ, and K are functions of x, y, z, and t. To get the total energy *within* the system, we again consider an arbitrary small element ΔV and add up the energy in all small elements to get the total energy in the system as illustrated in Fig. 2.4-1.

The energy in ΔV is approximately

$$(\hat{U} + \hat{\Phi} + \hat{K})\rho \,\Delta V \qquad (2.4\text{-}2)$$

where $\hat{U}$, $\hat{\Phi}$, and $\hat{K}$ are evaluated at any point in ΔV and Δt.

If we sum over all the small elemental volumes in V and take the limit as ΔV approaches zero, we have

$$\lim_{\Delta V \to 0} \sum_{V} (\hat{U} + \hat{\Phi} + \hat{K})\rho \,\Delta V = \int_{V} (\hat{U} + \hat{\Phi} + \hat{K})\rho \,dV \qquad (2.4\text{-}3)$$

Applying the definition of accumulation yields

$$\text{Accumulation} = \int_{V} (\hat{U} + \hat{\Phi} + \hat{K})\rho \,dV|_{t+\Delta t}$$
$$- \int_{V} (\hat{U} + \hat{\Phi} + \hat{K})\rho \,dV|_{t} \qquad (2.4\text{-}4)$$

Energy output minus energy input

We will first consider input and output energy *associated with mass*. Each increment of mass added to or removed from the system may carry with it kinetic, potential, and internal energy. In addition, energy is also transferred in the *process* of adding or removing mass. Consider, for ex-

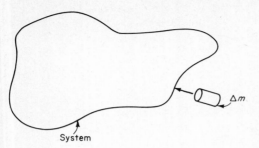

System   **Fig. 2.4-2** Addition of mass to system.

ample, the system shown in Fig. 2.4-2 and assume that we add a small mass Δm which we contain in a closed rigid container (a tin can, if you wish).

Note that to get the mass inside the system, we must push—that is, we must either compress some of the material in the system, thereby doing work on the mass already in the system and increasing its energy, or we must expand the system boundary, making the system do work on the surroundings and thus transfer energy across the boundary.

In the process of doing all this, we do not change the energy associated with Δm, but we cannot either add or remove mass without performing this so-called flow work. Thermodynamics teaches that it is convenient to combine this flow work with the energy associated with mass added or removed by defining the abstraction *enthalpy H*.

The flow work necessary to add a unit mass to our system is simply the pressure times the volume added:

$$\text{Flow work} = p \, \hat{V} \, \Delta m \tag{2.4-5}$$

Therefore, the energy associated with transferring a unit mass into the system becomes

$$(\hat{U} + \hat{\Phi} + \hat{K}) \, \Delta m + p \, \hat{V} \, \Delta m \tag{2.4-6}$$

but since we *define*

$$\hat{U} + p \, \hat{V} = \hat{H} \tag{2.4-7}$$

expression 2.4-6 becomes

$$(\hat{H} + \hat{\Phi} + \hat{K}) \, \Delta m \tag{2.4-8}$$

(This, of course, does not require that the material be encased in a container as above.)

The same argument holds for mass *removed* from the system. We can now combine the output and input terms for any small area ΔA and time Δt, writing the mass passing across the boundary as in Sec. 2.2:

$$\text{Output energy} - \text{input energy} = (\hat{H} + \hat{\Phi} + \hat{K})\rho(\mathbf{v} \cdot \mathbf{n}) \, \Delta A \, \Delta t$$
$$(2.4\text{-}9)$$

Adding the contributions of all the small areas to get the total for the whole surface and taking the limit as ΔA approaches zero:

$$\text{Output} - \text{input} \cong \lim_{\Delta A \to 0} \sum_A (\hat{H} + \hat{\Phi} + \hat{K})\rho(\mathbf{v} \cdot \mathbf{n}) \, \Delta A \, \Delta t$$

$$= \Delta t \int_A (\hat{H} + \hat{\Phi} + \hat{K})\rho(\mathbf{v} \cdot \mathbf{n}) \, dA \qquad (2.4\text{-}10)$$

We now have accounted for all energy *associated with mass* (both in the system and moving across the boundary). Next we must take into account energy which transfers across the boundary *not* associated with mass. We make an arbitrary division of this energy into two classes, (1) heat and (2) work.

By heat Q' we mean the *amount* of energy (for example, Btu) crossing the boundary (*a*) not associated with mass and (*b*) which flows as a result of a temperature gradient. [We add a prime (') to the symbol because we wish to reserve Q for *rate* of heat flow, for example, 1 Btu/hr.] This classification includes energy transferred by thermal radiation and conduction.[1] *Work* (*W'*—for example, ft lbf) is defined as all energy crossing the boundary (*a*) not associated with mass and (*b*) which does *not* transfer as a result of a temperature gradient. [We use the prime (') here for the same reason: we wish to reserve W for *rate* of work, for example, ft lbf/hr.] This energy is often called *shaft work* because it is commonly transmitted by a rotating shaft, but it also includes such things as energy transmitted by electrical leads crossing the boundary of the system (the electrons which transfer the energy are assumed to have a mass of zero in cases of interest to us).

By convention, heat *into* the system and work *out of* the system are regarded as positive (by analogy to a steam power plant, where one puts heat in and gets work out). Thermodynamics tells us that heat and work are not exact differentials—that is, they are not independent of path—but that their *difference* is. We therefore write small amounts of heat or work as $\delta Q'$ and $\delta W'$ to remind ourselves of this fact. If we add up the heat

[1] We use the common symbol Q for heat flow and volume flow. This notation is traditional.

and work for each differential area dA on the surface of our system we obtain:

$$\sum_A \delta Q' = Q' \tag{2.4-11}$$

$$\sum_A \delta W' = W' \tag{2.4-12}$$

If we now substitute in Eq. (2.4-1) for a small time Δt:

$$\Delta t \int_A (\hat{H} + \hat{\Phi} + \hat{K})\rho(\mathbf{v} \cdot \mathbf{n}) \, dA + \int_V (\hat{U} + \hat{\Phi} + \hat{K})\rho \, dV \Big|_{t+\Delta t}$$

$$- \int_V (\hat{U} + \hat{\Phi} + \hat{K})\rho \, dV \big|_t \cong Q' - W' \tag{2.4-13}$$

Dividing by Δt, taking the limit as Δt approaches zero, and defining

$$\lim_{\Delta t \to 0} \frac{Q'}{\Delta t} = Q \tag{2.4-14}$$

and

$$\lim_{\Delta t \to 0} \frac{W'}{\Delta t} = W \tag{2.4-15}$$

where Q and W are *rates*, we have

$$\int_A (\hat{H} + \hat{\Phi} + \hat{K})\rho(\mathbf{v} \cdot \mathbf{n}) \, dA + \frac{d}{dt} \int_V (\hat{U} + \hat{\Phi} + \hat{K})\rho \, dV = Q - W \tag{2.4-16}$$

which is the *macroscopic energy balance*.

Simplified forms of the macroscopic energy balance

We seldom use the overall energy balance in its form as shown above because it usually can be simplified for a particular situation. Many of these simplifications are so common that it is worth examining them in some detail. For example, we know that we can always rewrite the first term of the overall energy balance as

$$\int_A \hat{H} \rho(\mathbf{v} \cdot \mathbf{n}) \, dA + \int_A \hat{\Phi} \rho(\mathbf{v} \cdot \mathbf{n}) \, dA + \int_A \hat{K} \rho(\mathbf{v} \cdot \mathbf{n}) \, dA \tag{2.4-17}$$

We now proceed to examine ways of simplifying each of the above terms.

The potential energy term Let us first examine the second term in expression (2.4-17) to see what can be done to simplify things. First, we may observe that the imposed force fields of most interest to us as engineers are conservative; that is, the force may be expressed as the gradient of a *potential function*, defined up to an additive constant. This potential function gives the value of the potential energy *per unit mass* referred to some datum point (depending on the value of the additive constant).

For example, for the case illustrated in Fig. 2.4-3 where we have a gravitational field with reference point of zero energy at the center of the earth, the *potential function* Ω may be written as

$$\Omega = \frac{k}{r} \tag{2.4-18}$$

where k = an appropriate constant proportional to the mass of the earth
r = distance from the center of the earth
(This relation is valid only *outside* the earth's surface.)

The force produced by the field is the gradient of the potential, and in this case lies in the r direction (assuming the earth is a sphere, even though we know it is slightly pear shaped).

$$F_r = \frac{\partial \Omega}{\partial r} = -\frac{k}{r^2} = g \tag{2.4-19}$$

$(-k/r^2)$ is commonly called g, the acceleration of gravity. This says just what we expect; that is, the force is inversely proportional to the square of the distance from the center of the earth and directed toward the center. At sea level g is approximately 32.17 ft/sec².

The potential energy at a given point is the value of the gravitational potential at that point. Thus, we *should* evaluate changes in gravitational

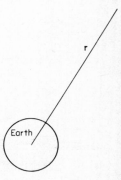

Fig. 2.4-3 Potential energy due to gravitational field.

potential energy by subtracting the values of the potential function at the two points; that is,

$$\text{Difference in } \hat{\Phi} \text{ at } r_1 \text{ and } r_2 = \Omega(r_2) - \Omega(r_1) = \frac{k}{r_2} - \frac{k}{r_1} \quad (2.4\text{-}20)$$

In practice, most processes take place over small *changes* in r, at large *values* of r, that is, at the surface of the earth, where $(r_2 - r_1)$ is of perhaps the order of 100 ft, while r_2 and r_1 themselves are of the order of 4,000 miles. Under such a circumstance Eq. (2.4-19) shows the acceleration of gravity to be nearly constant.

$$\frac{-k}{[(4,000 \text{ mile})(5,280 \text{ ft/mile})]^2} \cong \frac{-k}{[(4,000)(5,280 + 100)]^2}$$

Instead of taking differences in the potential function, therefore, we usually assume that g is constant and integrate force times distance using Eq. (2.4-19) to evaluate $\hat{\Phi}$:

$$\Delta\hat{\Phi} = \int_{r_{\text{ref}}}^{r} g \, dr = g(r - r_{\text{ref}}) \quad (2.4\text{-}21)$$

(For our problems, the reference radius is usually some radius near the surface rather than $r = 0$.)

This says $\hat{\Phi}$ is a *linear* function of r rather than the hyperbolic function equation (2.4-19), and simply means we are approximating the hyperbola over a short distance by a straight line, as shown in Fig. 2.4-4. Note that we must be careful in using this approximation for problems involving rockets, etc., where the changes in distance are large enough that g may change significantly.

What is the implication for the second term in expression (2.4-17)?

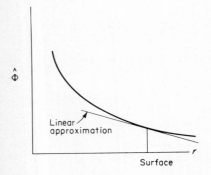

Fig. 2.4-4 Linear approximation for potential energy.

The assumption of constant g means that we can write, considering only gravitational fields,

$$\int_A \hat{\Phi}\rho(\mathbf{v} \cdot \mathbf{n}) \, dA = \int_A gr\rho(\mathbf{v} \cdot \mathbf{n}) \, dA$$

$$= g \int_{A_1} r\rho(\mathbf{v} \cdot \mathbf{n}) \, dA + g \int_{A_2} r\rho(\mathbf{v} \cdot \mathbf{n}) \, dA$$

$$= -gr_1 w_1 + gr_2 w_2 \qquad (2.4\text{-}22)$$

In addition, in processes of interest to us, we can often assume that ρ is constant across any single inlet or outlet (for example, across the cross section *individually* of each pipe crossing the system boundary). With this additional assumption, for each individual area,

$$gr \int_A \rho(\mathbf{v} \cdot \mathbf{n}) \, dA = gr\rho \int_A (\mathbf{v} \cdot \mathbf{n}) \, dA = gr\rho \langle v \rangle A \qquad (2.4\text{-}23)$$

where $\rho \langle v \rangle A$ gives the mass rate of flow through A. Note that r can be referred to any arbitrary point desired. We usually further replace r by the rectangular coordinate z.

The kinetic energy term   From physics we know that the kinetic energy of a unit mass is expressible as

$$\hat{K} = \frac{v^2}{2} \qquad (2.4\text{-}24)$$

The v which appears in this equation is with respect to some reference point. The v in our overall balances is the velocity *relative to the boundary of the system*[1] at the particular point. We are normally interested only in changes in kinetic energy, which are not affected by the reference point for velocity. We write the third term in expression (2.4-17) as

$$\int_A \hat{K}\rho(\mathbf{v} \cdot \mathbf{n}) \, dA = \int_A \frac{v^2}{2} \rho(\mathbf{v} \cdot \mathbf{n}) \, dA \qquad (2.4\text{-}25)$$

For a fluid where ρ and the angle between the velocity vector and the outward normal are constant across A we may write

$$\int_A \hat{K}\rho(\mathbf{v} \cdot \mathbf{n}) \, dA = \rho \int_A \frac{v^2}{2} (\mathbf{v} \cdot \mathbf{n}) \, dA = \rho \int_A \frac{v^2}{2} v(\cos \alpha) \, dA$$

$$(2.4\text{-}26)$$

---

[1] This is a somewhat awkward velocity to use unless the system boundary is stationary with respect to the coordinate system. Fortunately, in our examples we can almost always choose coordinate axes such that this is true.

Combining the velocities and removing $(\cos \alpha)/2$ from the integral

$$\int_A \hat{K} \rho (\mathbf{v} \cdot \mathbf{n}) \, dA = \rho \frac{\cos \alpha}{2} \int_A v^3 \, dA = \rho \frac{\cos \alpha}{2} \langle v^3 \rangle A \qquad (2.4\text{-}27)$$

(By definition the integral is what we have called the average times the area; see Appendix 2A.) Note that $\langle v^3 \rangle$ is not the same as $\langle v \rangle^3$ except for special cases:

$$\langle v^3 \rangle = \frac{\displaystyle\int_A v^3 \, dA}{\displaystyle\int_A dA} \neq \left(\frac{\displaystyle\int_A v \, dA}{\displaystyle\int_A dA} \right)^3 = \langle v \rangle^3 \qquad (2.4\text{-}28)$$

The enthalpy term   We now consider the first term in expression (2.4-17). If we apply to this term the assumption that enthalpy and density are constant across any single inlet or outlet area (such as the cross section of a pipe crossing the boundary), we may write, for each such single area,

$$\int_A \hat{H} \rho (\mathbf{v} \cdot \mathbf{n}) \, dA = \hat{H} \rho \int_A (\mathbf{v} \cdot \mathbf{n}) \, dA = \hat{H} \rho \langle v \rangle A \qquad (2.4\text{-}29)$$

To make it easier to relate temperature and enthalpy, we define a so-called mixing-cup or bulk temperature, which is an average for temperature *somewhat* analogous to the bulk velocity for velocity:

$$T_b = \frac{\displaystyle\int_A \rho T (\mathbf{v} \cdot \mathbf{n}) \, dA}{\displaystyle\int_A \rho (\mathbf{v} \cdot \mathbf{n}) \, dA} = \langle T \rangle \qquad (2.4\text{-}30)$$

The bulk temperature is *not* simply an area average, but is an average with respect to $\rho (\mathbf{v} \cdot \mathbf{n})$, which is the local mass flow rate.

$$T_b = \frac{\langle T \rho (\mathbf{v} \cdot \mathbf{n}) \rangle}{\langle \rho (\mathbf{v} \cdot \mathbf{n}) \rangle} \qquad (2.4\text{-}31)$$

For constant-density systems, which we frequently encounter, the density will cancel, and the denominator is simply $\langle v \rangle A$, so that we may write

$$T_b \langle v \rangle A = \int_A T (\mathbf{v} \cdot \mathbf{n}) \, dA \qquad (2.4\text{-}32)$$

The bulk temperature is that temperature one would obtain by catching the fluid in an insulated cup at the existing flow rate, mixing it perfectly, and measuring the resulting temperature.

Note: It is also possible to define an average which is *directly* analogous to the bulk velocity as

$$T_{\mathrm{avg}} = \frac{\displaystyle\int_A T \, dA}{\displaystyle\int_A dA} \tag{2.4-33}$$

This average temperature is the one obtained by *instantly* isolating some infinitesimally thin cross section of a homogeneous flowing fluid by two insulating baffles as shown in Fig. 2.4-5 and then mixing the fluid so isolated. This average does not take into account the variation of flow velocity across the pipe. A little reflection should make it clear that the bulk *temperature* is not so much a temperature average as an *energy flux* average.

A common special case of the macroscopic energy balance is that of a *steady-state* system with a single inlet, a single outlet, and no variation in $\hat{H}$ or $\hat{\Phi}$ across either inlet or outlet. For this case Eq. (2.4-16) reduces as follows:

$$\hat{H} \int_A \rho(\mathbf{v} \cdot \mathbf{n}) \, dA + \hat{\Phi} \int_A \rho(\mathbf{v} \cdot \mathbf{n}) \, dA + \int_A \frac{v^2}{2} \rho(\mathbf{v} \cdot \mathbf{n}) \, dA = Q - W \tag{2.4-34}$$

Recognizing the first two integrals as the mass flow rates:

$$\hat{H}_2 w_2 - \hat{H}_1 w_1 + \hat{\Phi}_2 w_2 - \hat{\Phi}_1 w_1 + \int_{A_1} \frac{v^2}{2} \rho(\mathbf{v} \cdot \mathbf{n}) \, dA$$
$$+ \int_{A_2} \frac{v^2}{2} \rho(\mathbf{v} \cdot \mathbf{n}) \, dA = Q - W \tag{2.4-35}$$

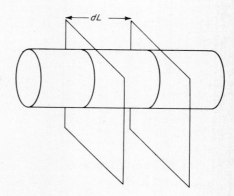

Fig. 2.4-5 Area average temperature of a flowing fluid.

But for steady state, a macroscopic mass balance tells us:

$$w_2 = \int_{A_2} \rho(\mathbf{v} \cdot \mathbf{n}) \, dA = -\int_{A_1} \rho(\mathbf{v} \cdot \mathbf{n}) \, dA = w_1 \qquad (2.4\text{-}36)$$

We may, therefore, base Eq. (2.4-35) on a unit mass flowing by dividing by Eq. (2.4-36).

$$\Delta\hat{H} + \Delta\hat{\Phi} - \frac{\displaystyle\int_{A_1} (v^2/2)\rho(\mathbf{v} \cdot \mathbf{n}) \, dA}{\displaystyle\int_{A_1} \rho(\mathbf{v} \cdot \mathbf{n}) \, dA} + \frac{\displaystyle\int_{A_2} (v^2/2)\rho(\mathbf{v} \cdot \mathbf{n}) \, dA}{\displaystyle\int_{A_2} \rho(\mathbf{v} \cdot \mathbf{n}) \, dA} = \hat{Q} - \hat{W}$$

$$(2.4\text{-}37)$$

or, multiplying and dividing the third and fourth terms on the left by $\displaystyle\int_{A} dA$ for the appropriate area,

$$\Delta\hat{H} + \Delta\hat{\Phi} + \frac{1}{2} \Delta \frac{\langle v^3 \rangle}{\langle v \rangle} = \hat{Q} - \hat{W} \qquad (2.4\text{-}38)$$

The kinetic energy term above is somewhat cumbersome to integrate. For turbulent flow, a velocity profile with constant value equal to $\langle v \rangle$ is often assumed (plug flow), and the term taken to be

$$\frac{1}{2} \Delta \frac{\langle v^3 \rangle}{\langle v \rangle} \cong \frac{1}{2} \Delta \langle v \rangle^2 \qquad (2.4\text{-}39)$$

The error involved in this assumption is examined in detail in Appendix 2A.

Example 2.4-1 Energy balance Water as liquid enters a boiler operating at a steady state and leaves as steam as shown. Calculate both the kinetic and potential energy changes in Btu/lb$m$. (See Fig. 2.4-6.)

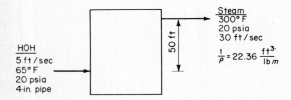

Fig. 2.4-6 Sketch for Example 2.4-1.

Solution

$$\hat{K}_1 = \frac{v^2}{2}\bigg|_1 = \left[\frac{(5)^2 \text{ ft}^2}{2 \text{ sec}^2}\right]\left(\frac{1 \text{b} f \text{ sec}^2}{32.2 \text{ 1b} m \text{ ft}}\right)\left(\frac{\text{Btu}}{778 \text{ ft lb} f}\right)$$
$$= 0.0005 \text{ Btu/lb} m$$

$$\hat{K}_2 = \frac{v^2}{2}\bigg|_2 = \frac{(30)^2}{2}\frac{1}{32.2}\frac{1}{778} = 0.018 \text{ Btu/lb} m$$

Note that this is quite a small amount of energy compared to ΔH_v, the latent heat of vaporization for water, which is of the order of 1000 Btu/lb$m$, or even compared to the heat required to raise the temperature 1° either for liquid water (about 1 Btu/lb$m$), or for steam (about 0.5 Btu/lb$m$).

$$\Delta\hat{\Phi} = g\,\Delta z = \left(32.2\,\frac{\text{ft}}{\text{sec}^2}\right)(50 \text{ ft})\left(\frac{\text{lb} f \text{ sec}^2}{32.2 \text{ lb} m \text{ ft}}\right)\left(\frac{\text{Btu}}{778 \text{ ft lb} f}\right)$$
$$= 0.0642 \text{ Btu/lb} m$$

Again note that thermally this is a very small quantity even for the considerable elevation change considered.

Figures 2.4-7 and 2.4-8 show the approximate magnitude of these quantities for ordinary velocity and height changes.

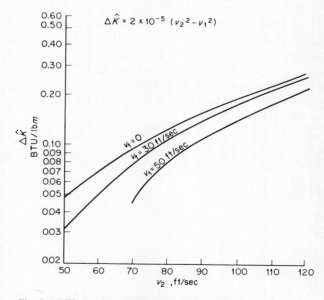

Fig. 2.4-7 Thermal equivalent of kinetic energy.

$\Delta\hat{\Phi} = 1.29 \times 10^{-3}\Delta z$

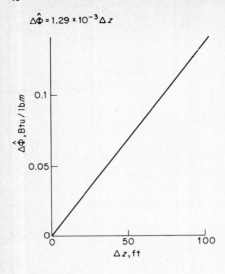

Fig. 2.4-8 Thermal equivalent of potential energy.

□ □ □

Example 2.4-2 Energy balance on a boiler Water enters a boiler at 5 psig, 65°F, through a 4-in. pipe at a bulk velocity of 5 ft/sec and steam leaves at 600°F and a pressure of 400 psig. At what rate must heat be supplied to the boiler under steady-state conditions?

Solution We calculate the heat load by applying the overall energy balance. A convenient system to choose is the one shown in Fig. 2.4-9. The integral over the surface vanishes everywhere except over the inlet and outlet because elsewhere there is no velocity component normal to the surface. At steady state there is no variation with time, so that the derivative of the volume integral vanishes. No shaft work is done, and so if we denote the inlet by (1) and the outlet by (2), the overall energy balance reduces to

$$\int_{A_1} (\hat{H} + \hat{\Phi} + \hat{K})\rho(\mathbf{v} \cdot \mathbf{n}) \, dA + \int_{A_2} (\hat{H} + \hat{\Phi} + \hat{K})\rho(\mathbf{v} \cdot \mathbf{n}) \, dA = Q$$

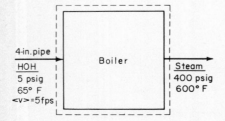

Fig. 2.4-9 Sketch for Example 2.4-2.

Neglecting changes in potential and kinetic energy and assuming that $\hat{H}$ and ρ are constant across the inlet and across the outlet, we have

$$(a) \qquad \hat{H}_1\rho_1 \int_{A_1} (\mathbf{v} \cdot \mathbf{n}) \, dA + \hat{H}_2\rho_2 \int_{A_2} (\mathbf{v} \cdot \mathbf{n}) \, dA = Q$$

From a table of thermodynamic properties of steam[1] we find that

$$\hat{H}_1 = 33 \text{ Btu/lb}m \qquad \rho_1 = \frac{1}{0.01604} \text{ lb}m/\text{ft}^3$$

$$\hat{H}_2 = 1307 \text{ Btu/lb}m \qquad \rho_2 = \frac{1}{1.4777} \text{ lb}m/\text{ft}^3$$

An overall mass balance yields

$$\rho_1 \int_{A_1} (\mathbf{v} \cdot \mathbf{n}) \, dA + \rho_2 \int_{A_2} (\mathbf{v} \cdot \mathbf{n}) \, dA = 0$$

but since the dot product is negative for A_1

$$\int_{A_1} (\mathbf{v} \cdot \mathbf{n}) \, dA = - \langle v_1 \rangle A_1$$

Thus

$$-\rho_1 \langle v_1 \rangle A_1 + \rho_2 \int_{A_2} (\mathbf{v} \cdot \mathbf{n}) \, dA = 0$$

But we know that

$$\langle v_1 \rangle = 5 \text{ ft/sec}$$

and

$$A_1 = \frac{\pi D_1^2}{4} = \frac{\pi}{4} \left(\frac{4.026}{12}\right)^2 = 0.08840 \text{ ft}^2$$

$$\rho_1 \langle v_1 \rangle A_1 = \rho_2 \int_{A_2} (\mathbf{v} \cdot \mathbf{n}) \, dA$$

$$= \left(\frac{1}{0.01604} \frac{\text{lb}m}{\text{ft}^3}\right) \left(5 \frac{\text{ft}}{\text{sec}}\right) (0.08840 \text{ ft}^2) = 27.5 \text{ lb}m/\text{sec}$$

[1] J. H. Keenan and F. G. Keyes, "Thermodynamic Properties of Steam," John Wiley & Sons, Inc., New York, 1961.

Substituting in equation (a)

$$Q = -\left(33 \; \frac{\text{Btu}}{\text{lb}m}\right)\left(27.5 \; \frac{\text{lb}m}{\text{sec}}\right) + \left(1307 \; \frac{\text{Btu}}{\text{lb}m}\right)\left(27.5 \; \frac{\text{lb}m}{\text{sec}}\right)$$

$$= 35{,}000 \; \text{Btu/sec}$$

□ □ □

Example 2.4-3 Energy balance on a flowing system Water is pumped through a "perfectly" insulated pipe from a tank on the ground to one of identical dimensions on the second floor (15 ft higher) of a building by a pump which supplies 5 hp to the fluid. The pipe is of uniform diameter, and the valve of the pump outlet is closed until a flow of 50 gal/min is obtained. What is the temperature rise of the water?

Solution Applying the overall energy balance with $Q = 0$, $W = -(5 \text{ hp}) \, 550 \, (\text{ft lb}f)/(\text{hp sec})$ and assuming $\rho = \text{constant}$

$$\hat{H}_2 w_2 + gz_2 w_2 - \hat{H}_1 w_1 - gz_1 w_1 = -W$$

Since the overall mass balance tells us that $w_2 = w_1$,

$$w_1[(\hat{H}_2 - \hat{H}_1) + g(z_2 - z_1)] = -W$$

From thermodynamics we know that $\Delta\hat{H} = \hat{C}_p \Delta T$, and the heat capacity of water is 1 Btu/ lb$m$ °F.

$$w_1(\hat{C}_p \, \Delta T + g \, \Delta z) = -W$$

Neglecting the effect of temperature on the density of water:

$$\Delta T = \left(-\frac{W}{w_1} - g \, \Delta z\right)\frac{1}{\hat{C}_p}$$

$$= \left[-5 \left(550 \; \frac{\text{ft lb}f}{\text{sec}}\right)\left(\frac{\text{min}}{50 \text{ gal}}\right)\left(\frac{\text{gal}}{8.33 \text{ lb}m}\right)\left(\frac{60 \text{ sec}}{\text{min}}\right)\right.$$

$$\left. - \left(\frac{32.2 \text{ ft}}{\text{sec}^2}\right)(15 \text{ ft})\left(\frac{\text{lb}f \text{ sec}^2}{32.2 \text{ lb}m \text{ ft}}\right)\right]\left(\frac{\text{lb}m}{1 \text{ Btu}}\right)(°\text{F})\left(\frac{\text{Btu}}{778 \text{ ft lb}f}\right)$$

$$= 0.91°\text{F}$$

Note the small ΔT; it takes an extremely large loss in *mechanical* energy to give a significant rise in temperature.

□ □ □

Example 2.4-4   Energy balance on a mixed tank   Figure 2.4-10 shows an insulated, perfectly mixed tank which contains a heating coil. Water flows into the tank at a rate of 10 ft³/min and out at the same rate. The volume of liquid in the tank is 100 ft and the initial temperature is 70°F. The water flowing into the tank is at 150°F. If the heater adds 5000 Btu/min to the tank and the horsepower actually added by the agitator (i.e., the power into the motor less losses in the motor, etc.) is 5 hp, calculate the tank temperature as a function of time.

Solution   The overall energy balance applied to the system shown gives, neglecting potential and kinetic energy changes,

$$(a) \qquad \int_{A_1} \hat{H} \rho (\mathbf{v} \cdot \mathbf{n}) \, dA + \int_{A_2} \hat{H} \rho (\mathbf{v} \cdot \mathbf{n}) \, dA$$
$$+ \frac{d}{dt} \int_V \hat{U} \rho \, dV = Q - W$$

Examining each term:

$$\int_{A_1} \hat{H} \rho (\mathbf{v} \cdot \mathbf{n}) \, dA = \rho \hat{H}_1 \int_{A_1} (\mathbf{v} \cdot \mathbf{n}) \, dA = - \hat{H}_1 \, w_1$$

$$\int_{A_2} \hat{H} \rho (\mathbf{v} \cdot \mathbf{n}) \, dA = \rho \hat{H}_2 \int_{A_2} (\mathbf{v} \cdot \mathbf{n}) \, dA = \hat{H}_2 \, w_2$$

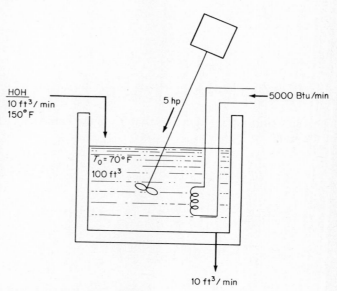

Fig. 2.4-10 Sketch for Example 2.4-4.

A mass balance shows $w_1 = w_2$ and $V = $ constant, so that

$$\frac{d}{dt} \int_V \hat{U} \rho \, dV = \rho \frac{d}{dt} \left[\hat{U} \int_V dV \right] = \rho \frac{d(\hat{U}V)}{dt} = \rho V \frac{d\hat{U}}{dt}$$

From thermodynamics $\hat{U} = \hat{C}_V (T - T_0)$, and since the fluid is essentially incompressible $\hat{C}_p \cong \hat{C}_V$

$$\frac{d\hat{U}}{dt} = \hat{C}_V \frac{dT}{dt} = \hat{C}_p \frac{dT}{dt}$$

Then

$$\frac{d}{dt} \int_V \hat{U} \rho \, dV = \rho V \hat{C}_p \frac{dT}{dt}$$

Substituting in equation (a)

$$-\hat{H}_1 w_1 + \hat{H}_2 w_1 + \rho V \hat{C}_p \frac{dT}{dt} = Q - W$$

We can write $\hat{H}$ as

$$\hat{H} = \hat{C}_p (T - T_0)$$

or

$$\hat{H}_1 = \hat{C}_p (T_1 - T_0)$$
$$\hat{H}_2 = \hat{C}_p (T_2 - T_0)$$

Then substituting, recognizing that T_2 is the same as T, the temperature of the tank:

$$w_1 \hat{C}_p [(T - T_0) - (T_1 - T_0)] + \rho V \hat{C}_p \frac{dT}{dt} = Q - W$$

$$\left(\frac{10 \text{ ft}^3}{\text{min}} \right) \left(\frac{62.4 \text{ lb}m}{\text{ft}^3} \right) \left(\frac{1 \text{ Btu}}{\text{lb}m \, °\text{F}} \right) (T - 150°\text{F})$$

$$+ \left(62.4 \, \frac{\text{lb}m}{\text{ft}^3} \right) (100 \text{ ft}^3) \left(\frac{1 \text{ Btu}}{\text{lb}m \, °\text{F}} \right) \left(\frac{dT}{dt} \right) \left(\frac{°\text{F}}{\text{sec}} \right)$$

$$= \left(5000 \, \frac{\text{Btu}}{\text{min}} \right) - (-5 \text{ hp}) \left(42.44 \, \frac{\text{Btu}}{\text{hp min}} \right)$$

Collecting:

$$624(T - 150) + 6,240 \frac{dT}{dt} = 5,000 + 212$$

[Note how small W is (212) compared with Q(5,000) in terms of equivalent thermal energy.] Integrating:

$$6,240 \frac{dT}{dt} = 624 \, (158 - T)$$

$$10 \int_{70}^{T_f} \frac{dT}{158 - T} = \int_0^{t_f} dt$$

$$-10 \ln \frac{158 - T_f}{158 - 70} = t_f$$

$$T_f = 158 - 88 \, e^{-t_f/10}$$

Notice that the steady-state temperature ($t \rightarrow \infty$) is 158°; in other words, the heat and work added are sufficient to raise the temperature by 8°F from inlet conditions.

□ □ □

The mechanical energy balance

Engineers are frequently concerned with problems in which particular forms of energy in the macroscopic energy balance are of special interest: the work term, the kinetic energy terms, the potential energy terms, and the flow work part of the enthalpy terms. These terms all represent a special type of energy—mechanical energy—that is, either work or a form that may be directly converted into work. The other terms in the energy balance (the internal energy and heat terms) do not permit simple conversion into *work*, as we know from thermodynamics.

To obtain *work* from the heat or internal energy forms we must go through some sort of heat engine which is subject to severe efficiency limitations from the second law of thermodynamics. These efficiency limits depend on the working temperatures of the heat engine. Conversely, the *mechanical* energy terms have no such limitation. We can convert, for example, a given change in potential energy very nearly completely into work (completely, if we could eliminate friction). We cannot accomplish the same goal with heat or internal energy because of the second law of thermodynamics. Heat and internal energy represent "lower-quality" energy, and so it is not surprising that we might wish to make a *mechanical* energy balance in order to see what is happening to this "high-quality" energy in our processes.

We make this balance the same way we made earlier balances, that is, by applying Eq. (2.1-1). We have already written down the output and the input terms of mechanical energy for an arbitrary system in our derivation of the macroscopic energy balance:

$$\begin{matrix} \text{Output} - \text{Input} \\ \text{of } \textit{mechanical} \\ \text{energy} \end{matrix} = \int_A (\hat{\Phi} + \hat{K} + p\hat{V})\rho(\mathbf{v} \cdot \mathbf{n}) \, dA - W \qquad (2.4\text{-}40)$$

We have also developed the accumulation terms:

$$\begin{matrix} \text{Accumulation} \\ \text{of } \textit{mechanical} \\ \text{energy} \end{matrix} = \frac{d}{dt}\left[\int_V (\hat{\Phi} + \hat{K})\rho \, dV \right] \qquad (2.4\text{-}41)$$

Mechanical energy is obviously *not* a conserved quantity.

If I apply the brakes while driving a car, the kinetic energy of the system (the car) gets converted to internal energy of the system at the point where the brake shoes rub the drums (or disks) and to heat where the tires rub the road. This energy can obviously not be recovered easily as work—say to raise the elevation of the car. If I had *not* applied the brakes, however, I could have used the kinetic energy to coast up a hill and thus accomplish an elevation (potential energy) change.

Energy which is converted to heat or internal energy we call *lost work*, because this represents a loss in *mechanical* energy. It would be more logical, perhaps, to call this sum *lost mechanical energy*, but we will not assail tradition. *The reader is cautioned not to confuse lost work with lost energy.* *Total* energy is still conserved—mechanical energy is not. The lost work, then, is our generation term (here a *negative* generation) in Eq. (2.1-1). Substituting in this equation, rearranging slightly, and denoting lost work by *lw*,

$$\int_A (\hat{\Phi} + \hat{K} + p\hat{V})\rho(\mathbf{v} \cdot \mathbf{n}) \, dA + \frac{d}{dt}\left[\int_V (\hat{\Phi} + \hat{K})\rho \, dV \right] = - \, lw - W$$

$$(2.4\text{-}42)$$

We have adopted the same sign convention for *lw* as for *W*: input is negative, output positive. The kinetic energy term may be written for the angle between the velocity and the outward normal either $0°$ or $180°$ and for an incompressible fluid:

$$\frac{\int_A \frac{1}{2} v^2 \rho (\mathbf{v} \cdot \mathbf{n}) \, dA}{\int_A \rho (\mathbf{v} \cdot \mathbf{n}) \, dA} = \frac{1}{2} \frac{\langle v^3 \rangle}{\langle v \rangle} \tag{2.4-43}$$

The denominator arises because we are discussing kinetic energy *per pound* of fluid $\hat{K}$, not total kinetic energy K. As was discussed earlier this whole term is frequently approximated by $\frac{1}{2} \langle v \rangle^2$ in turbulent flow. (See Appendix 2A for a discussion of the error involved.)

The most common form of the mechanical energy balance is for a steady-state system with no variation in $\hat{\Phi}$, $\hat{K}$ (flat velocity profiles), or $p \hat{V}$ across individual inlet and outlet areas, for which we obtain:

$$\sum_i (\hat{\Phi} + \hat{K} + p \hat{V}) \int_{A_i} \rho (\mathbf{v} \cdot \mathbf{n}) \, dA = -lw - W \tag{2.4-44}$$

or, for a single inlet and outlet,

$$(\hat{\Phi}_2 + \hat{K}_2 + p_2 \hat{V}_2) w_2 - (\hat{\Phi}_1 + \hat{K}_1 + p_1 \hat{V}_1) w_1 = -lw - W \tag{2.4-45}$$

which may be written (since $w_2 = w_1$, an overall mass balance at steady state):

$$\Delta \hat{\Phi} + \Delta \hat{K} + \Delta (p \hat{V}) = -\widehat{lw} - \widehat{W} \tag{2.4-46}$$

Since we have divided by the mass flow rate, the lost work and work terms are per pound of fluid flowing and the symbols therefore incorporate a caret ($\hat{}$). For an incompressible fluid with the only body force being gravity:

$$\Delta \left(\frac{\langle v \rangle^2}{2} \right) + g \, \Delta z + \frac{\Delta p}{\rho} = -\widehat{lw} - \widehat{W} \tag{2.4-47}$$

In the absence of friction and shaft work this is called the Bernoulli equation:

$$\Delta \left(\frac{\langle v \rangle^2}{2} \right) + g \, \Delta z + \frac{\Delta p}{\rho} = 0 \tag{2.4-48}$$

Example 2.4-5 Mechanical energy and pole vaulting The event of pole vaulting can be analyzed in terms of the mechanical energy balance. Part of the idea in this event is to convert the kinetic energy of a running man into potential energy (height above the ground). This, of course, is not the whole story, since the man also supplies additional energy via his legs, shoulders, and arms.

It is instructive to see how fast a man would have to run in order to pole-vault 17 ft high if he could not supply any energy other than his kinetic energy. We assume further that there is no dissipation of energy by his arms, the pole, wind resistance, the pole sliding into the socket, etc. We also assume that he converts *all* his kinetic energy to potential energy, which means that he would no longer really have the linear velocity to carry him *across* the bar, and we neglect angular rotation (pole vaulters rotate as they vault).

Solution   Take the man and his pole as the system. Applying the mechanical energy balance, we see that there are no convective terms, no heat, no work, and no change in internal energy of the system, leaving us with

$$\frac{d}{dt}\left[\int_V (\hat{\Phi} + \hat{K})\rho\, dV\right] = 0$$

Let C = constant and integrate with respect to t:

$$\int_V (\hat{\Phi} + \hat{K})\rho\, dV = C$$

Assume that all the man's mass m is concentrated at his center of gravity:

$$mgz + \frac{mv^2}{2} = C$$

To evaluate C let $v = v_1$ where $z = 0$ (on the ground)

$$C = mg(0) + \frac{mv_1^2}{2}$$

$$= \frac{mv_1^2}{2}$$

Thus,

$$mgz + \frac{mv^2}{2} = \frac{mv_1^2}{2}$$

But where $z = 17$ ft, $v = 0$, so that

$$\left(32.2 \frac{\text{ft}}{\text{sec}^2}\right)(17 \text{ ft}) + 0 = \frac{v_1{}^2}{2}$$

$$v_1 = 33 \text{ ft/sec} = 22.5 \text{ mph}$$

This is equivalent to running at a constant speed which would cover 100 yards in 9.3 sec—pretty fast. Obviously legs, arms, and shoulders are important.

The man's center of gravity ($z = 0$) is not at ground level but perhaps at 3 ft above ground, and his center of gravity need only clear the bar by perhaps 6 in., so that this corresponds to more than a 17-ft pole vault. We also should be able to add on the height which the man can raise his center of gravity by jumping; and so things are a little easier than we have calculated.

□ □ □

Example 2.4-6 Lost work calculation Water is flowing through a fouled 500 ft length of 3-in. schedule 40 pipe at a rate of 230 gal/min (see Fig. 2.4-11). The pipe expands to a 4-in. pipe as shown, and the gage pressures at points 1 and 2 are as shown. Calculate the lost work in the pipeline.

Solution Applying the mechanical energy balance in the form of Eq. (2.4-47), noting there is no shaft work:

$$(a) \qquad \Delta\left(\frac{\langle v \rangle^2}{2}\right) + g\,\Delta z + \frac{\Delta p}{\rho} = -\widehat{l w}$$

Using the density of water as 8.33 lb$m$/gal and the flow cross section as 0.0513 ft² at point 1 and 0.08840 ft² at point 2, we calculate the in-

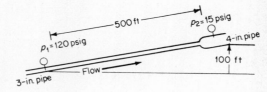

Fig. 2.4-11 Sketch for Example 2.4-6.

dividual terms as follows:

$$\langle v_1 \rangle = \frac{Q}{A} = \left(\frac{230 \text{ gal}}{\text{min}} \right) \left[\frac{\text{ft}^3}{7.48 \text{ gal}} \right] \left(\frac{1}{0.0513 \text{ ft}^2} \right) \left(\frac{\text{min}}{60 \text{ sec}} \right) = \frac{10 \text{ ft}}{\text{sec}}$$

$$\langle v_2 \rangle = \langle v_1 \rangle \frac{A_1}{A_2} = 10 \frac{0.0513}{0.08840} = 5.82 \text{ ft/sec}$$

$$\frac{\Delta \langle v \rangle^2}{2} = \left[\frac{(10^2 - 5.82^2) \text{ ft}^2}{2 \text{ sec}^2} \right] \left(\frac{\text{lbf sec}^2}{32.2 \text{ lbm ft}} \right) = -1.03 \text{ ft lbf/lbm}$$

$$g \, \Delta z = \left[(32.2) \frac{\text{ft}}{\text{sec}^2} (100 - 0) \text{ ft} \right] \left(\frac{\text{lbf sec}^2}{32.2 \text{ lbm ft}} \right) = 100 \text{ ft lbf/lbm}$$

$$\frac{\Delta p}{\rho} = \left[\frac{(15 - 120) \text{ lbf}}{8.33 \text{ lbm}} \right] \left(144 \frac{\text{ft}}{7.48} \right) = -243 \text{ ft lbf/lbm}$$

Substituting in equation (a)

$$-\widehat{lw} = -243 - 1.03 + 100 = -142 \text{ ft lbf/lbm}$$

Note that the kinetic energy term is negligible compared to the other terms.

This amount of lost work, if all dissipated in the fluid as internal energy, would raise the temperature:

$$\Delta T = \left(\frac{142 \text{ ft lbf}}{\text{lbm}} \right) \left(\frac{\text{Btu}}{778 \text{ ft lbf}} \right) \left(\frac{\text{lbm } ^\circ\text{F}}{1 \text{ Btu}} \right)$$

$$= 0.18 \, ^\circ\text{F}$$

If there were *no* flow, the Δp would be (since there would be no lost work or KE change):

$$\frac{\Delta p}{\rho} = -g \, \Delta z$$

$$\Delta p = -g \rho \, \Delta z$$

$$= -\left[(32.2) \frac{\text{ft}}{\text{sec}^2} (62.4) \frac{\text{lbm}}{\text{ft}^3} \right] [(100 - 0) \text{ ft]} \left(\frac{\text{lbf sec}_2}{32.2 \text{ lbm ft}} \right)$$

$$\left(\frac{\text{ft}^2}{144 \text{ in.}^2} \right)$$

$$= -43.3 \text{ lbf/in.}^2$$

If the pipe were mounted *horizontally* (assuming this does not change the lost work significantly) with flow still at 230 gal/min, the pressure drop would be

$$\frac{\Delta\langle v\rangle^2}{2} + \frac{\Delta p}{\rho} = -\widehat{lw}$$

$$\Delta p = \rho\left(\frac{-\Delta\langle v\rangle^2}{2} - \widehat{lw}\right)$$

Neglecting the small KE term:

$$\Delta p = \left(\frac{62.4\ \text{lbm}}{\text{ft}^3}\right)\left(-142\frac{\text{ft lbf}}{\text{lbm}}\right)\left(\frac{\text{ft}^3}{144\ \text{in.}^2}\right)$$

$$= -61.5\ \text{lbf/in.}^2$$

Since the pressure drop for *clean* horizontal 3-in. pipe at this flow of water is about (6.5 psi)/(100 ft), we would get about 32.5 psi pressure drop if our pipe were clean and not fouled. This is very bad fouling or more than one gets in the typical plant.

Note that if the flow were in the *opposite* direction, at *low* flow rates the flow would actually be from a low pressure to a high pressure because of the potential energy change.

□ □ □

2.5   THE DIFFERENTIAL ENERGY BALANCE[1]

In the following we discuss the transition from the integral equation for the energy balance (first law of thermodynamics) to the differential equation for the energy balance—from the macroscopic to the microscopic equation. The resulting differential balance is one of the starting equations for the study of transport phenomena.

The macroscopic energy balance was determined above to be

$$\int_A (\hat{H} + \hat{\Phi} + \hat{K})\rho(\mathbf{v}\cdot\mathbf{n})\,dA + \frac{d}{dt}\int_V (\hat{U} + \hat{\Phi} + \hat{K})\rho\,dV = Q - W$$

$$(2.5\text{-}1)$$

[1] This section provides a transition for subsequent courses in transport phenomena—the differential approach referred to in Sec. 1.4.

If we consider Q as the rate of heat transfer to the control volume expressed in terms of the heat flux $\mathbf{q}$, and W as the rate of surface work done on the control volume in terms of the varying stress vector $\mathbf{t}^n$ which acts at each point on the surface described by the outward normal $\mathbf{n}$, Eq. (2.5-1) may be rewritten:

$$\int_A (\hat{U} + \hat{\Phi} + \hat{K})\rho(\mathbf{v} \cdot \mathbf{n}) \, dA + \frac{d}{dt} \int_V (\hat{U} + \hat{\Phi} + \hat{K})\rho \, dV + \int_A (\mathbf{q} \cdot \mathbf{n}) \, dA$$
$$- \int_A (\mathbf{t}^n \cdot \mathbf{v}) \, dA = 0 \qquad (2.5\text{-}2)$$

where we have split the enthalpy into internal energy and flow work.

The last term in Eq. (2.5-2) represents the work done by surface forces.[1] Let $\hat{\Omega} = (\hat{U} + \hat{\Phi} + \hat{K})$ and if we also assume as in Sec. 2.3 a stationary volume, move the time derivative inside the integral, and apply the divergence theorem, we have

$$\int_V \left[\frac{\partial \rho \hat{\Omega}}{\partial t} + (\boldsymbol{\nabla} \cdot \rho \mathbf{v} \hat{\Omega}) + (\boldsymbol{\nabla} \cdot \mathbf{q}) - (\boldsymbol{\nabla} \cdot [\mathbf{T} \cdot \mathbf{v}]) \right] dV = 0 \quad (2.5\text{-}3)$$

Since the limits of integration are arbitrary the terms in brackets must be zero and the differential energy balance is

$$\frac{\partial \rho \hat{\Omega}}{\partial t} + (\boldsymbol{\nabla} \cdot \rho \mathbf{v} \, \hat{\Omega}) = -(\boldsymbol{\nabla} \cdot \mathbf{q}) + (\boldsymbol{\nabla} \cdot [\mathbf{T} \cdot \mathbf{v}]) \qquad (2.5\text{-}4)$$

$$(a) \qquad\qquad (b) \qquad\qquad\quad (c) \qquad\qquad\quad (d)$$

where terms (a), (b), (c), (d) are rates of change per unit volume representing

(a) Gain of energy
(b) Energy input by convection
(c) Energy input by conduction
(d) Surface work done on system

[1] This term involves a tensor

$$\mathbf{t}^n = [\mathbf{n} \cdot \mathbf{T}]$$

where $\mathbf{T}$ is the stress tensor. You are already familiar with two special types of tensors—zero order (scalars) and first order (vectors). The stress tensor is a second-order tensor. A second-order tensor simply associates a *vector* with each direction in space. The stress tensor is an entity that gives one back the stress vector at a point on a surface when it is supplied with the outward normal (direction in space).

For many engineering applications we prefer to have the energy equation in terms of *thermal* energy using the temperature and heat capacity of the fluid in question. This is accomplished by substituting for the mechanical energy in terms of thermal energy, writing internal energy as a total derivative, and substituting from thermodynamics in terms of T, p, and $\hat{C}_V$ such that

$$\rho\hat{C}_V\frac{\partial T}{\partial t} + \rho\hat{C}_V(\mathbf{v}\cdot\mathbf{\nabla})T = -(\mathbf{\nabla}\cdot\mathbf{q}) - T\left(\frac{\partial p}{\partial T}\right)_\rho(\mathbf{\nabla}\cdot\mathbf{v}) \qquad (2.5\text{-}5)$$

The thermal energy equation in component form is

$$\rho\hat{C}_V\left(\frac{\partial T}{\partial t} + v_x\frac{\partial T}{\partial x} + v_y\frac{\partial T}{\partial y} + v_z\frac{\partial T}{\partial z}\right) = -\left(\frac{\partial q_x}{\partial x} + \frac{\partial q_y}{\partial y} + \frac{\partial q_z}{\partial z}\right)$$
$$- T\left(\frac{\partial p}{\partial T}\right)_\rho\left(\frac{\partial v_x}{\partial x} + \frac{\partial v_y}{\partial y} + \frac{\partial v_z}{\partial z}\right)$$
$$(2.5\text{-}6)$$

For most problems of interest, we want to relate temperature profiles to heat fluxes, and so we combine Fourier's law with Eq. (2.5-6). Fourier's law, which is discussed in Chap. 8, relates the transport properties of the individual fluid to the heat flux produced by a temperature gradient. The investigation of solutions of the differential energy equation for various systems (boundary conditions) and its forms where energy is transferred by conduction and convection is the topic of energy transport as a transport phenomenon.[1] For example, the resulting equation of change for an incompressible fluid at constant pressure in one-dimensional flow is

$$\rho\hat{C}_p\frac{\partial T}{\partial t} + \rho\hat{C}_p v_x\frac{\partial T}{\partial x} = k\frac{\partial^2 T}{\partial y^2} \qquad (2.5\text{-}7)$$

where k is the coefficient of thermal conductivity. (Note that for a solid $v_x = 0$ and the second term drops out.)

2.6 THE MACROSCOPIC MOMENTUM BALANCE

We now examine another balance, that of momentum. Momentum, in contrast to mass and energy, is not in general a conserved quantity. It is, moreover, a vector quantity, and so we will now write vector equations.

[1] R. B. Bird, W. E. Stewart, and E. N. Lightfoot, "Transport Phenomena," John Wiley & Sons, Inc., New York, 1960.

The momentum vector of a mass m may be written as

$$\mathbf{P} = m\mathbf{v} \tag{2.6-1}$$

Newton's second law of motion[1] tells us that

$$\Sigma \mathbf{F} = (\text{output} - \text{input} + \text{accumulation}) \text{ rate of momentum} \tag{2.6-2}$$

which is just a form of our entity balance written in terms of rates with the sum of forces as the generation rate term. Here we are interested only in linear momentum, and so we will not consider changes in angular momentum. We accordingly restrict our mathematical model to situations where things are not changing in speed of rotation, that is, where there is no angular acceleration.

Remember that we now have a *vector* equation to treat. Equation 2.6-2 shows clearly that *momentum is not conserved*—it is generated by external forces imposed on the system. Momentum is conserved only in the absence of external forces.

Consider a system moving in space at some arbitrary velocity. We will restrict the origin of coordinates, as in Sec. 2.4, to being at rest with respect to the surface of our system (to avoid the problem of constructing the relative velocity between system and mass crossing the boundary).

The sum of external forces on the system, $\Sigma \mathbf{F}$, corresponds to the *rate* of generation of momentum, and so (output − input) *rate* is constructed by multiplying the mass crossing the boundary (a scalar) by its velocity (a vector). Considering, as before, a small element of area, ΔA, on the surface, the *rate* of momentum entering or leaving through ΔA is approximately:[2]

$$\frac{\partial}{\partial t} \begin{bmatrix} (\text{Momentum output} - \\ \text{momentum input}) \\ \text{through } \Delta A \end{bmatrix} \cong \underbrace{\mathbf{v}}_{\text{velocity}} \rho \underbrace{(\mathbf{v} \cdot \mathbf{n}) \Delta A}_{\substack{\text{rate of} \\ \text{mass flow}}} \tag{2.6-3}$$

---

[1] The form of this equation is valid for inertial coordinate frames. We must modify this form for use with accelerating coordinate frames, such as coordinate frames attached to accelerating rockets. The discussion of the case is given in Appendix 2B.

[2] Strictly speaking, we should add a term which accounts for momentum transfer through ΔA because of mass transfer. This term is seldom significant and will never be significant for our treatment here. The method of including this term is apparent from the material in Chap. 12.

Summing over the whole surface of our system and letting ΔA approach zero

$$\frac{\partial}{\partial t}\left[\begin{array}{l}\text{(Momentum output} -\\ \text{momentum input) total}\end{array}\right] = \lim_{\Delta A \to 0} \sum_A [\mathbf{v}\rho(\mathbf{v} \cdot \mathbf{n}) \Delta A]$$

$$= \int_A \mathbf{v}\rho(\mathbf{v} \cdot \mathbf{n}) \, dA \tag{2.6-4}$$

The rate-of-accumulation term is written by once again considering a small volume ΔV and writing its mass times its velocity:

$$\begin{array}{l}\text{Rate of momentum}\\ \text{accumulation of}\\ \Delta V\end{array} \cong \frac{d}{dt}(\mathbf{v}\rho\Delta V) \tag{2.6-5}$$

Adding up all ΔV's to get V and taking the limit as $\Delta V \to 0$ gives us

$$\begin{array}{l}\text{Rate of momentum}\\ \text{accumulation}\end{array} = \frac{d}{dt} \lim_{\Delta V \to 0} \sum_V \mathbf{v}\rho\Delta V = \frac{d}{dt}\int_V \mathbf{v}\rho \, dV \tag{2.6-6}$$

We can then construct the macroscopic momentum balance[1] as

$$\int_A \mathbf{v}\rho(\mathbf{v} \cdot \mathbf{n}) \, dA + \frac{d}{dt}\int_V \mathbf{v}\rho \, dV = \Sigma\mathbf{F} \tag{2.6-7}$$

Note that in general $\Sigma\mathbf{F}$ may have a component in any direction. Also notice that $\mathbf{F}$ is the force the surroundings exert *on* the control volume, *not* the force the control volume exerts on the surroundings.

There are two types of forces with which we must contend: body forces and surface forces. Body forces arise from the same fields that we considered when discussing potential energy—gravitational, electrostatic, etc. The change in energy from positional changes in a potential field can be accomplished by a force acting on the body during its displacement. This force—for example, gravity—does not necessarily disappear when the body is at rest. All potential fields do not produce what we call body forces, but all our body forces are produced by potential fields.

---

[1] Equation (2.6-7) can be written in terms of the substantial derivative by application of Gauss' theorem and the continuity equation. We do not use this form of the momentum balance. For such treatment see, for example, W. Prager, "Introduction to Mechanics of Continua," Ginn and Company, New York, 1961.

The forces we call body forces are those which are distributed throughout the volume of the system and are proportional somehow to an extensive property of the system, that is, *not* through direct traction on the surface of the system. Gravity will be the only body force of concern to us in this text.

The second type of force is the surface force. By surface force we mean the force exerted *by* the surroundings *on* the system by direct traction on the surface of the system. At any point in space-time we can define a unique force at each point of the system surface. Since the surface has no mass, this force will not include the body force. The surface traction at a point on the surface can be resolved into two components: one parallel to the surface (shear) and one perpendicular to the surface (normal).

As is shown in elementary physics, for a static fluid the normal force is the same in all directions, and is called the hydrostatic pressure. Further, there is no shear force in static fluid systems, because by definition a fluid deforms continuously under shear. The hydrostatic pressure is the same as the pressure used in a thermodynamic equation of state.

When a fluid is flowing, the normal stresses are not in general equal if shear is present (viscous fluid). If the fluid is incompressible, however, the pressure can be defined satisfactorily as the mean of the normal stresses in the three coordinate directions. For problems to be considered here, this definition will be satisfactory. For a full discussion of this topic, the reader can consult texts in continuum mechanics.

There is a useful observation we can make at this point with regard to the pressure force. Without going into a great deal of proof, we can observe that any object, no matter how irregularly shaped, will not move in a uniform static pressure field. For example, if we have an egg-shaped object of the same density as water and we immerse it in water, it will remain stationary. This shows that the integral of the pressure force over the entire exterior surface of this system results in a zero net force. If we replace the egg-shaped object by an object of some extremely irregular shape similar to a piece of cinder, which again has the same density as water, the object still does not move, which shows that the integral of the static pressure over a surface of even more irregular shape is still zero. The usefulness of the observation is as follows: in systems which involve some minimum external pressure (such as atmospheric pressure), and we work with pressures in *gage* pressure we automatically ignore the integral of the constant external pressure over the entire outer surface. This does not change the problem in any essential way but sometimes makes the calculations simpler in that we do not have to calculate integrals of pressure forces over, for example, the outside of pipes. This will become clear as we examine the example below.

Note that if above we had chosen an object which was of *different* density than the water or the air in which it was immersed, the body would, in fact, move. The motion, however, would *not* be a result of the pressure forces but rather a result of the *body* force—gravity—which acts upon the object. Body forces are forces which act proportional to the mass of the object, and arise from gradients in certain potential fields, e.g., gravitational fields, electromagnetic fields, electrostatic fields, etc. Pressure forces and drag forces, which we have already discussed, do not depend on the mass of the object, but rather are related to the surface area of the object and the velocity flow field external to the object. For example, the reason a balloon rises if filled with helium is that the body force which results from the gravitational acceleration on the mass of helium inside the balloon is insufficient to overcome the pressure-gradient force from top to bottom of the balloon. This pressure field is *not* a *uniform* static pressure field, but rather contains a gradient. This gives rise to the so-called buoyancy force which is, in essence, a force arising from a pressure gradient. In most of the problems that we treat this pressure gradient is so small that for all practical purposes we are not in a pressure-gradient field but rather in a uniform static field.

We may write Eq. (2.6-7) in component form for rectangular coordinates as

$$\int_A (v_x\mathbf{i} + v_y\mathbf{j} + v_z\mathbf{k})\rho v \cos \alpha \, dA + \frac{d}{dt}\int_V (v_x\mathbf{i} + v_y\mathbf{j} + v_z\mathbf{k})\rho \, dV$$
$$= \Sigma(F_x\mathbf{i} + F_y\mathbf{j} + F_z\mathbf{k})$$

$$(2.6\text{-}8)$$

Since Eq. (2.6-7) is a vector equation, the equality signifies that each component of the vector on the left must equal the corresponding component of the vector on the right. The same is true, obviously, of Eq. (2.6-8) since it is merely the same equation rewritten in different form. We may, therefore, equate the *components* and obtain the following set of scalar equations:

$$\int_A v_x\rho v \cos \alpha \, dA + \frac{d}{dt}\int_V v_x\rho \, dV = \Sigma F_x \qquad (2.6\text{-}9)$$

$$\int_A v_y\rho v \cos \alpha \, dA + \frac{d}{dt}\int_V v_y\rho \, dV = \Sigma F_y \qquad (2.6\text{-}10)$$

$$\int_A v_z\rho v \cos \alpha \, dA + \frac{d}{dt}\int_V v_z\rho \, dV = \Sigma F_z \qquad (2.6\text{-}11)$$

The momentum equation (since it is a vector equation) therefore yields *three* algebraic equations. Similar results are obtained in other coordinate systems (e.g., cylindrical and spherical).

Frequently we will be able to choose a coordinate system such that momentum is transferred in only one direction, so that there is only one equation to solve. This should be done whenever possible.

The reader is cautioned to remember that the components of **v** must have the appropriate sign, and that care must be taken not to confuse the magnitude of **v**, v, with one of the *components* of **v** in cases where $\cos \alpha$ is not $+1$ or -1.

We have a situation with average velocity in the momentum balance for pipe flow similar to that with the energy balance in Eq. (2.4-43), except that now we are faced with evaluating expressions of the form (for $\cos \alpha = \pm 1$ and $\rho = $ constant)

$$\int_A v^2 \, dA = \langle v^2 \rangle A \tag{2.6-12}$$

That is, we wish to replace an area integral by the area times an appropriate average of the integrand. It is common to replace $\langle v^2 \rangle$ by $\langle v \rangle^2$ for turbulent flow. The approximate error caused by replacing $\langle v^2 \rangle$ by $\langle v \rangle^2$ is calculated in Appendix 2A.

We now consider four examples. First, we do a fixed system with momentum transfer in one dimension. Next we consider a moving system with momentum transfer in one dimension which we reduce to the same problem as a stationary system by proper choice of coordinates. Third, we consider a stationary system with momentum transfer in two dimensions. Fourth, we consider a moving system with drag.

Example 2.6-1 Momentum transfer to a nozzle Figure 2.6-1 shows a nozzle which is attached to a fire hose. The hose carries water at 500 gal/min. What is the force in the x direction required of the threads to keep the nozzle attached to the hose? Frictionless flow may be assumed.

Solution First, choose a system bounded by the *outside* surface of the

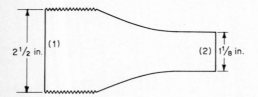

$2\frac{1}{2}$ in.   (1)   (2) $1\frac{1}{8}$ in.

Fig. 2.6-1 Nozzle of fire hose.

nozzle and two planes, (1) and (2), normal to the flow. Apply the overall momentum balance.

$$\int_A \mathbf{v}\rho(\mathbf{v}\cdot\mathbf{n})\,dA + \frac{d}{dt}\int_V \mathbf{v}\rho\,dV = \Sigma\mathbf{F}$$

$$\underset{\text{I}}{} \qquad \underset{\text{II}}{} \qquad \underset{\text{III}}{}$$

Term I may be written as the sum of three integrals: the integral over surface (1), the integral over surface (2), and the integral over the outside surface of the nozzle. The integral over the outside surface of the nozzle is zero because the velocity on the outside surface is everywhere zero. If we assume that the density of water is constant and that the velocity profile is flat, we may write
I:

$$\int_A \mathbf{v}\rho(\mathbf{v}\cdot\mathbf{n})\,dA = \rho\int_{A_1}\mathbf{v}(\mathbf{v}\cdot\mathbf{n})\,dA + \rho\int_{A_2}\mathbf{v}(\mathbf{v}\cdot\mathbf{n})\,dA$$

There is a component only for the x direction, which is

$$\left(\int_A \mathbf{v}\rho(\mathbf{v}\cdot\mathbf{n})\,dA\right)_x = \rho A_2\langle v_2\rangle^2 - \rho A_1\langle v_1\rangle^2$$

$$= w_2\langle v_2\rangle - w_1\langle v_1\rangle$$

II:

Because the system is at steady state

$$\frac{d}{dt}\int_V \mathbf{v}\rho\,dV = 0$$

III:

The external forces acting on the system in the x direction are the force in the threads and the pressure forces. (We assume that there is no force on the outside surface from, for example, air blowing past the nozzle giving a frictional drag at the surface. Even if such a force existed, it would normally be small compared to the momentum change.)

The force of gravity has a component only in the y direction, and so does not affect our x-directed force.

Substituting in (I) we have

$$w_2\langle v_2\rangle - w_1\langle v_1\rangle = \Sigma F_x$$

$$w_2\frac{Q_2}{A_2} - w_1\frac{Q_1}{A_1} = \Sigma F_x$$

But by an overall mass balance

$$w_2 = w_1 = w$$

and since ρ is constant

$$Q_2 = Q_1 = Q$$

Thus

$$wQ\left(\frac{1}{A_2} - \frac{1}{A_1}\right) = \Sigma F_x$$

$$= \left(500 \ \frac{\text{gal}}{\text{min}}\right)\left(8.33 \ \frac{\text{lb}m}{\text{gal}}\right)\left(500 \ \frac{\text{gal}}{\text{min}}\right)$$

$$\times \left(\frac{\text{ft}^3}{7.48 \ \text{gal}}\right)\left(\frac{\text{min}^2}{3,600 \ \text{sec}^2}\right) \times \left(\frac{\text{lb}f \ \text{sec}^2}{32.2 \ \text{lb}m \ \text{ft}}\right) \times$$

$$\left[\frac{4}{\pi \ (1.125/12)^2} - \frac{4}{\pi \ (2.5/12)^2}\right]\left(\frac{1}{\text{ft}^2}\right) = \Sigma F_x \ \text{lb}f$$

$$\Sigma F_x = 280 \ \text{lb}f$$

This net force is made up of two forces, the pressure force and the reaction force in the threads. We evaluate the pressure drop using the mechanical energy balance since the mechanical energy balance relates the pressure drop to the change in kinetic energy:

$$\frac{\Delta(\langle v \rangle^2)}{2} + g \ \Delta z + \frac{\Delta p}{\rho} = - \widehat{lw} - \widehat{W}$$

If $p_2 = 0$ psig:

$$p_1 = \rho \ \frac{\langle v_2 \rangle^2 - \langle v_1 \rangle^2}{2}$$

$$v_2 = \frac{Q}{A_2} = \left(500 \ \frac{\text{gal}}{\text{min}}\right)\left(\frac{\text{ft}^3}{7.48 \ \text{gal}}\right)\left[\frac{4}{\pi \ (1.125/12)^2}\right]\left(\frac{1}{\text{ft}^2}\right)$$

$$= 9,700 \ \text{ft/min}$$

$$v_1 = 9,700 \ \frac{A_2}{A_1} = 9,700 \ \frac{\pi D_2^2}{4} \ \frac{4}{\pi D_1^2} = 9,700 \ \frac{(1.125/12)^2}{(2.5/12)^2}$$

$$= 1,960 \ \text{ft/min}$$

$$p_1 = \left(62.4 \ \frac{\text{lb}m}{\text{ft}^3}\right)\left[\frac{9,700^2 - 1,960^2}{(2)(32.2)}\right]\left(\frac{\text{ft}^2 \ \text{lb}f \ \text{sec}^2}{\text{min}^2 \ \text{lb}m \ \text{ft}}\right)\left(\frac{\text{min}^2}{3,600 \ \text{sec}^2}\right)$$

$$= 24,200 \ \text{lb}f/\text{ft}^2 \quad (168 \ \text{lb}f/\text{in.}^2)$$

This is p_1 in *gage* pressure. If F_T = force on threads

$$280 = 24{,}200A_1 + F_T = 24{,}200\frac{\pi}{4}\left(\frac{2.5}{12}\right)^2 + F_T$$

$$F_T = -547 \ \text{lb}f \quad \text{(acts to } left \text{ on system)}$$

□ □ □

Example 2.6-2   Momentum balance on a jet plane   A plane is to fly at a speed of 600 mph. Each engine will take in 60 lb$m$ of air per lb$m$ of fuel, and each engine is to supply 10,000 lb$f$ of thrust. If the engine burns 3 lb$m$/sec of fuel, what must the outlet velocity be? Figure 2.6-2 shows the engine.

Solution   We choose coordinates attached to the plane as shown. Using the system indicated, we can see that mass enters and leaves only through areas 1, 2, and 3. The input and output terms of momentum balance become

$$\int_A \mathbf{v}\rho(\mathbf{v}\cdot\mathbf{n})\,dA = \int_{A_1} \mathbf{v}\rho(\mathbf{v}\cdot\mathbf{n})\,dA$$
$$+ \int_{A_2} \mathbf{v}\rho(\mathbf{v}\cdot\mathbf{n})\,dA + \int_{A_3} \mathbf{v}\rho(\mathbf{v}\cdot\mathbf{n})\,dA$$

The fuel enters at low velocity and flow rate, and so we neglect the momentum passing through A_3. We assume the velocity to be constant across A_1 and A_2 (although *different* at A_1 and A_2) since velocities are high. Therefore

$$\int_A \mathbf{v}\rho(\mathbf{v}\cdot\mathbf{n})\,dA = \mathbf{v}_1 \int_{A_1} \rho(\mathbf{v}\cdot\mathbf{n})\,dA$$
$$+ \mathbf{v}_2 \int_{A_2} \rho(\mathbf{v}\cdot\mathbf{n})\,dA = -w_1\mathbf{v}_1 + w_2\mathbf{v}_2$$

Note the differing signs from the cos α terms.

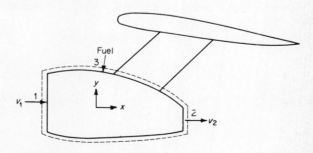

Fig. 2.6-2 Jet engine.

The system moves at constant velocity, and the flows are at steady state; therefore, the accumulation term in the momentum balance becomes

$$\frac{d}{dt} \int_V \mathbf{v} \rho \, dV = 0$$

The most important force involved is the force from the strut supporting the engine pod. We are given that this must be 10,000 lbf.

The force in the strut comes from the drag force on the fuselage, wings, etc., of the aircraft. In the usual case we would have to *calculate* this force from given conditions—we will see how to do this later in this text.

If the exhaust is sonic or supersonic, the pressure at 1 and 2 may differ, but still would probably be negligible compared to the strut force. The other force involved is the drag from the air and this also will be less important than the strut force.

If we substitute in the overall momentum balance we obtain

$$(a) \qquad -w_1 v_1 \mathbf{i} + w_2 v_2 \mathbf{i} = + 10,000 \, \mathbf{i} \ \text{lbf}$$

The sign on the right-hand side of the equation above is positive because the force *on the system* is in the positive x direction.

In the above equation we do not know the inlet velocity. This velocity can be taken to a good degree of approximation as the free-stream velocity, 600 mph, since the change in velocity from 1 to 2 is much more than the difference between v_1 and 600 mph.

We can relate w_1 and w_2 with a total mass balance

$$w_1 + w_3 = w_2$$

and since we are given that $60 \, w_3 = w_1$ and that $w_3 = 3$ lbm/sec:

$$w_2 = 61 \, w_3 = 183 \ \text{lbm/sec}$$

Substituting in equation (a) and equating components:

$$w_1 \left(-v_1 + 1.0167 \, v_2 \right) = 10,000 \ \text{lbf}$$

$$\left(180 \frac{\text{lbm}}{\text{sec}} \right) \left[\left(600 \, \frac{\text{mi}}{\text{hr}} \right) \left(\frac{\text{hr}}{3,600 \ \text{sec}} \right) \left(\frac{5,280 \ \text{ft}}{\text{mile}} \right) - \left(1.0167 \, v_2 \right) \right]$$

$$= - \left(10,000 \ \text{lbf} \right) \left(\frac{32.2 \ \text{lbm ft}}{\text{lbf sec}^2} \right)$$

$$v_2 = 890 \ \text{ft/sec}$$

□ □ □

Example 2.6-3 Momentum transfer to an elbow Figure 2.6-3 shows a horizontal pipeline elbow carrying seawater at a rate of 7,000 gal/min for cooling in an oil refinery. The line is 24-in. nominal-diameter schedule 40 steel pipe, and the pressure at both points 1 and 2 may be taken to be 50 psig. (Actually there will, of course, be a small pressure drop caused by friction. We will see how to calculate this in a later chapter.) Calculate the magnitude and direction of the resultant force in the pipe and on the pipe support.

For seawater:

$$\rho = 64 \ \frac{\text{lb}m}{\text{ft}^3}$$

For 24-in. nominal-diameter pipe:

$$\text{ID} = 22.626 \text{ in.}$$

$$A_{xs} = 2.792 \text{ ft}^2$$

where the subscript xs denotes cross-sectional area.

Solution Using the system indicated, the overall momentum balance may be written, term by term, as

$$\int_A \mathbf{v}\rho(\mathbf{v} \cdot \mathbf{n}) \, dA = \int_{A_1} \mathbf{v}\rho(\mathbf{v} \cdot \mathbf{n}) \, dA + \int_{A_2} \mathbf{v}\rho(\mathbf{v} \cdot \mathbf{n}) \, dA$$

$$= -w_1\mathbf{v}_1 + w_2\mathbf{v}_2$$

There is a sign difference because α is 180° at the inlet and 0° at the outlet. Because we are at steady state:

$$\frac{d}{dt}\int_V \mathbf{v}\rho \, dV = 0$$

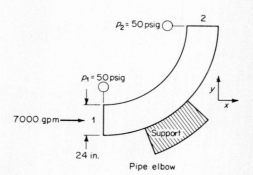

Fig. 2.6-3 Sketch for Example 2.6-3.

Substituting in the overall momentum balance and noting $w_1 = w_2 = w$

$$(a) \qquad w(\mathbf{v}_2 - \mathbf{v}_1) = \Sigma \mathbf{F}$$

where

$$\mathbf{v}_1 = \frac{w}{A_1 \rho_1} \mathbf{i} = \frac{Q}{A_1} \mathbf{i}$$
$$= \left(7{,}000 \, \frac{\text{gal}}{\text{min}}\right) \left(\frac{\text{ft}^3}{7.48 \, \text{gal}}\right) \left(\frac{1}{2.792 \, \text{ft}^2}\right) \mathbf{i} = \left(335 \, \frac{\text{ft}}{\text{min}}\right) \mathbf{i}$$

Therefore, since $A_2 = A_1$,

$$\mathbf{v}_2 = \left(335 \, \frac{\text{ft}}{\text{min}}\right) \mathbf{j}$$

Substituting in equation (a):

$$\left(7{,}000 \, \frac{\text{gal}}{\text{min}}\right) \left(\frac{\text{ft}^3}{7.48 \, \text{gal}}\right) \left(\frac{64 \, \text{lbm}}{\text{ft}^3}\right) \left(\frac{\text{lbf sec}^2}{32.2 \, \text{lbm ft}}\right) \left(\frac{\text{min}^2}{3{,}600 \, \text{sec}^2}\right)$$

$$(335 \, \mathbf{j} - 335 \, \mathbf{i}) \, \frac{\text{ft}}{\text{min}} = \Sigma \mathbf{F} \quad \text{lbf}$$

$$\mathbf{F} = -173 \mathbf{i} + 173 \mathbf{j} \quad \text{lbf}$$

This is the *total* external force and is made up of the pressure force, the force in the pipe wall, and the force from the pipe support.

$$\mathbf{F} = \mathbf{F}_{\text{pres}} + \mathbf{F}_{\text{pipe}} + \mathbf{F}_{\text{sup}}$$

We can calculate the pressure force as

$$\mathbf{F}_{\text{pres}} = p_1 A_1 \mathbf{i} - p_2 A_2 \mathbf{j}$$

The minus sign appears because the force at (2) is in the negative y direction.

$$\mathbf{F}_{\text{pres}} = \left(50 \, \frac{\text{lbf}}{\text{in.}^2}\right) \left(\frac{144 \, \text{in.}^2}{\text{ft}^2}\right) (2.792 \, \text{ft}^2) \mathbf{i}$$

$$- \left(50 \, \frac{\text{lbf}}{\text{in.}^2}\right) \left(\frac{144 \, \text{in.}^2}{\text{ft}^2}\right) (2.792 \, \text{ft}^2) \mathbf{j}$$

$$= 20{,}100 \mathbf{i} - 20{,}100 \mathbf{j}$$

Substituting for $\mathbf{F}$ and $\mathbf{F}_{\text{pres}}$ we have

$$\mathbf{F}_{\text{pipe}} + \mathbf{F}_{\text{sup}} = -173\mathbf{i} + 173\mathbf{j} - (20{,}100\mathbf{i} - 20{,}100\mathbf{j})$$
$$= -20{,}273\mathbf{i} + 20{,}273\mathbf{j}$$

The force in the x direction is exerted to the left, as intuitively expected, and the force in the y direction is in the $+y$ direction.

We cannot separate the forces into the forces at (1), (2), and the support without knowing more about the system.

□ □ □

Example 2.6-4 Momentum transfer in a jet boat We wish to build a boat that is propelled by a jet of water, as shown in Fig. 2.6-4. The inlet is to be 6-in. inside diameter and the outlet 3-in. inside diameter. The boat is supplied with a motor and pump powerful enough to transfer water at 300 gal/min, and the drag on the hull can be expressed as

$$F_{\text{drag}} = 0.025v^2$$

where $v = $ speed of boat, ft/sec
$\quad F = $ total drag, lbf

Note that the constant 0.025 has dimensions of lbf sec²/ft². Calculate the top speed of the boat.

Solution Using the system shown, and coordinates *moving with the boat*, the overall momentum may be considered term by term:

$$\int_A \mathbf{v}\rho(\mathbf{v} \cdot \mathbf{n})\, dA = \int_{A_1} \mathbf{v}\rho(\mathbf{v} \cdot \mathbf{n})\, dA$$
$$+ \int_{A_2} \mathbf{v}\rho(\mathbf{v}\cdot\mathbf{n})\, dA = -\mathbf{v}_1 w_1 + \mathbf{v}_2 w_2$$

assuming plug flow for $\mathbf{v}_1$ and $\mathbf{v}_2$.

Since the system is at steady state

$$\frac{d}{dt}\int_V \mathbf{v}\rho\, dV = 0$$

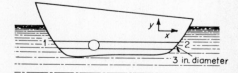

Fig. 2.6-4 Jet boat for Example 2.6-4.

The sum of forces is

$$\mathbf{F} = 0.025\,v^2\mathbf{i} \ \ \text{lbf}$$

Substituting in the overall momentum balance

$$(a) \qquad -w_1\mathbf{v}_1 + w_2\mathbf{v}_2 = 0.025\,v^2\mathbf{i} \ \ \text{lbf}$$

but an overall total mass balance shows

$$-w_1 + w_2 = 0$$

or

$$w_1 = w_2$$

Thus equation (a) becomes (in component form)

$$w_1(\mathbf{v}_2 - \mathbf{v}_1) = 0.025\,v^2\mathbf{i} \ \ \text{lbf}$$

But

$$Q_1 = Q_2$$

$$v_2 = \frac{Q_2}{A_2} = \frac{Q_1}{A_2}$$

$$v_1 = \frac{Q_1}{A_1}$$

$$w_1 = \rho_1 Q_1$$

So

$$\rho_1 Q_1{}^2 \left(\frac{1}{A_2} - \frac{1}{A_1}\right) = 0.025\,v^2 \ \ \text{lbf}$$

Substituting known data:

$$\left(\frac{\text{lbf sec}^2}{32.2 \ \text{lbm ft}}\right)\left(62.4\,\frac{\text{lbm}}{\text{ft}^3}\right)\left[\left(300\,\frac{\text{gal}}{\text{min}}\right)\left(\frac{\text{min}}{60 \ \text{sec}}\right)\left(\frac{\text{ft}^3}{7.48 \ \text{gal}}\right)\right]^2$$

$$\left\{\left[\frac{(4)(144)}{\pi(3)^2 \ \text{ft}^2}\right] - \left[\frac{(4)(144)}{\pi(6)^2 \ \text{ft}^2}\right]\right\}$$

$$= 0.025\,v^2 \ \ \text{lbf}$$

Therefore

$$v = 23 \text{ ft/sec}$$

or about 15 to 16 mph.

□ □ □

A possible question in the reader's mind at this point might be "Why aren't we finished with the course?" The answer, of course, is that we have considered problems involving the macroscopic balances in which we always know every term except one. Unfortunately, this is seldom the case. For example, consider the case of the macroscopic energy balance. The usual way in which we use this balance is that we are interested in designing something like a heat exchanger. We know the energy *input* and energy *output* in the stream that we wish to heat or cool and we also know only the energy *input* in the stream that we use for the heating or cooling. We do not, however, know *either* the heat transfer area *or* the amount of energy exiting in the stream we are using to accomplish our heating or cooling. The energy exiting depends on the area. To determine the relationship between the area and the energy of the exiting stream, we use a rate equation. For this specific example, we would use the rate equation which related heat transfer to the temperature differences which exist in the system, and then knowing the heat transfer rate we could find the output temperature of one of our streams. We could also work the problem in the reverse fashion and by specifying the output temperature of the stream calculate the desired heat transfer rate and from this determine the temperature differences required in the system or the area required to accomplish this heat transfer. The objective of much of the remainder of the book is to introduce the rate expressions for momentum, heat, and mass transfer and to illustrate their usefulness in design calculations.

2.7 THE DIFFERENTIAL MOMENTUM BALANCE[1]

In the following we discuss the transition from the integral equation for the momentum balance (Newton's second law of motion) to the differential equation for the momentum balance—from the macroscopic to the microscopic equation. The resulting differential balance is one of the starting equations for the study of transport phenomena.

[1]This section provides a transition for subsequent courses in transport phenomena—the differential approach referred to in Sec. 1.4.

The macroscopic momentum balance was determined above to be

$$\int_A \mathbf{v} \rho (\mathbf{v} \cdot \mathbf{n}) \, dA + \frac{d}{dt} \int_V \mathbf{v} \rho \, dV = \Sigma \mathbf{F} \tag{2.7-1}$$

Normally we are concerned with the body force due to gravity, expressed in terms of $\mathbf{g}$, and the surface force, which may be expressed in terms of the stress vector $\mathbf{t}^n$ discussed in Sec. 2.5. For these forces Eq. (2.7-1) becomes

$$\int_A \mathbf{v} \rho (\mathbf{v} \cdot \mathbf{n}) \, dA + \frac{d}{dt} \int_V \mathbf{v} \rho \, dV = \int_V \rho \, \mathbf{g} \, dV + \int_A \mathbf{t}^n \, dA \tag{2.7-2}$$

To parallel the previous discussion of scalar equations (in Secs. 2.3 and 2.5) we could consider only the x motion for the above equation or form the scalar product with an arbitrary vector $\mathbf{b}$ which transforms Eq. (2.7-2) into a scalar equation.[1] If we perform the latter step we obtain the scalar equation

$$\int_A \rho (\mathbf{v} \cdot \mathbf{b})(\mathbf{v} \cdot \mathbf{n}) \, dA + \frac{d}{dt} \int_V (\mathbf{v} \cdot \mathbf{b}) \, \rho \, dV - \int_V \rho (\mathbf{g} \cdot \mathbf{b}) \, dV$$

$$+ \int_A (\mathbf{b} \cdot \mathbf{t}^n) \, dA = 0 \tag{2.7-3}$$

At this point we proceed in a parallel way to Secs. 2.3 and 2.5, but the form of the area integrals on the left-hand side of Eq. (2.7-3) leads us to forms involving second-order tensors, not vectors. Since we are concerned with illustration only, let us write the result (having assumed, as before, stationary volume) after applying the divergence theorem for *tensors* and separating the pressure from the viscous stress tensor:

$$\frac{\partial \rho \mathbf{v}}{\partial t} + (\nabla \cdot \rho \mathbf{v} \mathbf{v}) = -\nabla p - [\nabla \cdot \mathbf{T}] + \rho \mathbf{g} \tag{2.7-4}$$

$$(a) \qquad (b) \qquad (c) \qquad (d) \qquad (e)$$

where (a), (b), (c), (d), (e) are rate of change of momentum (or force on

[1] S. Whitaker, "Introduction to Fluid Mechanics," Prentice-Hall, Inc., Englewood Cliffs, N. J., 1968.

an element) per unit volume representing

(a) Increase of momentum
(b) Momentum input by convection
(c) Pressure force
(d) Momentum input by viscous forces
(e) Gravitational force

This is an intermediate step which roughly parallels Eqs. (2.3-12) and (2.5-4).

We are again faced with the problem of relating the transport properties of the fluid to the flux for a given driving force. Here the flux is the momentum flux and the driving force the velocity gradient. This topic is discussed in Chap. 5.

For a particularly simple fluid (an incompressible *Newtonian* fluid) whose flux/driving force relationship is of the same form as Fourier's law and Fick's law, substitution in Eq. (2.7-4) yields

$$\rho \frac{D\mathbf{v}}{Dt} = -\nabla p + \mu \nabla^2 \mathbf{v} + \rho \mathbf{g} \qquad (2.7\text{-}5)$$

which is called the Navier-Stokes equation. The operators are defined for rectangular Cartesian coordinates as

$$\nabla^2 = \frac{\partial^2}{\partial x^2} + \frac{\partial^2}{\partial y^2} + \frac{\partial^2}{\partial z^2} \qquad (2.7\text{-}6)$$

$$\frac{D}{Dt} = \frac{\partial}{\partial t} + v_x \frac{\partial}{\partial x} + v_y \frac{\partial}{\partial y} + v_z \frac{\partial}{\partial z} \qquad (2.7\text{-}7)$$

The investigation of solutions of this equation for various systems (boundary conditions) where momentum is transferred by viscous motion is the topic of momentum transport as a transport phenomenon.[1] For example, the axial flow of an incompressible Newtonian fluid in a horizontal pipe is described by (in *cylindrical* coordinates)

$$\rho \frac{\partial v_x}{\partial t} = -\frac{\partial p}{\partial x} + \mu \frac{1}{r} \frac{\partial}{\partial r} \left(r \frac{\partial v_x}{\partial r} \right) \qquad (2.7\text{-}8)$$

where μ is the coefficient of viscosity.

[1] R. B. Bird, W. E. Stewart, and E. N. Lightfoot, "Transport Phenomena," John Wiley & Sons, Inc., New York, 1960.

APPENDIX 2A: APPROXIMATIONS TO AVERAGE VELOCITY

Let us see how much error the assumption that the velocity profile is constant at the bulk velocity gives by looking at the ratio of the right to the wrong expression and integrating for a known profile.

The velocity distribution for turbulent flow in pipes is fairly well represented by the following empirical profile:

$$v = v_{max} \left(1 - \frac{r}{R}\right)^{1/7}$$

Energy balance approximation—turbulent flow

The ratio we wish to consider with respect to the kinetic energy term in the energy balance is

$$\frac{\langle v^3 \rangle / \langle v \rangle}{\langle v \rangle^2} = \frac{\langle v^3 \rangle}{\langle v \rangle^3} = \frac{\int_0^R [v_{max}(1 - r/R)^{1/7}]^3 \, 2\pi r \, dr}{\int_0^R 2\pi r \, dr}$$

$$\frac{\left(\int_0^R 2\pi r \, dr\right)^3}{\left\{\int_0^R [v_{max}(1 - r/R)^{1/7}] \, 2\pi r \, dr\right\}^3}$$

Defining an auxiliary variable, $t = 1 - r/R$, we have

$$dt = -\frac{dr}{R}$$

$$r = (1 - t)R$$

Substituting:

$$\frac{\langle v^3 \rangle}{\langle v \rangle^3} = \frac{\int_1^0 t^{3/7}[(1-t)R](-R\,dt)}{R^2/2} \quad \frac{(R^2/2)^3}{\left\{\int_1^0 t^{1/7}[(1-t)R](-R\,dt)\right\}^3}$$

$$= \frac{-(7/10)t^{10/7} + (7/17)t^{17/7} \Big|_1^0}{4\left\{\left[-(7/8)t^{8/7} + (7/15)t^{15/7}\right]\Big|_1^0\right\}^3} = \frac{7/17 - 7/10}{4(7/15 - 1/8)^3}$$

$$\tag{2A-1}$$

$$\cong 1.06$$

Therefore, the assumption creates an error of about

$$\% \text{ error} = 100 \frac{\text{approximate quantity} - \text{correct quantity}}{\text{correct quantity}}$$

$$= 100 \frac{1 - 1.06}{1.06} = -5.7\% \tag{2A-2}$$

We could, of course, apply this correction, but it is seldom worthwhile because of the approximate nature of the velocity distribution used and other uncertainties that are usually present in real problems.

Energy balance approximation—laminar flow

As a matter of interest, let us see what the error would be if we made this approximation in the energy balance for *laminar* flow, where

$$v = v_{\max}\left[1 - \left(\frac{r}{R}\right)^2\right] \tag{2A-3}$$

Then $\dfrac{\langle v^3 \rangle / \langle v \rangle}{\langle v \rangle^2} = \dfrac{\langle v^3 \rangle}{\langle v \rangle^3}$

$$= \frac{\int_0^R \{v_{\max}[1 - (r/R)^2]\}^3 \, 2\pi r \, dr}{\int_0^R 2\pi r \, dr} \frac{\left(\int_0^R 2\pi r \, dr\right)^3}{\left\{\int_0^R v_{\max}[1 - (r/R)^2] 2\pi r \, dr\right\}^3}$$

This may be written as

$$\frac{\langle v^3 \rangle}{\langle v \rangle^3} = \frac{(R^2/2)\int_0^1 [1 - (r/R)^2] (2r/R) \, d(r/R)}{R^2/2}$$

$$\frac{(R^2/2)^3}{\left\{(R^2/2)\int_0^1 [1 - (r/R)^2] (2r/R) \, d(r/R)\right\}^3}$$

This time we define $t = (r/R)^2$ so that $dt = 2(r/R) \, d(r/R)$. Substituting

$$\frac{\langle v^3 \rangle}{\langle v \rangle^3} = \frac{\int_0^1 (1-t)^3 \, dt}{1} \frac{1}{\left[\int_0^1 (1-t) dt\right]^3} = \frac{(1-t)^4/4 \Big|_1^0}{\left[(1-t)^2/2 \Big|_1^0\right]^3}$$

$$\frac{\langle v^3 \rangle}{\langle v \rangle^3} = \frac{1/4}{(1/8)} = 2$$

Therefore, our quantity using the approximation would be *half* the correct answer.

Momentum balance approximation—turbulent flow

In using the momentum balance, we often replace $\langle v^2 \rangle$ by $\langle v \rangle^2$. Let us investigate the corresponding errors in the same fashion as above. For turbulent flow:

$$\frac{\langle v^2 \rangle}{\langle v \rangle^2}$$

$$= \frac{\int_0^R [v_{max}(1 - r/R)^{1/7}]^2 2\pi r \, dr}{\int_0^R 2\pi r \, dr} \frac{\left(\int_0^R 2\pi r \, dr\right)^2}{\left[\int_0^R v_{max}(1 - r/R)^{1/7} 2\pi r \, dr\right]^2}$$

$$(2A-4)$$

Defining t as $t = 1 - r/R$

$$\frac{\langle v^2 \rangle}{\langle v \rangle^2} = \frac{\int_1^0 t^{2/7}(1 - t)R(- R \, dt)}{R^2/2} \frac{(R^2/2)^2}{\left[\int_1^0 t^{1/7}(1 - t)R(-R \, dt)\right]^2}$$

$$(2A-5)$$

$$= \frac{1}{2} \frac{[(7/9)t^{9/7} - (7/16)t^{16/7}] \Big|_0^1}{[(7/8)t^{8/7} - (7/18)t^{15/7}]^2 \Big|_0^1} = 1.02$$

Thus

$$\text{Percent error} = (100)\left(\frac{1 - 1.02}{1.02}\right) \cong - 2 \text{ percent} \qquad (2A-6)$$

Momentum balance approximation—laminar flow

For laminar flow:

$$\frac{\langle v^2 \rangle}{\langle v \rangle^2}$$

$$= \frac{\int_0^R \{v_{max}[1 - (r/R)^2]\}^2 2\pi r \, dr}{\int_0^R 2\pi r \, dr} \frac{\left(\int_0^R 2\pi r \, dr\right)^2}{\left\{\int_0^R v_{max}[1 - (r/R)^2] 2\pi r \, dr\right\}^2}$$

Rearranging and defining $t = (r/R)^2$ so that $dt = (2r/R)d(r/R)$ as in the previous laminar flow case:

$$\frac{\langle v^2 \rangle}{\langle v \rangle^2} = \frac{\int_0^1 (1-t)^2 dt}{\left[\int_0^1 (1-t)dt \right]^2} = \frac{(1-t)^3/3 \Big|_1^0}{[(1-t)^2/2]^2 \Big|_1^0} = \frac{4}{3} \tag{2A-7}$$

Therefore, if the approximation is used, a result 75 percent of the correct result will be obtained.

APPENDIX 2B: MOMENTUM EQUATION FOR NONINERTIAL REFERENCE FRAMES

The momentum equation as we have written it in Chap. 2 applies to an *inertial* reference frame system. By an *inertial* frame we mean a frame which is not *accelerated*.[1] (In our use of *unaccelerated* we exclude rotation and permit only uniform translation.) These frames are also referred to as Galilean or Newtonian frames. The acceleration of frames attached to the earth's surface, however, is slight enough that our equations give good answers for most engineering problems—the error is of the order of one-third of 1 percent.[2] For problems involving relatively large acceleration of the reference frame, however, Newtonian mechanics must be reformulated. An example of such a problem is the use of a reference frame attached to a rocket as it accelerates from zero velocity on the pad to some orbital velocity. For such a problem phenomena observed from an inertial frame and from the accelerating frame can differ greatly.

For example, if an astronaut in zero gravity but in an accelerating rocket drops his toothbrush, he sees it apparently accelerate toward the "floor" (wherever that direction might be at the time). His observation is with respect to an accelerating frame (his spaceship). To an observer in an inertial frame, however, the toothbrush simply continues at the constant velocity it had when released, and the spacecraft accelerates past the toothbrush.

---

[1] For a more extended discussion of this material, the reader is referred to C. G. Fanger, "Engineering Mechanics: Statics and Dynamics," Charles E. Merrill Books, Inc., Columbus, Ohio, 1970.

[2] Fanger, *ibid.*

All this is a complicated way of saying that *observed acceleration depends on the motion of the reference frame*. If this is true, it is obvious that Eq. (2.6-2) (Newton's second law) must change in form when written with respect to an accelerating reference frame. Here we will consider only a simple example of this—that of a system of *constant* mass.

Return to the astronaut's toothbrush. With respect to an *inertial* reference frame, acceleration is zero. Newton's law is

$$\Sigma \mathbf{F} = m\mathbf{a} = 0$$

With respect to a reference frame attached to the spacecraft, however, there is an *apparent acceleration*. This apparent acceleration is equal to the relative acceleration of the two coordinate frames. Newton's law will appear valid if we *add an additional force* to the left-hand side and interpret the acceleration on the right-hand side as the object acceleration relative to the accelerating frame. The inertial body force is the negative of the product of the mass and the *relative acceleration* of inertial and accelerating frames:[1]

$$\Sigma \mathbf{F} + \text{(inertial body force)} = m\mathbf{a}$$

An example is the object sitting on the ledge above the rear seat of your car. You slam on the brakes and decelerate from 60 mph. The object flies forward. To the hitchhiker standing at the side of the road, the sum of forces on the object is zero and therefore its acceleration is zero—it continues at 60 mph until the presence of a force (perhaps the windshield). To your accelerating (decelerating) reference frame, there is no external force on the object, but there is an inertial body force of $m\mathbf{a}$, where $\mathbf{a}$ is the relative acceleration of car and ground. This is balanced on the right-hand side of the equation by the same mass times the relative acceleration of the object and the windshield, which is the same in magnitude and direction as that of the car and the ground.

In the case of rotating frames things become more complex, and the reader is referred to texts such as Fanger.[2]

---

[1] A. G. Hansen, "Fluid Mechanics," John Wiley & Sons, Inc., New York, 1967.
[2] Fanger, *op. cit.*

PROBLEMS

2.1 For the following situation:

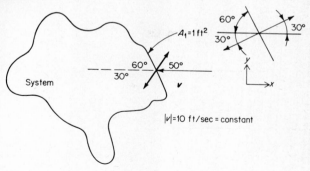

(*a*) Calculate $(\mathbf{v} \cdot \mathbf{n})$.
(*b*) Calculate v_x.
(*c*) Calculate w.

Use coordinates shown.

2.2 Calculate $\displaystyle\int_{A_1} \rho(\mathbf{v} \cdot \mathbf{n}) \; dA$ for the following situation

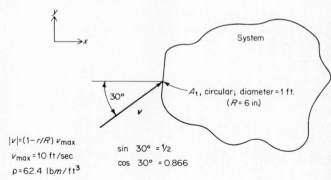

Use coordinates shown.

2.3 A pipe system sketched below is carrying water through section 1 at a velocity of 3 ft/sec. The diameter at section 1 is 2 ft. The same flow passes through section 2 where the diameter is 3 ft. Find the discharge and bulk velocity at section 2.

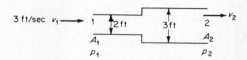

2.4 Consider the steady flow of water at 70°F through the piping system shown below where the velocity distribution at station 1 may be expressed by

$$v(r) = 9.0\left(1 - \frac{r^2}{R_1{}^2}\right) \qquad \text{ft/hr}$$

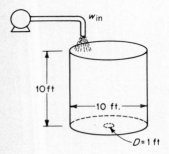

Find the average velocity at 2.

2.5 The velocity profile for turbulent flow of a fluid in a circular smooth tube follows a power law such that

$$v = v_{\text{маx}} \left(\frac{r_i - r}{r_i}\right)^{1/n}$$

Find the value of $\langle v\rangle / v_{\text{max}}$ for this situation. If $n = 7$ this is the result from the Blasius resistance formula.

(Let $y = r_i - r$.)

2.6 Oil of specific gravity 0.8 is pumped to a large open tank with a 1.0-ft hole in the bottom. Find the maximum *steady* flow rate Q_{in}, ft³/sec, which can be pumped to the tank without its overflowing.

2.7 Shown is a perfectly mixed tank containing 100 ft³ of an aqueous solution of 0.1 salt by weight. At time $t = 0$, pumps 1 and 2 are turned on. Pump 1 is pumping pure water into the tank at a rate of 10 ft^3/hr and pump 2 pumps solution from the tank at 15 ft^3/hr. Calculate the concentration in the outlet line as an explicit function of time.

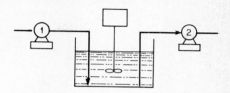

2.8 Water is flowing into an open tank at the rate of 50 ft^3/min. There is an opening at the bottom where water flows out at a rate proportional to

$$w = 0.1\rho\sqrt{2gh} \qquad lbm/sec$$

where h = height of the liquid above the bottom of the tank. Set up the equation for height versus time; separate the variables and integrate. The area of the bottom of the tank is 100 ft^2.

2.9 Salt solution is flowing into and out of a stirred tank at a rate of 5 gal solution/min. The input solution contains 2 lbm salt/gal solution. The volume of the tank is 100 gal. If the initial salt concentration in the tank was 1 lbm salt/gal, what is the outlet concentration as a function of time?

2.10 A batch of fuel-cell electrolyte is to be made by mixing two streams in a large stirred tank. Stream 1 contains a water solution of H_2SO_4 and K_2SO_4 in mass fractions x_{A1} and x_{B1} lbm/lbm of solution, respectively ($A = H_2SO_4$, $B = K_2SO_4$). Stream 2 is a solution of H_2SO_4 in water having a mass fraction x_{A2}. If the tank is initially empty, the drain is closed, and the streams are fed at steady rates w_1 and w_2 lbm of solution/hr, show that x_A is independent of time and find the total mass m at any time. Hint: Do not expand the derivative of the product in the species balance, but integrate directly.

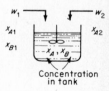

Concentration
in tank

2.11 You have been asked to design a stirring tank for salt solutions. The tank volume is 126 gal and a salt solution flows into and out of the tank at 7 gal of solution/min. The input salt concentration contains 3 lbm of salt/gal of solution. If the initial salt concentration in the tank is $\frac{1}{2}$ lbm salt/gal, find:

   (a) The output concentration of salt as a function of time assuming that the output concentration is the same as that within the tank.

   (b) Plot the output salt concentration as a function of time and find the asymptotic value of the concentration $m(t)$ as $t \to \infty$.

   (c) Find the response time of the stirring tank, i.e., the time at which the output concentration $m(t)$ will equal 62.3 percent of the change in the concentration from time $t = 0$ to time $t = \infty$.

2.12 Initially a mixing tank has fluid flowing in at a steady rate w_1 and temperature T_1. The level remains constant and the exit temperature is $T_2 = T_1$. Then at time zero, Q Btu/sec flows through the tank walls and heats the fluid (Q-constant). Find T_2 as a function of time. Use $dH = \hat{C}_p\, dT$ and $dU = \hat{C}_v\, dT$ where $\hat{C}_p = \hat{C}_v$.

2.13 Two streams of medium oil are to be mixed and heated in a steady process. For the conditions shown in the sketch, calculate the outlet temperature T. The specific heat at constant pressure is 0.52 Btu/lbm °F and the base temperature for zero enthalpy is 32°F.

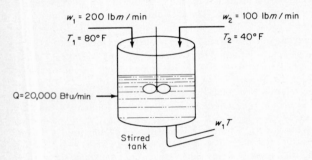

2.14 An organic liquid is being evaporated in a still using hot water flowing in a cooling coil. Water enters the coil at 130°F at the rate of 1,000 1bm/min, and exits at 140°F. The organic liquid enters the still at the same temperature as the exiting vapor. Calculate the steady-state vapor flow rate in 1bm/sec.

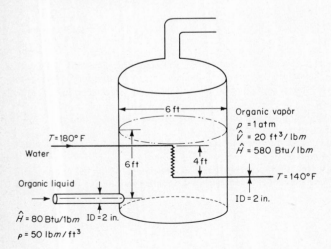

2.15 Steam at 200 psia, 600°F is flowing in a pipe. Connected to this pipe through a valve is an evacuated tank. The valve is suddenly opened and the tank fills with steam until the pressure is 200 psia and the valve is closed. If the whole process is insulated (no heat transfer) and $\hat{K}$ and $\hat{\phi}$ are negligible, determine the final internal energy of the steam.

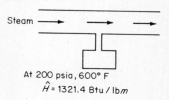

Steam ⟶

At 200 psia, 600° F
$\hat{H}$ = 1321.4 Btu / lbm

2.16 Oil (sp. gr. = 0.8) is draining from the tank as shown

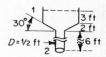

30° | 3 ft
2 ft
D = ½ ft → | 6 ft
2

Find: (a) efflux velocity as a function of height, (b) initial volumetric flow rate (gal/min), and (c) mass flow rate (1bm/sec). (d) How long will it take to empty the upper 5 ft of the tank?

2.17

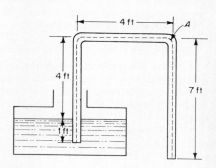

|— 4 ft —| A

4 ft

7 ft

1 ft

A siphon as shown in the sketch is used to drain gasoline (density = 50 lbm/ft³) from a tank open to the atmosphere. The liquid is flowing through the tube at a rate of 5.67 ft/sec.

(a) What is the lost work $\widehat{lw}$?

(b) If the $\widehat{lw}$ term is assumed to be linear with tube length, what is the pressure inside the bend at point A?

2.18 Cooling water for a heat exchanger is pumped from a lake through an insulated 3-in.-diameter schedule 40 pipe. The lake temperature is 50°F and the vertical distance from the lake to the exchanger is 200 ft. The rate of pumping is 150 gal/min. The electric power input to the pump is 10 hp.

(a) Draw the flow system showing the control surfaces and control volume with reference to a zero energy reference plane.

(b) Find the temperature at the heat exchanger inlet where the pressure is 1 psig.

2.19 It is proposed to generate hydroelectric power from a dammed river as shown below.

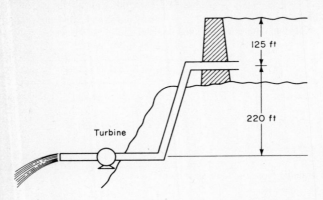

The diameter of the conduit is 25 ft, and the bulk velocity of the water is 10 mph at the outlet.

(a) Draw a control volume for the system and label all control surfaces.

(b) Find the power generated by a 90 percent efficient water turbine if $\widehat{lw} =$ 90 ft lbf/lbm.

2.20 River water at 70 °F flowing at 10 mph is diverted into a $30' \times 30'$ channel which is connected to a turbine operated 500 ft below the channel entrance. The water pressure at the channel inlet is 25 psia and the pressure 20 ft below the turbine is atmospheric. An estimate of the lost work in the channel is 90 ft lbf/lbm.

(a) Draw a flow diagram showing all control volumes and control surfaces with reference to a zero energy plane.

(b) Calculate the horsepower from a 90 percent efficient water turbine.

2.21 Water flows from the bottom of a large tank where the pressure is 90 psia to a turbine which produces 30 hp. The turbine is located 90 ft below the bottom of the tank. The pressure at the turbine exit is 30 psia and the water velocity is 30 ft/sec. The flow rate in the system is 120 lbm/sec. If the frictional loss in the system, not including the turbine, is 32 ft lbf/lbm,

(a) Draw a flow system showing the control volume and control surfaces with reference to a zero energy plane.

(b) Find the efficiency of the turbine.

2.22 A pump operating at 80 percent efficiency delivers 30 gal/min of water from a reservoir to a chemical plant 1 mile away. A schedule 40 3-in. pipe is used to transport the water and the lost work due to pipe friction is 200 ft lbf/lbm. At the plant the fluid passes through a reactor to cool the reacting chemicals and 800,000 Btu/hr are transferred before the water is discharged to the drain. The elevation of the plant is 873 ft above sea level; the elevation of the reservoir is 928 ft above sea level.

(a) What is the minimum horsepower required for this pump?

(b) The pump is removed from the system; will water flow?

Can you calculate the flow rate? Do so or state why not.

2.23 Nitrogen gas is flowing isothermally at 77°F through a horizontal 3-in. inside-diameter pipe at a rate of 0.28 lbm/sec. The absolute pressures at the inlet and outlet of the pipe are 2 atm and 1 atm respectively. Determine the lost work $\widehat{lw}$ in Btu/lbm. The gas constant is 0.7302 ft³ atm/lb mole °R. Molecular weight of N_2 is 28.

2.24 Water is pumped from a lake to a standpipe. The input pipe to the pump extends 10 ft below the level of the lake; the pump is 15 ft above lake level and the water in the standpipe is almost constant at 310 ft above the pump inlet. The lost work due to friction is 140 ft lbf/lbm for pumping water through the 6,000 ft of 4-in. schedule 40 pipe used. If the pump capacity is 6,000 gal/hr, what is the work required from the pump in ft lbf/lbm?

2.25 Water flows from the bottom of a large tank where the pressure is 100 psig through a pipe to a turbine which produces 5.82 hp. The pipe leading to the turbine is 60 ft below the bottom of the tank. In this pipe (at the turbine) the pressure is 50 psig, and the velocity is 70 ft/sec. The flow rate is 100 lbm/sec. If the friction loss in the system, not including the turbine, is 40 ft lbf/lbm, find the efficiency of the turbine.

2.26 Water flowing in a 1-in.-ID pipe is suddenly expanded to a 9-in.-ID pipe by means of an expanding conical section similar to a nozzle. Heat is added at 3 Btu/ lbm. Find the outlet temperature if the inlet temperature is 80°F, inlet pressure is 40 psig, outlet pressure is 30 psig, and the outlet velocity is 5 ft/sec. Assume

$$\Delta \widehat{H} = \left(1 \; \frac{\text{Btu}}{\text{lbm °F}}\right) \Delta T + \frac{\Delta p}{\rho}$$

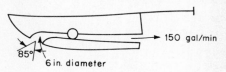

with constant density $\rho = 62.4$ lbm/ft³.

2.27 A fire hose has an inside diameter of 1-½ in. and a nozzle diameter of 0.5 in. The inlet pressure to the hose is 40 psig at a water flow of 8 ft³/min. If a flat velocity profile is assumed what is the value of and the direction of the force exerted by the fireman holding the hose?

2.28 Your company is planning to manufacture a boat propelled by a jet of water. The boat must be capable of exerting a 200-lbf pull on a line which holds it motionless. The intake to the boat is inclined at an angle of 85° and is a pipe with a 6-in. inside diameter. The motor and pump are capable of delivering 150 gal/min through the horizontal outlet. Calculate the outlet pipe size which gives the required thrust. Calculate the minimum horsepower output required of the motor if its efficiency is 100 percent. List all your assumptions.

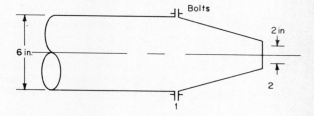

2.29 Water is flowing out of the frictionless nozzle shown below where $p_1 = 120$ psig and the nozzle discharges to the atmosphere.

(a) Show a control volume and all control surfaces with all forces labeled.

(b) Find the resultant force in the bolts at (1) which is necessary to hold the nozzle on the pipe.

(c) If mercury, sp. gr. = 13.6, were flowing through the nozzle instead of water at the same conditions, what would be the new resultant force on the bolts?

2.30 Consider the straight piece of pipe

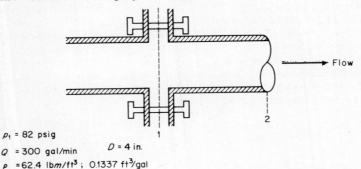

p_1 = 82 psig

Q = 300 gal/min D = 4 in.

ρ = 62.4 lbm/ft^3 ; 0.1337 ft^3/gal

(a) Draw the control volume and label the control surfaces.

(b) Calculate the force on the bolts at point 1. Are the bolts in tension or compression?

2.31 Consider steady, constant-temperature flow of water through the 45° reducing elbow shown below. The volume of the elbow is 28.3 in.^{3}.

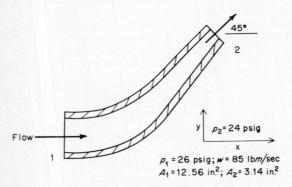

p_2 = 24 psig

p_1 = 26 psig; w = 85 lbm/sec

A_1 = 12.56 in.^{2}; A_2 = 3.14 in.^{2}

Draw a control volume and label all control surfaces and forces on the control volume. Calculate the *total* resultant force acting on the elbow.

2.32 Two streams of water join together at a reducing tee where the upstream pressure of both is 40 psig. Stream 1 is flowing in a 1-in. pipe at 50 lbm/min, stream 2 in a 2-in. pipe at 100 lbm/min, and stream 3 exits from the tee in a 3-in. pipe. The straight run flow is from stream 2 to stream 3. Calculate the force of the pipe threads on the tee if the downstream pressure is 35 psig. Use positive coordinates in the direction of flow as shown in the following figure.

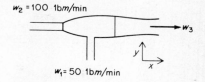

$w_2 = 100$ lb$m$/min

w_3

$w_1 = 50$ lb$m$/min

2.33 Consider the horizontal lawn sprinkler shown. Observe that if the sprinkler is split in half as shown, the two halves are identical and thus only one half need be considered. For one-half of the sprinkler:

(a) Use a mass balance to obtain an expression for v_0 in terms of v_1.
(b) Use the frictionless mechanical energy balance to find v_1.
(c) Find the value of F_R that will just prevent the sprinkler from rotating.

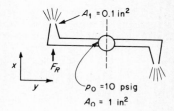

$A_1 = 0.1$ in^2

F_R

$p_0 = 10$ psig
$A_0 = 1$ in^2

2.34 A stand is to be built to hold a rocket ship stationary while a lateral or side thrust engine is being fired. Twenty lb$m$/sec of fuel is consumed and ejected only out of the side engine at a velocity of 4,000 ft/sec. The direction of flow is as shown in the diagram. Find the x and y components of the restraining force required of the stand.

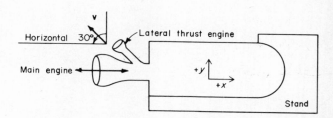

Horizontal 30°

Lateral thrust engine

Main engine

$+y$
$+x$

Stand

2.35 The tank shown below is secured to a concrete slab by bolts and receives an organic liquid (sp. gr. 0.72) from a filler line which enters the tank 1 ft above the tank floor. The liquid enters at a rate of 4,500 gal/min. What are the forces exerted on the restraining bolts? Assume that the filler line transmits no force to the tank (due to expansion joints).

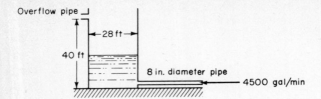

Overflow pipe ⌐

40 ft ←28 ft→

8 in. diameter pipe

4500 gal/min

2.36 A cylindrical storage tank is connected to a nozzle by a short length of horizontal pipe as shown in the diagram. At time equal to zero, the height of liquid in the tank is H ft above the axis of the discharge pipe. The liquid discharges through the nozzle into the atmosphere as the liquid level $h(t)$ changes.

(a) Using a mass balance, show that the liquid level $h(t)$ is related to the nozzle discharge velocity v_2 by

$$\frac{dh}{dt} = -\left(\frac{D_2}{D_T}\right)^2 v_2$$

(b) Using a mechanical energy balance on the discharge pipe and the results of part (a), show that the nozzle discharge velocity v_2 and the gage pressure at the pipe entrance $(p_1 - p_2)$ are given as functions of time by

$$v_2(t) = \left(\frac{2gH}{\rho(1-\beta^2)}\right)^{1/2} - \frac{g}{\rho(1-\beta^2)}\left(\frac{D_2}{D_T}\right)^2 t$$

where $\beta = (D_2/D_1)^2$

$$p_1 - p_2 = \left[H^{1/2} - \frac{1}{2}\left(\frac{D_2}{D_T}\right)^2\left(\frac{2g}{\rho(1-\beta^2)}\right)^{1/2} t\right]^2$$

(c) Using a momentum balance on the nozzle and the results of part (b), show that the force exerted on the threads of the nozzle F_r is given as a function of time by

$$F_R = \left[\rho D_2^2 (1-\beta^2)\left(\frac{2g}{\rho(1-\beta^2)}\right)^{1/2} - D_1^2\right]\frac{\pi}{4}\rho D_2^2\left[H^{1/2} - \frac{1}{2}\left(\frac{D_2}{D_T}\right)^2\left(\frac{2g}{1-\beta^2}\right)^{1/2} t\right]^2$$

Assume there is no pressure drop across the short section of pipe between the tank and the nozzle.

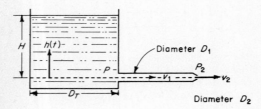

H $h(t)$

Diameter D_1

P P_2

v_1 v_2

D_T

Diameter D_2

3
Dimensional Analysis

This chapter is an introduction to the fundamentals of dimensional analysis.[1-4] Dimensional analysis is defined; systems of measurement and dimensions are discussed. Buckingham's theorem is introduced and two methods are presented for dimensional analysis of data. Examples of the use of these methods are given. The concepts of similarity and scaling are introduced.

[1] P. W. Bridgman, "Dimensional Analysis," Yale University Press, New Haven, Conn., 1931.

[2] W.J. Duncan, "Physical Similarity and Dimensional Analysis," Edward Arnold (Publishers) Ltd., London, 1953.

[3] R.E. Johnstone and M.W. Thring, "Pilot Plants, Models and Scale-up Methods in Chemical Engineering," McGraw-Hill Book Company, New York, 1957.

[4] H.L. Langhaar, "Dimensional Analysis and Theory of Models," John Wiley & Sons, Inc., New York, 1951.

Most problems in transport of momentum, heat, and mass are too complicated to solve by integrating the rate equations. In many instances the correct form of the equations is not known. Dimensional analysis is a means of analyzing data by assuming the data are described by a dimensionally homogeneous equation. The dimensionally homogeneous equation of which we are speaking is normally either the rate equation for the process or its integrated form. The advantage of the dimensional analysis is that one does not need to know the explicit functional form of either of these two equations in order to obtain the dimensionless groups which, when correlated from laboratory data, will, in fact, give the simplest form for the fundamental equation.

Dimensions are a means of telling how the numerical value of a given quantity changes when its unit of measurement changes. For example, 1.0 kg is equivalent to 2.2 lb. An equation is dimensionally homogeneous if the form of the equation does not depend on the fundamental units of measurement. An equation of the form $y = a + b + c \ldots$ is dimensionally homogeneous if and only if y, a, b, $c \ldots$ have the same dimensions. The equation for distance traveled starting from rest with uniform acceleration, $s = \frac{1}{2}gt^2$, is valid whether s is measured in centimeters, feet, etc. and time is in seconds, minutes or days, etc. The equation is dimensionally homogeneous as defined. If 32.2 ft/sec^2 is substituted for g the equation becomes $s = 16.1\ t^2$. This equation is correct on the earth at sea level but it is no longer dimensionally homogeneous since the factor 16.1 applies only if s is in feet and time is in seconds.

The variables in dimensional analysis problems are combined in such a way that the resulting variables are fewer and dimensionless. Listed below is an example for pipe flow in which the significant variables are the pressure difference, length of the conduit, velocity of the fluid, density of the fluid, the viscosity of the fluid, speed of sound in the fluid, surface tension of the fluid, and acceleration of gravity; but we reduce these eight variables by means of dimensional analysis to only five variables, each of which is dimensionless. The net result of such combinations is a reduction in the number of variables that must be considered. This reduction results in a simplification in the treatment of data and in understanding the meaning of data. Dimensional analysis of variables is important in all phases of analysis since it allows generalization of results. Further, it should be noted that this generalization is the basis for scale-up of process equipment.

The idea of a dimensionless group is familiar to all of us in the common concept of percentage. By use of percentage, we can describe widely varying things in a way which is independent of the units in which they are measured. For example, we can talk about the percent increase in population or the percent interest on a bank loan or the percent of time worked during the year. If I plan to take an automobile trip, and a certain time after leaving I tell you I have gone 60 percent of the distance, this gives you certain information which is independent of the total distance of the trip. Dimensional analysis is basically an expression of something with which we are all familiar—similarity. Within reason, if we know how long an elephant's trunk is we have some idea of how long his ear is. We know this because baby elephants are basically similar to big elephants; that is, they are small-scale replicas of big elephants. Therefore, the dimensionless ratio of ear length to trunk length will, within reason, remain constant. (Zoologists will be somewhat unhappy with this statement since, of course, the ratio will change slightly, but the basic idea is contained within the analogy.)

The objective of this chapter is to introduce the concept of dimensional analysis as a means of systematically reducing the number of variables used to express data by reducing the variables to dimensionless form. In this way the amount of data to be manipulated is reduced, making it easier to understand and to use. Since the same mathematical model often describes different phenomena (consider the analogies of mass, momentum, and heat transfer[1]), treating data for similar phenomena by making the data dimensionless not only reduces the total number of variables to be considered (thus reducing the total amount of data), but allows similar analysis of data accumulated from *different* experiments or processes. The chapter introduces the necessary fundamentals and methods that can be used to help analyze data and provides the background for the design equations presented in later chapters.

3.1 SYSTEMS OF MEASUREMENT

Conventional systems of measurement conform to Newton's second law (force = mass × acceleration) in that the weight of a unit mass is equal to g units of force. Table 3.1-1 shows the units of length, time, mass, force, work, and weight for six common systems of measurement. Notice that length and time are always fundamental units in these six systems. In the centimeter-gram-second (cgs) *mass* system the gram is the unit of mass, the dyne (the force that gives a gram mass an acceleration of one centimeter per second squared) is the unit of force, and the weight (force exerted by) a gram mass is $w = mg = 981$ dynes. The unit of work is the erg (dyne cm).

---

[1] R. B. Bird, W. E. Stewart, and E.N. Lightfoot, "Transport Phenomena," John Wiley & Sons, Inc., New York, 1960.

Table 3.1-1 Units for six common systems of measurement

Unit of system	Cgs (mass)	mks (mass)	mks (force)	British (mass)	American Engineering	$(LtMF)^a$
Length	centimeter (cm)	meter (m)	meter (m)	foot (ft)	foot (ft)	foot (ft)
Time	second (sec)	second (sec)	second (sec)	second (sec)	second (sec)	second (sec)
Mass ($F = ma$)	gram (g)	kilogram (kg$m$)	kilogram second²/meter (kg$f$ sec²/m)	pound (lb$m$)	slug (lb$f$ sec²/ft)	pound mass (lb$m$)
Force ($F = ma$)	dyne (g cm/sec²)	newton (kg$m$/sec²)	kilogram (kg$f$)	poundal (lb$m$ ft/sec²)	pound (lb$f$)	pound force (lb$f$)
Weight ($w = mg$)	981 dynes	9.81 newtons	9.81 kg$f$	32.2 poundals	32.2 lb$f$	pound force $\dfrac{32.17/\text{m}}{g_c} = \text{lb}f$
Work ($W = Fd$)	erg (dyne cm)	joule (newton m)	kilogram meter (kg m)	foot poundal	foot pound (ft lb$f$)	foot pound (ft lb$f$)

[a] In this system Newton's second law of motion must be written containing a dimensional constant g_c, since a unit of mass no longer weighs g units of weight at the earth's surface. $F = (m/g_c)a$.

In the meter-kilogram-second (mks) *mass* system the kilogram is the unit of mass, the newton (the force that gives a kilogram mass an acceleration of one meter per second squared) is the unit of force, and the unit of weight is 9.81 newtons. The joule (newton meter) is the unit of work (a joule is 10^7 ergs).

In the mks *force* system the kilogram is the unit of force (weight of a kilogram of mass under standard pull of gravity—980,655 dynes) and the unit of *mass* is (kilogram force) (seconds)^2/meter. A kg$f$ sec^2/m (a unit mass) weighs 9.81 kg$f$.

The unit of mass in the British mass system is the pound and the unit of force is that force that gives a pound mass an acceleration of one foot per second squared. This force unit is called the poundal. The weight of a pound mass in the British mass system is 32.2 poundals.

In the American engineering system the pound is the unit of force and the mass unit is (pound force) (seconds)^2/feet which is called the slug. A slug weighs 32.2 lb$f$.

One of the most common systems of measurement in the United States contains four fundamental units rather than three; these units are the pound force lb$f$, the pound mass lb$m$, the length L, and the time t. The pound force is defined as the force which will give one pound mass an acceleration of 32.174 ft/sec^2. For this system Newton's second law must be defined with a dimensional constant to make it correspond to observation and $F = ma/g_c$ where $g_c = 32.174$. (In the above systems this constant is 1.) The pound force in this system does not depend on the local value of g.

3.2 BUCKINGHAM'S THEOREM

Dimensionally homogeneous functions are a special class of functions and the theory of dimensional analysis is the mathematical theory of this class of functions. Dimensional analysis is based on the hypothesis that the mathematical model for a given problem is a dimensionally homogeneous equation. We assume nature works this way. A set of dimensionless products obtained from a set of variables described by a dimensionally homogeneous equation is *complete* if each product in the set is independent of all the others and every other dimensionless product of the variables is not independent but is a product of powers of dimensionless products in the set.

For example, if we have two dimensionless products, the Reynolds number Re $= vL\rho/\mu$ and the Prandtl number Pr $= \widehat{C_p}\mu/k$ in a complete set, the dimensionless product found by multiplying these two together, PrRe, is not an independent product. Buckingham's theorem states:

If an equation is dimensionally homogeneous, it can be reduced to a relationship among a *complete set* of dimensionless products. One can show if the variables (quantities) are related by

$$f(Q_1,Q_2,Q_3,\ldots,Q_n) = 0 \tag{3.2-1}$$

Where Q are the variables and there are n of them, the independent dimensionless products in a complete set are related by

$$g(\pi_1,\pi_2,\pi_3,\ldots,\pi_{n-r}) = 0 \tag{3.2-2}$$

where π_i are the dimensionless variables and r is the rank of the dimensional matrix.[1] Usually $r = k$, the number of fundamental dimensions. Therefore in most applications the number of dimensionless products is equal to the number of variables minus the number of dimensions. Example 3.2-3 illustrates the relatively rare case where $r \neq k$.

Example 3.2-1 A complete set of dimensionless variables for flow through a pipe The common variables involved when a fluid flows through a conduit such as a pipe are the pressure difference Δp, which is a force per unit area causing flow (usually the pressure drop across the conduit); the length of the conduit, L; the velocity of the flowing fluid, v; the density of the fluid, ρ; the fluid viscosity μ; the speed of sound in the fluid, c; the surface tension of the fluid, σ; the diameter of the conduit, D; and the acceleration of gravity, g. If we express the variables in an MLt (mass, length, time) system of measurement there will be six dimensionless products (number of variables—number of dimensions $= 9 - 3 = 6$). Since at this point we have not yet considered a detailed procedure for developing the dimensionless groups, we will simply write these down and proceed. The point here is to illustrate what these groups are. A complete set of dimensionless products for these variables is

Reynolds number $\qquad \mathrm{Re} = \dfrac{Dv\rho}{\mu} = \dfrac{\text{inertial force}}{\text{viscous force}}$ $\qquad$ (3.2-3)

Euler number $\qquad \mathrm{Eu} = \dfrac{\Delta p}{\rho v^2} = \dfrac{\text{pressure force}}{\text{inertial force}}$ $\qquad$ (3.2-4)

Froude number $\qquad \mathrm{Fr} = \dfrac{v^2}{Lg} = \dfrac{\text{inertial force}}{\text{gravity force}}$ $\qquad$ (3.2-5)

---

[1]The dimensional matrix is constructed by writing the variables as column headings and the fundamental dimensions as row headings and filling in the matrix with the power of the dimension for each variable. The rank is the order of the highest-order nonzero determinant of the matrix.

Mach number $\qquad\qquad \text{Ma} = \dfrac{v}{c} = \dfrac{\text{velocity}}{\text{sonic velocity}}$ (3.2-6)

Weber number $\qquad \text{We} = \dfrac{Lv^2\rho}{\sigma} = \dfrac{\text{inertial forces}}{\text{interfacial forces}}$ (3.2-7)

Dimensionless length $\qquad \dfrac{L}{D}$ (3.2-8)

Note that each product has a variable which the others do not contain; therefore it can be seen that they are independent. Further, it is not possible to find another dimensionless product using the variables in this set which is not a combination of these six products. There are *only* six independent products.

□ □ □

As we shall continue to mention, these dimensionless groups have physical significance. They are the parameters in the dimensionless differential equations which are the mathematical model for the system involved.[1] (As such the products may, for example, represent the ratios of forces acting on the system.) Example 3.2-1 is a practical illustration of Buckingham's theorem. If n variables are related by an unknown dimensionally homogeneous equation (the differential equation describing the process), the solution of this equation can be expressed as a relationship among $n - r$ dimensionless products (a complete set). The value of r is the rank of the dimensional matrix and is usually equal to k, the number of fundamental dimensions of the dimension system used.

Example 3.2-2   Drag force on a ship[2]   The drag force water exerts on the hull of a ship depends on the shape of the hull. Dimensional analysis can seldom be used to predict the ways in which phenomena are affected by varying shapes. Therefore, when applying dimensional analysis to a problem it is usually convenient to eliminate the consideration of shape effects by considering bodies of the same shape, i.e., bodies that are geometrically similar.[3] If we limit this example to consider only hulls of similar shape, the hull is specified by a single characteristic dimension such as its length L. If hulls are considered to have the same relative waterline, L also takes care of the draft. Then, the drag force F depends on the length L, the speed of the ship, v, the viscosity of the water, μ, the density of the

---

[1] See, for example, D. V. Boucher, and G. E. Alves, *Chem. Eng. Progr.*, **1959**:55, **1963**:75.

[2] H.L. Langhaar, "Dimensional Analysis and Theory of Models," John Wiley & Sons, Inc., New York, 1951.

[3] Geometrically similar bodies have corresponding lengths in a constant ratio.

water, ρ, and the acceleration of gravity, g. Gravity is important where flow is around floating objects (since energy used to propel the ship is dissipated by waves and the energy dissipation of the waves depends on g). There is an equation of the form

$$(a) \qquad f(F, v, L, \mu, \rho, g) = 0$$

Again, we skip the intermediate steps for simplicity and simply write down the result. The reader can, in fact, check to see that the dimensionless groups below are independent. A complete set of dimensionless products for this situation consists of the pressure coefficient (Euler number) $F/\rho v^2 L^2$, the Reynolds number $vL\rho/\mu$, and the Froude number v^2/Lg. Therefore equation (a) reduces to the form

$$g(\mathrm{Eu}, \mathrm{Re}, \mathrm{Fr}) = 0$$

or

$$(b) \qquad \mathrm{Eu} = h\,(\mathrm{Re}, \mathrm{Fr})$$

Equation (b) would be represented graphically by plotting Eu versus Re with a curve for each value of Fr. Note that for our original problem which contained six variables, plotting the data would require almost a library of charts. By doing the dimensional analysis, however, we have reduced this functional relationship to one among only three dimensionless groups, which can be plotted with one on the ordinate, one on the abscissa, and the third as a parameter. Since what we are normally interested in is the drag force, we substitute for the Euler number and rearrange to obtain

$$F = \rho v^2 L^2 h(\mathrm{Re}, \mathrm{Fr})$$

The cross-sectional area for that portion of the hull below the waterline is proportional to L^2, with a constant of proportionality (determined by the shape of the hull) which may be defined such that $L^2 = \frac{1}{2}kA$. Substituting this in equation (b) and defining

$$C_{\mathrm{drag}} = kh(\mathrm{Re}, \mathrm{Fr})$$

where C_{drag} is called the drag coefficient, gives

$$(c) \qquad F = \frac{1}{2}\,C_{\mathrm{drag}}\,\rho v^2 A$$

This is the commonly used expression for drag on the hull of a ship, and contains the kinetic energy $\rho v^2/2$, in which v is the velocity of the ship through the water, and A, which is the cross section of the hull normal to the flow. Equation (c) is of the same form as the formula for the drag on a *totally* immersed body, but here the drag coefficient C_{drag} depends on both the Reynolds number *and* the Froude number.

□ □ □

Example 3.2-3 Rank of matrix different from number of fundamental dimensions The important variables for determining the maximum pressure p_{max} resulting when flow of a compressible fluid is stopped instantly (shutting a valve) are velocity v of the fluid at shutoff, density of the fluid ρ, and the bulk modulus of the fluid $\beta = \rho(\partial p/\partial \rho)_T$. Functionally

$$p_{max} = p_{max}(v,\rho,\beta)$$

Find a complete set of dimensionless groups for this problem.
Solution The dimensional matrix is (*MLt* system)

	p_{max}	v	ρ	β
M	1	0	1	1
L	-1	1	-3	-1
t	-2	-1	0	-2

With $n = 4$, $k = 3$, $n - k = 1$

$$\pi_1 = \rho^a v^b \beta^c p_{max}$$

By inspection $\pi_1 = p/\beta$. This does not look reasonable since v and ρ are not used directly. Check the rank of the dimensional matrix.

3×3 DETERMINANTS

$$\begin{vmatrix} 1 & 0 & 1 \\ -1 & 1 & -3 \\ 2 & -1 & 0 \end{vmatrix} = 0$$

$$\begin{vmatrix} 0 & 1 & 1 \\ 1 & -3 & -1 \\ -1 & 0 & -2 \end{vmatrix} = 0$$

$$\begin{vmatrix} 1 & 1 & 1 \\ -1 & -3 & -1 \\ -2 & 0 & -2 \end{vmatrix} = 0$$

$$\begin{vmatrix} 1 & 0 & 1 \\ -1 & 1 & -1 \\ -2 & -1 & -2 \end{vmatrix} = 0$$

2 × 2 DETERMINANTS

$$\begin{vmatrix} 1 & 0 \\ -1 & 1 \end{vmatrix} \neq 0$$

Therefore $r = 2$.
Note: for this problem $k \neq r$

$$n - r = 2$$

$$\pi_1 = \beta^a v^b p_{max} \qquad \pi_2 = \beta^c v^d \rho$$

$$\pi_1 = \frac{p_{max}}{\beta} \qquad \pi_2 = \frac{\rho v^2}{\beta}$$

□ □ □

3.3 SYSTEMATIC ANALYSIS OF VARIABLES

We will now consider a systematic way of obtaining dimensionless groups from the variables in a given problem. The method we consider below involves a fair amount of algebraic manipulation, but as an initial approach gives one a better feel for what is happening. The whole procedure below can be done very concisely using matrix methods, which, in fact, almost *must* be used for problems involving many variables. Appendix 3A demonstrates this. Buckingham's theorem states for $r = k$ that the general form of the functional relationship among n independent quantities Q_1, $Q_2, \ldots, Q_n$[1]

$$f(Q_1, Q_2, \ldots, Q_n) = 0 \qquad (3.3\text{-}1)$$

which involves k fundamental units can be reduced to a relationship

---

[1] For most cases of interest to us, r, the rank of the dimensional matrix, will be equal to k, the number of fundamental dimensions. We restrict ourselves to such cases in the remaining discussion.

among $n - k$ independent dimensionless products, the π's, thus:

$$g(\pi_1, \pi_2, \pi_3, \ldots, \pi_{n-k}) = 0 \tag{3.3-2}$$

Any of the π's will not depend on more than $k + 1$ of the Q's because a complete set is independent. To find the π's:

1. List the variables and their dimensions.
2. Find the number of fundamental dimensions.
3. Select a set of variables equal to the number of fundamental dimensions and designate them by overscoring them with a tilde (e.g., $\tilde{Q}$) such that none of these quantities ($\tilde{Q}$'s) is dimensionless, the set of $\tilde{Q}$'s includes all the fundamental dimensions, and no two $\tilde{Q}$'s of the set have the same dimensions.
4. The dimensionless products (π's) are found one at a time by finding the product of the $\tilde{Q}$'s raised to an unknown power times one of the remaining variables (Q's) to a known power. This process is repeated $n - k$ times, thus using all the variables and obtaining $n - k$ dimensionless products (π's).
5. The exponents of the $\tilde{Q}$'s are found by the principle of dimensional homogeneity (i.e., that the dimensions will cancel in each product).

In general, any variable that is dimensionless but is deemed important may be considered one of the π's. Also, if any two variables have the same dimensions their ratio may be one of the π's. Any algebraic manipulation of the dimensionless products that does not change the number of dimensionless products is allowed. (That is, a π may be replaced by any power of itself, by its product with any other π raised to any power, or by its product with a numerical constant.)

Example 3.3-1 Equivalence of dimensional analysis for different choices of fundamental dimensions (flow in a pipe)[1] Below is a list of the significant variables for flow in a pipe. Apply the systematic procedure for dimensional analysis using the fundamental systems of units LtM, LMF, LFt, tMF, and $LtMF$ and show that the results obtained are equivalent.

[1]We proceed formally in systems where (1) length, time, force, and mass are all taken as fundamental units and (2) length, time, force, and temperature are all taken as fundamental units by including an appropriate *constant* or *constants* in the list of *variables*. These constants are usually g_c and the mechanical equivalent of thermal energy as measured by temperature. This is a purely formal procedure which gives satisfactory results. The reason for doing this is that there are additional constraints on the system which are *physical* (not mathematical) in the cases above. In (1), Newton's law must be satisfied, and in (2) the mechanical equivalent of thermal energy which is also purely a physical relationship also constrains the system. This point is discussed at some length in Appendix 3A.

Solution

	(I) MLt	(II) FML	(III) FLt	(IV) FMLt	(V) tMF
$Q_1 = \Delta p$	$M/t^2 L$	F/L^2	F/L^2	F/L^2	$M^2/t^4 F$
$Q_2 = L$	L	L	L	L	Ft^2/M
$Q_3 = D$	L	L	L	L	Ft^2/M
$Q_4 = v$	L/t	$L/(ML/F)^{1/2}$	L/t	L/t	Ft/M
$Q_5 = \mu$	M/Lt	$M/L(ML/F)^{1/2}$	Ft/L	M/Lt	$M/(Ft^2/M)t$
$Q_6 = \rho$	M/L	M/L^3	Ft^2/L^4	M/L^3	$M/(Ft^2/M)^3$
$Q_7 = k$	L	L	L	L	Ft^2/M
$Q_8 = g_c$	$\uparrow$	$\uparrow$	$\uparrow$	ML/Ft^2	$\uparrow$
	$F = ML/t^2$	$t = (ML/F)^{1/2}$	$M = Ft^2/L$		$L = Ft^2/M$

I. $\tilde{Q}_1 = \Delta p,\ \tilde{Q}_3 = D,\ \tilde{Q}_4 = v$

π_1: $L \Delta p^a D^b v^c$

$$L \left(\frac{M}{t^2 L}\right)^a L^b \left(\frac{L}{t}\right)^c$$

$1 - a + b + c = 0 \qquad a = 0$
$a = 0 \qquad c = 0$
$-2a - c = 0 \qquad b = -1$

Therefore
$$\pi_1 = \frac{L}{D}$$

π_2: $\mu \Delta p^a D^b v^c$

$$\frac{M}{Lt} \left(\frac{M}{t^2 L}\right)^a L^b \left(\frac{L}{t}\right)^c$$

$-1 - a + b + c = 0 \qquad a = -1$
$1 + a = 0 \qquad b = -1$
$-1 - 2a - c = 0 \qquad c = 1$

Therefore
$$\pi_2 = \frac{\mu v}{\Delta p D}$$

π_3: $\rho \Delta p^a D^b v^c$

$$\frac{M}{L^3} \left(\frac{M}{t^2 L}\right)^a L^b \left(\frac{L}{t}\right)^c$$

$-3 - a + b + c = 0 \qquad a = -1$
$1 + a = 0 \qquad b = 0$
$-2a - c = 0 \qquad c = 2$

Therefore
$$\pi_3 = \frac{\rho v^2}{\Delta p}$$

π_4: $k \Delta p^a D^b v^c$

$$L \left(\frac{M}{t^2 L}\right)^a L^b \left(\frac{L}{t}\right)^c$$

$+1 - a + b + c = 0 \qquad a = 0$
$a = 0 \qquad b = -1$
$-2a - c = 0 \qquad c = 0$

Therefore

$$\pi_4 = \frac{k}{D}$$

II. $\tilde{Q}_1 = \Delta p,\ \tilde{Q}_3 = D,\ \tilde{Q}_6 = \rho$

$\pi_1: L(\Delta p)^a D^b \rho^c$

$$L\left(\frac{F}{L^2}\right)^a L^b \left(\frac{M}{L^3}\right)^c$$

$1 - 2a + b - 3c = 0 \qquad a = 0$
$a = 0 \qquad b = -1$
$c = 0 \qquad c = 0$

Therefore

$$\pi_1 = \frac{L}{D}$$

$\pi_2: v^2(\Delta p)^a D^b \rho^c$

$$\frac{L^2 F}{ML}\left(\frac{F}{L^2}\right)^a L^b \left(\frac{M}{L^3}\right)^c$$

$1 - 2a + b - 3c = 0 \qquad a = -1$
$1 + a = 0 \qquad b = 0$
$-1 + c = 0 \qquad c = 1$

Therefore

$$\pi_2 = \frac{v^2 \rho}{\Delta p}$$

$\pi_3: \mu^2(\Delta p)^a (D)^b (\rho)^c$

$$\frac{M^2 F}{L^2 ML}\left(\frac{F}{L^2}\right)^a L^b \left(\frac{M}{L^3}\right)^c$$

$-3 - 2a + b - 3c = 0 \qquad a = -1$
$1 + a = 0 \qquad b = -2$
$1 + c = 0 \qquad c = -1$

Therefore

$$\pi_3 = \frac{\mu^2}{\Delta p D^2 \rho}$$

From I, $\left(\dfrac{\pi_2^2}{\pi_3}\right)_I = (\pi_3)_{II}$

$\pi_4: k(\Delta p)^a (D)^b (\rho)^c$

$$L\left(\frac{F}{L^2}\right)^a L^b \left(\frac{M}{L^3}\right)^c$$

$1 - 2a + b - 3c = 0 \qquad a = 0$
$a = 0 \qquad b = -1$
$c = 0 \qquad c = 0$

Therefore

$$\pi_4 = \frac{k}{D}$$

III. $\tilde{Q}_1 = \Delta p$, $\tilde{Q}_3 = D$, $\tilde{Q}_4 = v$

π_1: $L(\Delta p)^a(D)^b(v)^c$

$$L\left(\frac{F}{L^2}\right)^a L^b \left(\frac{L}{t}\right)^c$$

$1 - 2a + b + c = 0$ $\qquad a = 0$
$a = 0$ $\qquad b = -1$
$c = 0$ $\qquad c = 0$

Therefore

$$\pi_1 = \frac{L}{D}$$

π_2: $\mu(\Delta p)^a(D)^b(v)^c$

$$\frac{Ft}{L^2}\left(\frac{F}{L^2}\right)^a L^b \left(\frac{L}{t}\right)^c$$

$-2 - 2a + b + c = 0$ $\qquad a = -1$
$1 + a = 0$ $\qquad b = -1$
$1 - c = 0$ $\qquad c = 1$

Therefore

$$\pi_2 = \frac{\mu v}{\Delta p \, D}$$

π_3: $\rho(\Delta p)^a(D)^b(v)^c$

$$\frac{Ft^2}{L^4}\left(\frac{F}{L^2}\right)^a L^b \left(\frac{L}{t}\right)^c$$

$-4 - 2a + b + c = 0$ $\qquad a = -1$
$1 + a = 0$ $\qquad b = 0$
$2 - c = 0$ $\qquad c = 2$

Therefore

$$\pi_3 = \frac{\rho v^2}{\Delta p}$$

π_4: $k(\Delta p)^a(D)^b(v)^c$

$$L\left(\frac{F}{L^2}\right)^a L^b \left(\frac{L}{t}\right)^c$$

$-1 - 2a + b + c = 0$ $\qquad a = 0$
$a = 0$ $\qquad b = 1$
$c = 0$ $\qquad c = 0$

Therefore

$$\pi_4 = \frac{k}{D}$$

IV. $\tilde{Q}_1 = \Delta p$, $\tilde{Q}_3 = D$, $\tilde{Q}_4 = v$, $\tilde{Q}_6 = \rho$

π_1: $L(\Delta p)^a D^b v^c \rho^d$

$$L\left(\frac{F}{L^2}\right)^a L^b \left(\frac{L}{t}\right)^c \left(\frac{M}{L^3}\right)^d$$

$1 - 2a + b + c - 3d = 0$ $\qquad a = 0$
$a = 0$ $\qquad b = -1$
$c = 0$ $\qquad c = 0$
$d = 0$ $\qquad d = 0$

Therefore

$$\pi_1 = \frac{L}{D}$$

π_2: $\mu(\Delta p)^a D^b v^c \rho^d$

$$\frac{M}{Lt} \left(\frac{F}{L^2}\right)^a L^b \left(\frac{L}{t}\right)^c \left(\frac{M}{L^3}\right)^d$$

$-1 - 2a + b + c - 3d = 0$	$a = 0$
$1 + d = 0$	$b = -1$
$a = 0$	$c = -1$
$-1 - c = 0$	$d = -1$

Therefore

$$\pi_2 = \frac{\mu}{Dv\,\rho}$$

From I, $\left(\dfrac{\pi_3}{\pi_2}\right)_I = (\pi_2)_{IV}$

π_3: $k(\Delta p)^a D^b v^c \rho^d$

$$L \left(\frac{F}{L^2}\right)^a L^b \left(\frac{L}{t}\right)^c \left(\frac{M}{L^3}\right)^d$$

$1 - 2a + b + c - 3d = 0$	$a = 0$
$a = 0$	$b = -1$
$c = 0$	$c = 0$
$d = 0$	$d = 0$

Therefore

$$\pi_3 = \frac{k}{D}$$

π_4: $g_c(\Delta p)^a D^b v^c \rho^d$

$$\frac{ML}{Ft^2} \left(\frac{F}{L^2}\right)^a L^b \left(\frac{L}{t}\right)^c \left(\frac{M}{L^3}\right)^d$$

$1 - 2a + b + c - 3d = 0$	$a = 1$
$1 + d = 0$	$b = 0$
$-2 - c = 0$	$c = -2$
$-1 + a = 0$	$d = -1$

Therefore

$$\pi_4 = \frac{g_c \Delta p}{v^2 \rho}$$

V. $\tilde{Q}_1 = \Delta p$, $\tilde{Q}_3 = D$, $\tilde{Q}_4 = v$

π_1: $L(\Delta p)^a D^b v^c$

$$\frac{Ft^2}{M} \left(\frac{M^2}{Ft^4}\right)^a \left(\frac{Ft^2}{M}\right)^b \left(\frac{Ft}{M}\right)^c$$

$1 - a + b + c = 0$	$a = 0$
$-1 + 2a - b - c = 0$	$b = -1$
$2 - 4a + 2b + c = 0$	$c = 0$

Therefore

$$\pi_1 = \frac{L}{D}$$

π_2: $\mu(\Delta p)^a D^b v^c$ $\quad\quad\quad\quad\quad\quad 2 + 2a - b - c = 0 \quad\quad a = -1$

$$\frac{M^2}{F t^3}\left(\frac{M^2}{F t^4}\right)^a \left(\frac{F t^2}{M}\right)^b \left(\frac{F t}{M}\right)^c \quad \begin{array}{l} -1 - a + b + c = 0 \\ -3 - 4a + 2b + c = 0 \end{array} \quad \begin{array}{l} b = -1 \\ c = 1 \end{array}$$

Therefore

$$\pi_2 = \frac{\mu v}{\Delta p\, D}$$

π_3: $\rho(\Delta p)^a D^b v^c$ $\quad\quad\quad\quad\quad\quad 4 + 2a - b - c = 0 \quad\quad a = -1$

$$\left(\frac{M^4}{F^3 t^6}\right)\left(\frac{M^2}{F t^4}\right)^a \left(\frac{F t^2}{M}\right)^b \left(\frac{F t}{M}\right)^c \quad \begin{array}{l} -3 - a + b + c = 0 \\ -6 - 4a + 2b + c = 0 \end{array} \quad \begin{array}{l} b = 0 \\ c = 2 \end{array}$$

Therefore

$$\pi_3 = \frac{\rho v^2}{\Delta p}$$

π_4: $k(\Delta p)^a D^b v^c$ $\quad\quad\quad\quad\quad\quad 1 - a + b + c = 0 \quad\quad a = 0$

$$\left(\frac{F t^2}{M}\right)\left(\frac{M^2}{F t^4}\right)^a \left(\frac{F t^2}{M}\right)^b \left(\frac{F t}{M}\right)^c \quad \begin{array}{l} -1 + 2a - b - c = 0 \\ 2 - 4a + 2b + c = 0 \end{array} \quad \begin{array}{l} b = -1 \\ c = 0 \end{array}$$

Therefore

$$\pi_4 = \frac{k}{D}$$

A comparison of these five sets of groups shows them to be equivalent.

□ □ □

Example 3.3-2 Corresponding states of gases According to a particular three-parameter equation of state, the necessary variables for studying the pressure p, volume V, and temperature T behavior of gases and liquids are given by the functional equation:

$$f(p, V, T_m, p_c, V_c, T_{mc}) = 0$$

The subscript c refers to the critical point and the subscript m refers to the molecular property. (The "molecular temperature" is the product kT where k is Boltzmann's constant). The number of variables can be reduced by applying the theory of dimensional analysis, in the force F, length L system. We follow the systematic procedure described above.

1. $Q_1 = p = [FL^{-2}]$
   $Q_2 = V = [L^3]$
   $Q_3 = T_m = [FL]$
   $Q_4 = p_c = [FL^{-2}]$
   $Q_5 = V_c = [L^3]$
   $Q_6 = T_{mc} = [FL]$
2. $k = 2$, $n = 6$, number of dimensionless products $n - k = 6 - 2 = 4$.
3. Let $\tilde{Q}_1 = Q_1$
   $\tilde{Q}_3 = Q_3$
4. $\pi_1 = p^a (T_m)^b V$
   $\pi_2 = p^c (T_m)^d p_c$
   $\pi_3 = p^e (T_m)^f V_c$
   $\pi_4 = p^g (T_m)^h T_{mc}$
5. π_1 for F: $a + b = 0$ $\qquad\qquad a = 1$

$$L: -2a + b + 3 = 0 \qquad b = -1 \qquad \pi_1 = \frac{pV}{T_m}$$

$$= z \text{ (compressibility}$$

π_2 for F: $c + d + 1 = 0 \qquad c = -1 \qquad\qquad$ factor$)$

$$L: -2c + d - 2 = 0 \qquad d = 0 \qquad \pi_2 = \frac{p_c}{p}$$

π_3 for F: $e + f = 0 \qquad\qquad e = 1$

$$L: -2a + f + 3 = 0 \qquad f = -1 \qquad \pi_3 = \frac{pV_c}{T_m}$$

π_4 for F: $g + h + 1 = 0 \qquad g = 0$

$$L: -2g + h + 1 = 0 \qquad h = -1 \qquad \pi_4 = \frac{T_{mc}}{T_m}$$

Since

$$g(\pi_1, \pi_2, \pi_3, \pi_4) = 0$$

then

$$g\left(\frac{pV}{T_m}, \frac{p_c}{p}, \frac{pV_c}{T_m}, \frac{T_{mc}}{T_m}\right) = 0$$

Rearranging and reorganizing the π's as follows to conform to the form of the van der Waals equation

$$\pi_1' = \pi_1 = \frac{pV}{T_m} = z$$

$$\pi_2' = \frac{1}{\pi_2} = \frac{p}{p_c} = p_r$$

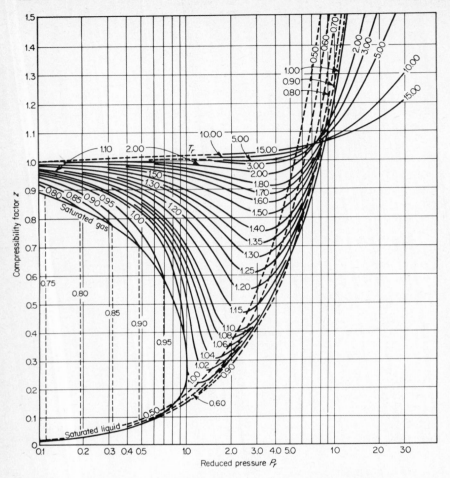

Fig. 3.3-1   Generalized compressibility factors $(z_c = 0.27)$. *O. A. Hougen, K. M. Watson, and R. A. Ragatz, "Chemical Process Principles Charts," John Wiley & Sons, Inc., New York, 1960.*)

$$\pi_3' = \frac{1}{\pi_4} = \frac{T_m}{T_{mc}} = T_r$$

$$\pi_4' = \frac{\pi_2 \pi_3}{\pi_4} = \frac{p_c}{p} \frac{p V_c}{T_m} \frac{T_m}{T_{mc}} = \frac{p_c V_c}{T_m} = z_c$$

gives

(a) $z = z(p_r, T_r, z_c)$

Thus, the six variables are reduced to four, and the last equation describes the data. Equation (a) is the form of the principle of corresponding states with a third parameter. In general all pure fluid pVT data can be described within reasonable limits by equation (a) as shown in Fig. 3.3-1.[1] (The lines represent "best fit" of data—within ± 1 percent.)

□ □ □

Example 3.3-3 Flow of a single fluid in a porous medium[2] Consider the variables involved in flow of a fluid through a porous medium, such as a sand bed, limestone, filter cake, glass beads, etc.

$$f(L,t,v,\mu,p,\mathscr{D},k,g,v_\infty,D,\rho) = 0$$

where the symbols are given in the nomenclature.

The systematic procedure is applied as follows, where the system of fundamental dimensions is the mass M, length L, and time t system:

1. $L = [L]$ $p = [M/Lt^2]$ $D = [L]$

  . $t = [t]$ $\mathscr{D} = [L^2/t]$ $\rho = [M/L^3]$

   $v = [L/t]$ $k = [L^2]$ $g = [L/t^2]$

   $\mu = [M/Lt]$ $v_\infty = [L/t]$

2. The number of variables, n, equals 11; number of fundamental dimensions, k, equals 3. There should be $n - k = 11 - 3 = 8$ dimensionless products.

3. Let $\tilde{Q}_1 = v_\infty$, $\tilde{Q}_2 = D$, $\tilde{Q}_3 = \rho$.

4. $\pi_1 = v_\infty{}^a D^b \rho^a L$ $\pi_5 = v_\infty{}^m D^n \rho^o \mathscr{D}$
   $\pi_2 = v_\infty{}^d D^e \rho^f t$ $\pi_6 = v_\infty{}^p D^q \rho^r k$
   $\pi_3 = v_\infty{}^g D^h \rho^i \mu$ $\pi_7 = v_\infty{}^s D^t \rho^u g$
   $\pi_4 = v_\infty{}^j D^k \rho^l p$ $\pi_8 = v_\infty{}^v D^w \rho^x v$

5. π_1 for M: $c = 0$ $a = 0$
      L: $a + b - 3c + 1 = 0$ $b = -1$ $\pi_1 = \dfrac{L}{D}$
      t: $-a = 0$ $c = 0$

[1] O.A. Hougen, K.M. Watson, and R. A. Ragatz, "Chemical Process Principles Charts," fig. 140a, John Wiley & Sons, Inc., New York, 1960.
[2] R. A. Greenkorn, Flow Models and Scaling Laws for Flow through Porous Media, *Ind. Eng. Chem.*, **56**:32, 1964.

π_2 for M: $\qquad\qquad\qquad f = 0 \qquad\qquad d = 0$

$\qquad L: \qquad d + e - 3f = 0 \qquad e = -1 \qquad \pi_2 = \dfrac{tv}{D}$

$\qquad t: \qquad\qquad -d + 1 = 0 \qquad f = 0$

π_3 for M: $\qquad\qquad\quad i + 1 = 0 \qquad g = -1$

$\qquad L: \quad g + h - 3i - 1 = 0 \qquad h = -1 \qquad \pi_3 = \dfrac{\mu}{v_\infty D\rho}$

$\qquad t: \qquad\qquad -g - 1 = 0 \qquad i = -1$

π_4 for M: $\qquad\qquad\quad l + 1 = 0 \qquad j = -2$

$\qquad L: \quad j + k - 3l - 1 = 0 \qquad k = 0 \qquad \pi_4 = \dfrac{p}{\rho v_\infty^2}$

$\qquad t: \qquad\qquad -j - 2 = 0 \qquad l = -1$

π_5 for M: $\qquad\qquad\qquad o = 0 \qquad o = 0$

$\qquad L: \; m + n - 3o + 2 = 0 \qquad n = -1 \qquad \pi_5 = \dfrac{\mathscr{D}}{v_\infty D}$

$\qquad t: \qquad\quad t - m - 1 = 0 \qquad m = -1$

π_6 for M: $\qquad\qquad\qquad r = 0 \qquad p = 0$

$\qquad L: \quad p + q - 3r + 2 = 0 \qquad q = -2 \qquad \pi_6 = \dfrac{k}{D^2}$

$\qquad t: \qquad\qquad\quad -p = 0 \qquad r = 0$

π_7 for M: $\qquad\qquad\qquad u = 0 \qquad s = -2$

$\qquad L: \quad s + t - 3u + 1 = 0 \qquad t = 1 \qquad \pi_7 = \dfrac{Dg}{v_\infty^2}$

$\qquad t: \qquad\qquad -s - 2 = 0 \qquad u = 0$

π_8 for M: $\qquad\qquad\qquad x = 0 \qquad v = -1$

$\qquad L: \quad v + w - 3x + 1 = 0 \qquad w = 0 \qquad \pi_8 = \dfrac{v}{v_\infty}$

$\qquad t: \qquad\qquad -v - 1 = 0 \qquad x = 0$

The result of this analysis is

$$(a) \qquad G\left(\frac{L}{D}, \frac{tv_\infty}{D}, \frac{v}{v_\infty}, \frac{v_\infty D\rho}{\mu}, \frac{D}{\rho v_\infty^2}, \frac{\mu}{\rho\mathscr{D}}, \frac{k}{D^2}, \frac{v_\infty^2}{Dg}\right) = 0$$

And the number of variables is reduced by three. We will use this result and the similar results of the next section to determine scaling laws for miscible flow in porous media in Sec. 3.6.

□ □ □

3.4 INSPECTIONAL ANALYSIS

It is interesting that the dimensionless groups obtained by dimensional analysis of the significant variables for a problem turn out, in fact, to be the variables and coefficients in the de-dimensionalized differential equation. We will illustrate this fact by considering examples of flow through a pipe, heat transfer by conduction, mass transfer by diffusion, and a more complex example of flow in a porous medium.

Example 3.4-1   Flow of fluid in a pipe[1]   The differential equation describing flow of an incompressible fluid of constant viscosity in the axial direction in a pipe, Eq. (2.7-8), is

$$\rho \frac{\partial v_x}{\partial t} = - \frac{\partial p}{\partial x} + \mu \frac{1}{r} \frac{\partial}{\partial r} \left(r \frac{\partial v_x}{\partial r} \right)$$

Make the equation dimensionless.

Solution   Let

$$v_x^* = \frac{v_x}{\langle v \rangle} \qquad p^* = \frac{p}{\rho \langle v \rangle^2} \qquad t^* = \frac{t \langle v \rangle}{D} \qquad x^* = \frac{x}{D} \qquad r^* = \frac{r}{D}$$

The difference between p^* at inlet and outlet is the Euler number. Substitute in the above equation to obtain

$$\frac{\rho \langle v \rangle^2}{D} \frac{\partial v_x^*}{\partial t^*} = - \frac{\rho \langle v \rangle^2}{D} \frac{\partial p^*}{\partial x^*} + \frac{\mu \langle v \rangle}{D^2} \frac{1}{r^*} \frac{\partial}{\partial r^*} \left(r^* \frac{\partial v_x^*}{\partial r^*} \right)$$

Multiply by $D/\rho \langle v \rangle^2$

$$\frac{\partial v_x^*}{\partial t^*} = - \frac{\partial p^*}{\partial x^*} + \left[\frac{\mu}{D \langle v \rangle \rho} \right] \frac{1}{r^*} \frac{\partial}{\partial r^*} \left(r^* \frac{\partial v_x^*}{\partial r^*} \right)$$

If two systems are described by this equation where they differ only in scale, the Reynolds number $D \langle v \rangle \rho / \mu$ must be the same for both. If the systems are geometrically similar (that is, dimensionless boundary and initial conditions the same) then

$$v^* = v^* (x^*, t^*)$$

$$p^* = p^* (x^*, t^*)$$

are the same in both systems and these systems are dynamically similar.

□ □ □

---

[1] This example provides a transition for subsequent courses in transport phenomena—the differential approach referred to in Sec. 1.4.

Example 3.4-2 Heat transfer by conduction[1] The differential equation
describing energy transport for a fluid at constant pressure in one dimension, Eq. (2.5-7), is

$$\rho \hat{C}_p \frac{\partial T}{\partial t} + \rho \hat{C}_p v_x \frac{\partial T}{\partial x} = k \frac{\partial^2 T}{\partial x^2}$$

Make this equation dimensionless.

Solution Let

$$v_x^* = \frac{v_x}{\langle v \rangle} \qquad t^* = \frac{t \langle v \rangle}{D} \qquad T^* = \frac{T - T_0}{T_1 - T_0} \qquad x^* = \frac{x}{D}$$

$$\frac{\rho \hat{C}_p (T_1 - T_0)\langle v \rangle}{D} \frac{\partial T^*}{\partial t^*} + \frac{\rho \hat{C}_p (T_1 - T_0)\langle v \rangle}{D} v_x^* \frac{\partial T^*}{\partial x^*}$$
$$= \frac{k(T_1 - T_0)}{D^2} \frac{\partial^2 T^*}{\partial x^{*2}}$$

Multiply the equation by $D / \rho \hat{C}_p (T_1 - T_0) \langle v \rangle$

$$\frac{\partial T^*}{\partial t^*} + v_x^* \frac{\partial T^*}{\partial x^*} = \left[\frac{k}{D \rho \hat{C}_p \langle v \rangle} \right] \frac{\partial^2 T^*}{\partial x^{*2}}$$

If we multiply the term in brackets by μ/μ we obtain

$$\frac{\partial T^*}{\partial t^*} + v_x^* \frac{\partial T^*}{\partial x^*} = \left[\frac{k}{\hat{C}_p \mu} \right] \left[\frac{\mu}{D \langle v \rangle \rho} \right] \frac{\partial^2 T^*}{\partial x^{*2}}$$

where the first term in brackets is the reciprocal of the Prandtl number
and the second term in brackets is the reciprocal of the Reynolds number.
If two systems are described by this equation where they differ only in
scale, the product 1/PrRe must be the same for both. If the systems are
geometrically similar and dimensionless boundary and initial conditions
the same then

$$T^* = T^*(x^*, t^*)$$

are the same in both systems and these systems are thermally similar.

□ □ □

[1] This example provides a transition for subsequent courses in transport phenomena—the
differential approach referred to in Sec. 1.4.

Example 3.4-3 Mass transfer by diffusion[1] The differential equation describing mass transfer of component A in a binary mixture of A and B in one dimension, Eq. (2.3-13), is

$$\frac{\partial x_A}{\partial t} + v_x \frac{\partial x_A}{\partial x} = \mathscr{D}_{AB} \frac{\partial^2 x_A}{\partial x^2}$$

(Note: This differential equation is of the same form as the previous example.) Make this equation dimensionless.

Solution Let

$$v_x^* = \frac{v_x}{\langle v \rangle} \qquad t^* = \frac{t\langle v \rangle}{D} \qquad x_A^* = \frac{x_A - x_{A0}}{x_{A1} - x_{A0}} \qquad x^* = \frac{x}{D}$$

$$\frac{(x_{A1} - x_{A0})\langle v \rangle}{D} \frac{\partial x_A^*}{\partial t^*} + \frac{(x_{A1} - x_{A0})\langle v \rangle}{D} v_x^* \frac{\partial x_A^*}{\partial x^*}$$
$$= \frac{\mathscr{D}_{AB}(x_{A1} - x_{A0})}{D^2} \frac{\partial^2 x_A^*}{\partial x^{*2}}$$

Multiply the equation by $D/(x_{A1} - x_{A0})\langle v \rangle$

$$\frac{\partial x_A^*}{\partial t^*} + v_x^* \frac{\partial x_A^*}{\partial x^*} = \left[\frac{\mathscr{D}_{AB}}{\langle v \rangle D} \right] \frac{\partial^2 x_A^*}{\partial x^{*2}}$$

If we multiply the term in brackets by $\rho\mu/\rho\mu$

$$\frac{\partial x_A^*}{\partial t^*} + v_x^* \frac{\partial x_A^*}{\partial x^*} = \left[\frac{\mathscr{D}_{AB}\rho}{\mu} \right] \left[\frac{\mu}{D\langle v \rangle \rho} \right] \frac{\partial^2 x_A^*}{\partial x^{*2}}$$

where the first term in brackets is the reciprocal of the Schmidt number and the second term in brackets is the reciprocal of the Reynolds number. If two systems are described by this equation where they differ only in scale the product 1/ScRe must be the same for both. If the systems are geometrically similar and have the same dimensionless boundary and initial conditions then

$$x_A^* = x_A^* (x^*, t^*)$$

is the same for both systems and they are similar in their diffusional behavior.

□ □ □

---

[1] This example provides a transition for subsequent courses in transport phenomena—the differential approach referred to in Sec. 1.4.

Example 3.4-4   Flow of a single fluid in a porous medium[1]   The equation
of continuity for flow through a porous medium is

$$(a) \qquad -\phi \frac{\partial c}{\partial t} - \frac{\partial}{\partial x_i} cv_i + \mathscr{D} \frac{\partial^2 c}{\partial x_i^2} = 0$$

where symbols are given in the nomenclature and the index indicates
summation over $i = 1, 2, 3$ (the three coordinate directions). The equation of motion for flow through a porous medium is given by Darcy's law:

$$(b) \qquad v_i = -\frac{k}{\mu} \frac{\partial p}{\partial x_i} - \delta_3 \frac{k\rho g}{\mu}$$

δ_3 is the magnitude of the unit vector in the z direction. Let $t^* = tv_\infty/D$,
$x_i^* = x_i/D, v_i^* = v_i/v_\infty$, and multiply equation (a) by D/v_∞ and equation (b)
by $1/v_\infty$. Then

$$(c) \qquad -\phi \frac{\partial c}{\partial t^*} - \frac{\partial}{\partial x_i^*} cv_i^* + \phi \left[\frac{\mathscr{D}}{v_\infty D} \right] \frac{\partial^2 c}{\partial x_i^{*2}} = 0$$

and

$$(d) \qquad v_i^* = \frac{\partial}{\partial x_i^*} \left[\frac{kp}{\mu v_\infty D} \right] - \delta_3 \left[\frac{k\rho g}{\mu v_\infty} \right]$$

where ϕ, x_i^*, c, and the groups in brackets are dimensionless. The dimensionless functional form for flow of a fluid through a porous medium according to equations (c) and (d) is

$$(e) \qquad H\left(\frac{L}{D}, \frac{tv_\infty}{D}, \frac{v}{v_\infty}, \frac{kp}{\mu v_\infty D}, \frac{k\rho g}{\mu v_\infty}, \frac{\mathscr{D}}{v_\infty D} \right) = 0$$

□ □ □

Equation (a) in Example 3.3-3 and equation (e) above differ in the
number of dimensionless groups and in their form. By applying the rule
that dimensionless products can be multiplied and divided by other
dimensionless groups it can be shown that the group $p/\rho v_\infty^2$ in equation (a)
of Example 3.3-3 can be transformed to $kp/\mu Dv_\infty$; that the group $\mu/\rho\mathscr{D}$
can be transformed to $\mathscr{D}/Dv_\infty$; and that the group $v_\infty^2/\rho g$ can be transformed to $k\rho g/\mu v_\infty$. Thus the equations are the same except that the

[1] R. A. Greenkorn, Flow Models and Scaling Laws for Flow through Porous Media, *Ind.
Eng. Chem.*, **56**:32 (1964).

groups $v_\infty D\rho/\mu$ and k/D^2 do not occur in equation (e) above. They are missing because the given equation of motion for flow in porous media, equation (d), is incomplete. According to the analysis of Example 3.3-3 there should be two more terms, one to take into account the effect of boundaries enclosing the medium ($v_\infty D\rho/\mu$) and the other to take into account the relationship between microscopic and macroscopic flow (k/D^2).

The difference between equation (a) and equation (e) points out an advantage to using both approaches in deriving a set of dimensionless products. The approach used in Example 3.3-3 gives the correct number of dimensionless groups for the postulated variables, but does not relate them physically. The method of Example 3.4-4, which requires a mathematical model, does give more physical sense, but it does not contain all the terms. It should be pointed out that equation (d) above, Darcy's law, is the one usually used to calculate flow in porous media. Even though we model using Darcy's law we know from Example 3.3-3 that other variables exist. It may prove, as it does in this case, that these additional terms need not be taken into account if the physical models are constructed and operated in such a manner that contributions of these terms are minimized. (That is, the particle diameter must be small compared to the distance between the containing walls, and the porous medium must be uniform.)

A possible further discrepancy in the analysis is that there may be more important variables than the ones shown in Example 3.3-3. Thus, it would seem more reasonable, especially in cases where little is known about the correct form of the equations of change, to get an exhaustive set of variables and dimensionless products and then eliminate dimensionless products rather than variables by interpreting | them physically. In this way, one will gain a little insight into the physics of the problem simply by knowing what has been ignored.

3.5 APPLICATIONS

In this section we discuss applications of dimensional analysis to the transfer of momentum, energy, and mass. These applications are extremely important as they will be referred to in later chapters in the book.

Example 3.5-1 Friction factor correlation In Example 3.2-1 we discussed many dimensionless groups involved for flow in conduits. In the usual flow problem, we are not concerned with all the possible variables and normally we will be most concerned with flow of a fluid in a pipe. The pressure drop Δp for fluid flowing in a horizontal uniform pipe depends on the length L in which the pressure drop occurs, the diameter D of the pipe, the average velocity $\langle v \rangle$ of the fluid, the viscosity μ, the density ρ, and the roughness height k of the surface. (The interpretation of roughness height k is a statistical problem[1] that has not been solved satisfactorily; we will not consider the statistics here but will use a single roughness height.) The relationship among the variables is

$$g(\Delta p, L, D, \langle v \rangle, \mu, \rho, k) = 0$$

By Buckingham's theorem, a complete set of dimensionless products for these variables is, by inspection,

$$G\left(\text{Eu}, \text{Re}, \frac{L}{D}, \frac{k}{D}\right) = 0$$

The pressure drop in a uniform horizontal pipe is related to the shear stress at the wall, which affects the velocity profile. The shear stress in fully developed steady flow is constant down the pipe; therefore, it follows that the pressure drop is proportional to the length of the pipe and

$$\Delta p = \frac{1}{2} \rho \langle v \rangle^2 \frac{L}{D} G\left(\text{Re}, \frac{k}{D}\right)$$

This equation is called the Darcy equation. The function $G(\text{Re}, k/D)$ is called the pipe friction factor f. A chart that presents the friction factor is known as a Stanton or Moody diagram. Figure 3.5-1 is such a diagram.[2]

□ □ □

As an extension of this example, one can show that the Reynolds number is one of the criteria that determine whether or not flow in a pipe is laminar (dominated by viscous forces, e.g., shear at the wall) or turbulent (dominated by inertia forces). If there is a single critical velocity v_{cr} for any single pipe at which the transition from laminar to turbulent flow occurs, and this velocity depends on the pipe diameter D, the viscosity μ, and the density ρ, there is a relationship

$$g(v_{cr}, D, \rho, \mu) = 0 \tag{3.5-1}$$

[1] J. Nikuradse, *VDI-Forschungsh.*, **361**:1 (1933).
[2] L. F. Moody, *Trans. ASME*, **66**:671 (1944).

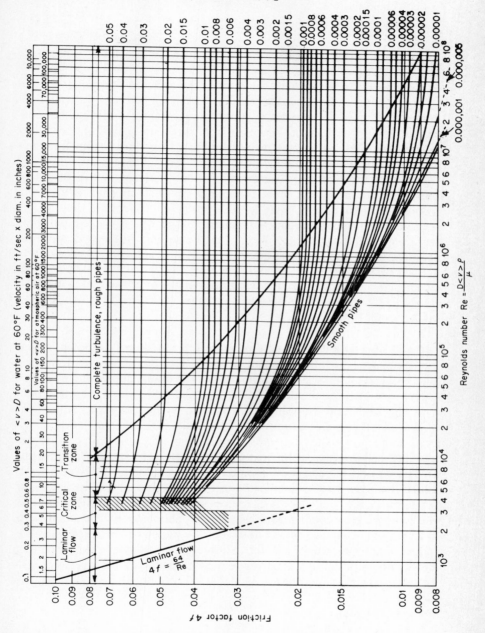

Fig. 3.5-1 Friction factor chart. *(Reprinted from the "Pipe Friction Manual," 3d ed., Copyright (c) 1961 by the Hydraulic Institute, 122 East 42d St., New York, N.Y. 10017.)*

By Buckingham's theorem

$$G(\text{Re}_{cr}) = 0 \qquad\qquad (3.5\text{-}2)$$

The solution of this equation is of the form $\text{Re}_{cr} = $ constant. It has been shown experimentally that transition does occur at about $\text{Re} = 2,100$. Actually, this critical Reynolds number is not a well-defined quantity but rather there is a rather broad transition region. In fact, one can maintain laminar flow up to perhaps $\text{Re} = 10,000$ under rigidly controlled boundary conditions; one cannot, however, retain turbulent flow much below $\text{Re} = 2,100$. The reason we do not obtain a transition region from our analysis is that we have not included in our original list some of the variables which are significant during transition from laminar to turbulent flow. For our purposes, flow is laminar for Re less than 2,100, and turbulent for flow greater than 2,100.

Note that the statements and equations apply to any size pipe. This gives us a basis for scale-up. We can take flow data in the laboratory on small-size pipes and then use these data to construct dimensionless plots such as the Stanton diagram from which we may calculate flow on a large scale as long as the flow regime remains constant over the region of scale-up. (Here, that changing size does not cause transition between laminar and turbulent flow.)

Example 3.5-2 Heat transfer coefficients for a fluid flowing in a pipe
Consider a pipe with fluid flowing inside in turbulent flow, where heat is transferred to the fluid through the wall of the pipe. The variables in this problem are the pipe diameter D, the average fluid velocity $\langle v \rangle$, the density of the fluid ρ, the viscosity of the fluid μ, the heat capacity of the fluid $\hat{C}_p$, and the thermal conductivity of the fluid k. The energy flux transferred to the fluid is the heat transfer coefficient times the temperature difference $h \triangle T$ where h is the heat transfer coefficient and $\triangle T$ is the temperature difference between the pipe wall and the fluid. We have not yet defined the heat transfer coefficient at this point, but an understanding of the heat transfer coefficient is not necessary to understand this example. The heat transfer coefficient is simply a phenomenological coefficient which is used in a model which describes the transfer of heat. What is the functional relationship for the dimensionless heat transfer coefficient? In other words, how could one correlate heat transfer data for this situation? The relationship among the variables is

$$g(D, \langle v \rangle, \rho, \mu, \hat{C}_p, k, h) = 0$$

There are seven variables and four dimensions (L, t, M, T); therefore, there are three dimensionless products in the complete set. By inspection

$$G\left(\frac{hD}{k}, \frac{D\langle v\rangle\rho}{\mu}, \frac{\hat{C}_p\mu}{k}\right) = 0$$

or

(a) $\text{Nu} = G'(\text{Re}, \text{Pr})$

where Nu is the Nusselt number, Re is Reynolds number, and Pr is the Prandtl number. Most of the correlations and equations for predicting the heat transfer coefficient for forced convection are of the form of equation (a). Figure 3.5-2 is a correlation of heat transfer data plotted in the general form of equation (a).[1] Since μ changes with temperature this chart has *an additional viscosity correction factor.*

□ □ □

Example 3.5-3 Convective mass transfer coefficients for flow in a pipe   If fluid is flowing turbulently through a pipe and mass is transferred between the fluid and the pipe wall as a result of a concentration difference of the component being transferred, the variables in the problem are the diameter of the pipe D, the fluid velocity $\langle v\rangle$, the density of the fluid ρ, the viscosity of the fluid μ, the mass transfer coefficient k_ρ, and the molecular diffusivity $\mathscr{D}_{AB}$. (For example, the pipe wall might be cooled with a volatile solid which evaporates with flow through the pipe.) The mass flux is the product of the mass transfer coefficient and the density difference. What is the functional relationship for the mass transfer coefficient? In other words, how would you correlate mass transfer data for this situation? The relationship among the variables is

$$f(D, \langle v\rangle, \rho, \mu, k_\rho, \mathscr{D}_{AB}) = 0$$

There are six variables, and using dimensions (M, L, t) there are three dimensionless products in a complete set. By inspection

$$G\left(\frac{k_\rho D}{\mathscr{D}_{AB}}, \frac{D\langle v\rangle\rho}{\mu}, \frac{\mu}{\rho\mathscr{D}_{AB}}\right) = 0$$

or

(a) $\text{Sh} = G'(\text{Re}, \text{Sc})$

[1] G. N. Sieder and G. E. Tate, *Ind. Eng. Chem.*, **28**:1429 (1936), Fig. 5.

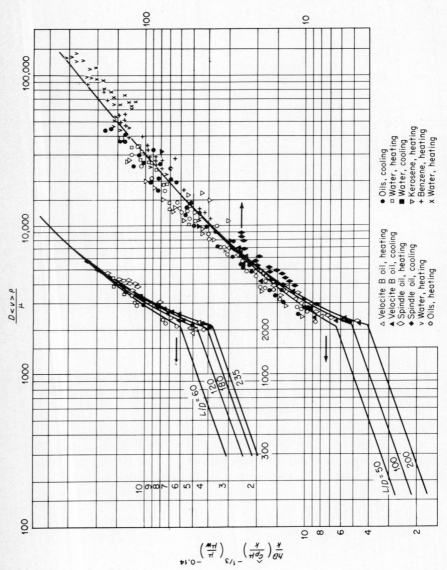

Fig. 3.5-2 Heat transfer coefficients for developed flow. [(G. N. Sieder and G. E. Tate, *Ind. Eng. Chem.*, 28:1429 (1936).]

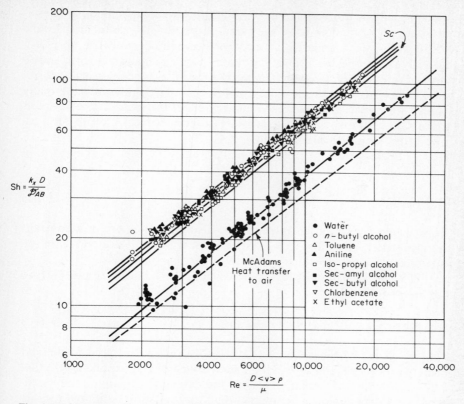

$$Sh = \frac{k_x D}{\mathscr{D}_{AB}}$$

$$Re = \frac{D \langle v \rangle \rho}{\mu}$$

Fig. 3.5-3 Mass transfer for vaporization in turbulent flow. [E. R. Gilliland and T. K. Sherwood, *Ind. Eng. Chem.*, **26**:516 (*1934*).]

where Sh is the Sherwood number, Re is the Reynolds number, and Sc is the Schmidt number. Many mass transfer correlations and equations for predicting mass transfer coefficients are of the form of equation (*a*). Typical results plotted in the form of equation (*a*) are shown in Fig. 3.5-3.[1]

□ □ □

3.6 SIMILARITY, MODELS, AND SCALING

The use of scaling laws in designing physical models is based on the concept of similarity. In geometry two plane figures are similar when corre-

[1] E. R. Gilliland and T. K. Sherwood, *Ind. Eng. Chem.*, **26**:516 (1934).

sponding angles are equal and when the ratio of corresponding sides is constant. Such figures have the same shape but may differ in size and position. Physical similarity is more than geometrical similarity. A physical body has properties other than shape—such as mass, velocity, and so forth. For example, dynamic similarity exists if two geometrically corresponding points experience similar net forces. Two bodies are thermally similar at geometrically corresponding points if the ratio of temperatures is the same. We say two systems are similar if their relevant physical properties are in a constant ratio.

The classical principle of similarity (from the Buckingham pi theorem) is

$$\pi_1 = f(\pi_2, \pi_3, \ldots, \pi_{n-r}) \tag{3.6-1}$$

Two systems in which all relevant physical quantities are in a constant ratio will each satisfy Eq. (3.6-1). The π's (dimensionless groups) are the criteria for geometric, dynamic, thermal, etc., similarity.

If we arbitrarily substitute a power function (product of powers) for f in Eq. (3.6-1) then

$$\pi_1 = C\,(\pi_2)^{x_2}(\pi_3)^{x_3} \cdots (\pi_{n-r})^{x_{n-r}} \tag{3.6-2}$$

where C is called the shape factor, since it has been found to depend primarily on shape, and $x_2, x_3, \ldots, x_{n-r}$, are empirical exponents. In this specific principle of similarity (called the extended principle), the shape factor C is dependent on geometry; thus this equation is limited to systems that have the same shape.

The shape factor can be eliminated if systems of similar geometric form are compared:

$$\frac{\pi'_1}{\pi_1} = \left(\frac{\pi'_2}{\pi_2}\right)^{x_2}\left(\frac{\pi'_3}{\pi_3}\right)^{x_3} \cdots \left(\frac{\pi'_{n-r}}{\pi_{n-r}}\right)^{x_{n-r}} \tag{3.6-3}$$

where the prime refers to the second system. Use of the equation in this manner (that is, deriving a relationship between two geometrically similar systems in terms of a power function relating dimensionless ratios of relevant physical quantities) is called extrapolation. The exponents $x_2, x_3, \ldots, x_{n-r}$ are said to be extrapolated from one system to the other, since they are assumed to be the same in each system.

It is possible to eliminate the exponents by comparing geometrically similar systems at the same values of the dimensionless groups. Thus

$$\pi'_1 = \pi_1$$

if $\pi'_2 = \pi_2, \pi'_3 = \pi_3, \ldots, = \pi_{n-r}$ \hfill (3.6-4)

In general, when two physical systems are similar, knowledge about one system provides knowledge about the other. Equation (3.6-4) is called the principle of corresponding states.

Normally, most physical models are designed to be geometrically similar and experiments are designed such that the important dimensionless ratios of relevant physical quantities are the same for both model and prototype.[1] One attempts to make the models similar in every important physical quantity. In practice, one usually cannot ensure similarity in all respects; it is usually necessary to assume that only a limited number of dimensionless groups is important, and the remainder can be ignored in actual experiment. (The dimensionless groups characterize the regimes.)

Physical models are used to interpret practical problems. The disadvantage of using physical models is that in order to build meaningful models we must have a certain amount of detailed information about the process. Normally we are interested in scaling in a direction to obtain information in a shorter time and on a scale that can be observed. Models of a process can be used to locate significant variables and to compare effects of changing variables.

A major problem in the use of models is the possibility of a scale effect. For example, in the study of the force of waves on a harbor breakwater, the model will have a scale effect if it is made so small that the surface tension of the water in the model affects the wave formation since this force is unimportant in the full-scale system. In scaling porous media, neither the distribution of particle and pore sizes (the nonuniformity, i.e., the distribution of permeability) nor even the average is scaled. If we attempted to scale these variables, the model would offer such resistance to flow that the time required to run the model studies would be the same as that in the prototype, defeating the purpose of the model study because pores are too small.

Points in the model and prototype that correspond to each other are called homologous points. Phenomena occurring in the model and prototype at homologous points are said to occur at homologous times, even though these events may not occur at the same real times. An equivalent definition of similarity is that two systems are similar if they behave the same at homologous points and times.

Langhaar[2] defines similarity abstractly by considering two functions $f(x,y,z,t)$ and $f'(x',y',z',t')$ where (x,y,z) and (x',y',z') are two homologous coordinate systems for prototype and model and t and t' are homologous times. "The function f' is similar to the function f provided

[1] Prototype—the system to be modeled.

[2] H. L. Langhaar, "Dimensional Analysis and Theory of Models," John Wiley & Sons, Inc., New York, 1951.

that the ratio f'/f is constant when the functions are evaluated for homologous points and homologous times. The constant ratio $f'/f = K_f$ is called the scale factor for the function f."

Consider a prototype and a model such that there are two homologous coordinate systems (x,y,z,t) and (x',y',z',t') where the prime refers to the model. Homologous points and times are defined by

$$x' = K_x x$$
$$y' = K_y y$$
$$z' = K_z z$$
$$t' = K_t t$$

$$(3.6\text{-}5)$$

where the constant K_x, K_y, K_z, and K_t are scale factors for length and time. For geometrically similar models

$$K_x = K_y = K_z \tag{3.6-6}$$

A distorted model might have the following relation between scale factors

$$K_x = K_y \neq K_z \tag{3.6-7}$$

where K_z/K_x is a distortion factor.

Let us use the principle of corresponding states, Eq. (3.6-4), to find the model or scaling laws for a system.

Example 3.6-1 Scaling drag on a body immersed in a fluid[1] Consider scaling the drag on a body immersed in a stream of incompressible fluid. The following equation describes the flow:

$$(a) \qquad \frac{F}{\rho v^2 D^2} = f\left(\frac{v D \rho}{\mu}\right)$$

If the unknown function f is to have the same value for the model and prototype, the Reynolds number for the two systems must be equal.

$$(b) \qquad \frac{v D \rho}{\mu} = \left(\frac{v D \rho}{\mu}\right)'$$

[1]H. L. Langhaar, "Dimensional Analysis and Theory of Models," John Wiley & Sons, Inc., New York, 1951.

Writing equations (*b*) and (*a*) in terms of scale factors

$$K_v K_D K_\rho = K_\mu$$

and

$$(c) \qquad K_F = K_\rho K_v{}^2 K_D{}^2 = \frac{K_\mu{}^2}{K_\rho}$$

Equation (*c*) is the model or scaling law of the system. In words: The dimensionless drag force on the model equals the dimensionless drag force on the prototype if the two bodies are tested in the same fluid.

□ □ □

Example 3.6-2 Scaling of miscible flow of two fluids in a porous medium[1] Example 3.3-3 gives the functional law for flow of a single fluid through porous media. To obtain scaling laws for flow of two miscible fluids such as oil and solvent, any further terms due to two-phase flow and interaction with the medium must be included. Thus, any terms which include fluid properties must be duplicated and we need three more coefficients in terms of the ratios of the two fluid properties:

$$G\left(\frac{L}{D}, \frac{tv_\infty}{D}, \frac{v}{v_\infty}, \frac{v_\infty D\rho}{\mu}, \frac{p}{\rho v_\infty{}^2}, \frac{v_\infty{}^2}{Dg}, \frac{\mu}{\rho\mathscr{D}}, \frac{D^2}{k}, \frac{\mu}{\mu_s}, \frac{\rho}{\rho_s}, \frac{\mathscr{D}}{\mathscr{D}_s}\right) = 0$$

where the subscript *s* refers to the solvent phase.

Since we are interested in gross effects, and since it is very difficult to scale permeability, the groups $(v_\infty D\rho)/\mu$ and k/D^2 will not be scaled and the final equation is

$$G'\left(\frac{L}{D}, \frac{tv_\infty}{D}, \frac{v}{v_\infty}, \frac{kp}{\mu Dv_\infty}, \frac{k\rho g}{\mu v_\infty}, \frac{\mathscr{D}}{Dv_\infty}, \frac{\mu}{\mu_s}, \frac{\rho}{\rho_s}, \frac{\mathscr{D}}{\mathscr{D}_s}\right) = 0$$

We may choose the dependent variables as v/v_∞ and $kp/(\mu Dv_\infty)$, and the independent variables are L/D and tv_∞/D, and the remaining groups are parameters.

---

[1] R. A. Greenkorn, Flow Models and Scaling Laws for Flow through Porous Media, *Ind. Eng. Chem.*, **56**:32 (1964).

The scaling is geometric and down in scale by the quantity a; thus $K_L = K_D = 1/a$ and length scales down by a. If the same fluids are used in the model and prototype:

$$K_\mu = K_{\mu s} = K_\rho = K_{\rho s} = K_\mathscr{D} = K_{\mathscr{D} s} = 1$$

Since $K_\mathscr{D} = K_D K_{v_\infty}$ and $K_\mathscr{D} = 1$, then $K_{v_\infty} = 1/K_D = a$, and velocity scales up by a. From $K_k K_\rho K_g = K_\mu K_{v_\infty}$, then $K_k = K_{v_\infty}$ and permeability scales up by a. $K_t K_{v_\infty} = K_D$; thus $K_t = K_D/K_{v_\infty}$ and time scales down by a^2. And, since $K_k K_p = K_\mu K_D K_{v_\infty}$, $K_p = 1/K_k$, pressure scales down by a.

The scaling law in words: The dimensionless pressure and dimensionless velocity are the same functions of dimensionless time in geometrically scaled models if the same fluids are used in the model and prototype.

□ □ □

APPENDIX 3A: ALGEBRAIC METHOD FOR DETERMINING DIMENSIONLESS GROUPS

The proof of Buckingham's theorem is based on a set of algebraic theorems concerned with the class of functions that are dimensionally homogeneous.[1] A systematic approach to finding independent dimensionless products can be based on this algebraic theory and this approach is described in the following.

In general, the number of dimensionless products in a complete set is equal to the total number of variables less the maximum number of these variables that will not form a dimensionless product. As we shall use the theorem the number of dimensionless products in a complete set is equal to the number of variables minus the rank of the dimensional matrix.

Suppose the variables we wish to consider are velocity v, length L, force F, mass density ρ, viscosity μ, and the acceleration due to gravity, g. We can display the exponents in the dimensional expression for each variable as the following, which is the dimensional matrix in the LtM system of units.

[1] L. Brand, *Arch. Rat. Mech. Anal.*, 1:35 (1957).

	v	L	F	ρ	μ	g
M	0	0	1	1	1	0
L	1	1	1	-3	-1	1
t	-1	0	-2	0	-1	-2

$$(3A\text{-}1)$$

where each column gives the exponent in the dimensional expression corresponding to the variable. For example, the dimension of v is $M^0 L^1 t^{-1}$, of L is $M^0 L^1 t^0$, of F is $M^1 L^1 t^{-2}$, etc. We can construct square matrices from this dimensional matrix by deleting various rows or columns. The determinants of these square matrices are the determinants of the original matrix and if the original matrix contains a nonzero determinant of order r, and if all determinants of order greater than r contained in the matrix have a value of zero, the rank of the original matrix is r. For example, the determinant formed from the first three columns in the dimensional matrix (3A-1) is

$$\begin{vmatrix} 0 & 0 & 1 \\ 1 & 1 & 1 \\ -1 & 1 & 2 \end{vmatrix} = (0)\begin{vmatrix} 1 & 1 \\ 1 & -2 \end{vmatrix} - (1)\begin{vmatrix} 0 & 1 \\ 1 & -2 \end{vmatrix} + (-1)\begin{vmatrix} 0 & 1 \\ 1 & 1 \end{vmatrix}$$
$$= 0 + (-1)(-1) + (-1)(-1) = 2$$

Since the largest-order determinant we can construct is a third-order determinant, and the above third-order determinant is different from zero, the rank of the dimensional matrix is 3. Since the rank is 3, the number of dimensionless products in a complete set is $N = n - r = 6 - 3 = 3$. The complete set of dimensionless products is a Reynolds number $vL\rho/\mu$, a Froude number v^2/Lg, and a Euler number $F/\rho v^2 L^2$. The rows of a matrix are linearly independent if and only if the rank of the matrix is equal to or less than the number of rows.

Let's digress a moment and consider an example where the rank of the matrix is *not* equal to the number of rows (for dimensional analysis, the fundamental dimensions). If we have the matrix

$$\begin{pmatrix} 1 & 2 & 3 & 4 \\ 5 & 6 & 7 & 8 \\ 6 & 8 & 10 & 12 \end{pmatrix}$$

the third row is the sum of the first two rows, that is, a linear combination of the other two rows. Therefore, the rank of the matrix is less than three. Since it is easy to find a nonzero *second*-order determinant, the rank is two. If this were the dimensional matrix for a problem with three "fundamental" dimensions, only two of the fundamental dimensions would be independent and we would get two dimensionless products $(n - r = 2)$ rather than one if we had used the number of fundamental dimensions as the rank.

Now returning to our *original* dimensional matrix (3A-1), any product of the variables has the form

$$\pi = v^{k_1} L^{k_2} F^{k_3} \rho^{k_4} \mu^{k_5} g^{k_6} \tag{3A-2}$$

or for the corresponding dimensions of π

$$\pi = [Lt^{-1}]^{k_1} [L]^{k_2} [MLt^{-2}]^{k_3} [ML^{-3}]^{k_4} [ML^{-1}t^{-1}]^{k_5} [Lt^{-2}]^{k_6} \tag{3A-3}$$

or rewritten:

$$\pi = M^{(k_3+k_4+k_5)} L^{(k_1+k_2+k_3-3k_4-k_5+k_6)} t^{(-k_1-2k_3-k_5-2k_6)} \tag{3A-4}$$

In order for π to be dimensionless the exponents of M, L, and t must all be zero:

$$k_3 + k_4 + k_5 = 0$$
$$k_1 + k_2 + k_3 - 3k_4 - k_5 + k_6 = 0 \tag{3A-5}$$
$$-k_1 - 2k_3 - k_5 - 2k_6 = 0$$

Any solution of these equations results in a set of exponents for the dimensionless product π. *Notice that the coefficients in each equation are a row of numbers in the dimensional matrix.* Equation (3A-5) is a system of three equations in six unknowns and is *undetermined*, therefore, since there is an infinite set of solutions. For our purposes we can assign any values to three of the unknowns, say k_1, k_2, and k_3, and solve for the other three. From this we would get a dimensionless product with, for example, $k_1 = 1$, $k_2 = 1$, $k_3 = 0$, and solving for the remaining unknowns $k_4 = 1$, $k_5 = -1$, $k_6 = 0$, so that

$$\pi = \frac{vL\rho}{\mu} = \text{Re} \tag{3A-6}$$

Other choices of three k's will lead to other products.

We wish to know the maximum number of *linearly independent* solutions of Eq. (3A-5). This is, of course, why we must know the rank of the dimensional matrix [the matrix of coefficients of Eq. (3A-5)]. Ignoring the trivial solution $k_i = 0$, Eq. (3A-5) has $(n - r)$ linearly independent solutions. Thus, there are three linearly independent solutions of Eq. (3A-5) (all other solutions are linear combinations of these three)—the number of independent dimensionless products will be three.

Our set of equations is a set of linear algebraic equations. For this example, we had only three equations, but in other situations, we may have many more. This presents no real complication as matrix methods allow us to solve large systems of linear algebraic equations conveniently.

Let's determine the dimensionless products for a set of variables associated with flow in porous media, using the algebraic method.

Example 3A-1   Dimensional analysis by the algebraic method for flow with heat transfer in a porous medium[1]   The variables involved for flow with heat transfer in a porous medium are given by the functional equation

$$f(L,t,\mu,p,g,c,\sigma,\mathscr{D},k,k_m,\beta,\widehat{C}_p,T,v,D,\rho) = 0$$

where in this case k_m is used for thermal conductivity to differentiate it from the permeability k, β is the coefficient of thermal expansion, $\widehat{C}_p$ is heat capacity, and T is the temperature. Determine a complete set of dimensionless products by the algebraic method.

Solution   We select the $LtMT$ system of fundamental units.

DIMENSIONAL MATRIX FOR THE VARIABLES

	1 L	2 t	3 μ	4 p	5 g	6 c	7 σ	8 $\mathscr{D}$	9 k	10 k_m	11 β	12 $\widehat{C}_p$	13 T	14 v	15 D	16 ρ
M	0	0	1	1	0	0	1	0	0	1	0	0	0	0	0	1
L	1	0	-1	-1	1	1	0	2	2	1	0	2	0	1	1	-3
t	0	1	-1	-2	-2	-1	-2	-1	0	-3	0	-2	0	-1	0	0
T	0	0	0	0	0	0	0	0	0	-1	-1	-1	1	0	0	0

In order to determine the number of independent dimensionless products, we must determine the rank of the dimensional matrix

[1] R. A. Greenkorn, Flow Models and Scaling Laws for Flow through Porous Media, *Ind. Eng. Chem.*, **56**:32 (1964).

EVALUATION OF RANK OF DIMENSIONAL MATRIX USING DETER-
MINANT TO EXTREME RIGHT

$$
D = \begin{array}{cccc}
13 & 14 & 15 & 16 \\
T & v & D & \rho \\
\end{array}
\begin{vmatrix}
0 & 0 & 0 & 1 \\
0 & 1 & 1 & -3 \\
0 & -1 & 0 & 0 \\
1 & 0 & 0 & 0 \\
\end{vmatrix} = -1
$$

$D \neq 0$:   rank $= 4$

Therefore

NUMBER OF DIMENSIONLESS PRODUCTS

$$N = n - 4 = 16 - 4 = 12$$

The coefficient equations (four of them—since the rank of the dimensional matrix is 4—that is, the four fundamental dimensions are independent) are found by writing out the exponent equations for each dimension; the coefficients are the rows of the dimensional matrix:

COEFFICIENT EQUATIONS

$$k_3 + k_4 + k_7 + k_{10} + k_{16} = 0$$

$$k_1 - k_3 - k_4 + k_5 + k_6 + 2k_8 + 2k_9 + k_{10} + 2k_{12}$$
$$+ k_{14} + k_{15} - 3k_{16} = 0$$

$$k_2 - k_3 - 2k_4 - 2k_5 - k_6 - 2k_7 - k_8 - 3k_{10} - 2k_{12} - k_{14} = 0$$

$$-k_{10} - k_{11} - k_{12} + k_{13} = 0$$

We can decide to write all of the dimensionless products such that they contain a maximum of the same four variables plus one additional variable since there are four coefficient equations. In this case, we decide to always use T, v, D, and ρ, and thus we solve the coefficient equations in terms of k_{13}, k_{14}, k_{15}, and k_{16}.

SOLUTION OF COEFFICIENT EQUATIONS

$$k_{13} = k_{10} + k_{11} + k_{12}$$

$$k_{14} = k_2 - k_3 - 2k_4 - 2k_5 - k_6 - 2k_7 - k_8 - 3k_{10} - 2k_{12}$$

$$k_{15} = -k_1 - k_2 - k_3 + k_5 - k_7 - k_8 - 2k_9 - k_{10}$$

$$k_{16} = -k_3 - k_4 - k_7 - k_{10}$$

To obtain numerical values from this set of equations, we must pick values for all the variables but the four $k_{13}, k_{14}, k_{15}, k_{16}$. The most convenient sets of values to assign the other k's are $k_1 = 1, k_2, \ldots, k_{12} = 0$; $k_1 = 0, k_2 = 1, k_3, \ldots, k_{12} = 0$, etc., and then determine the value of $k_{13}, k_{14}, k_{15},$ and k_{16} for each set. We will obtain by this procedure one set of the total number of sets of solutions. All the other possible solutions may be generated as linear combinations of the elements of this set.

MATRIX OF SOLUTIONS

	k_1 L	k_2 t	k_3 μ	k_4 p	k_5 g	k_6 c	k_7 σ	k_8 $\mathscr{D}$	k_9 k	k_{10} k_m	k_{11} β	k_{12} $\hat{C}_p$	k_{13} T	k_{14} v	k_{15} D	k_{16} ρ
1	1	0	0	0	0	0	0	0	0	0	0	0	0	0	-3	0
2	0	1	0	0	0	0	0	0	0	0	0	0	0	1	-1	0
3	0	0	1	0	0	0	0	0	0	0	0	0	0	-1	-1	-1
4	0	0	0	1	0	0	0	0	0	0	0	0	0	-2	0	-1
5	0	0	0	0	1	0	0	0	0	0	0	0	0	-2	1	0
6	0	0	0	0	0	1	0	0	0	0	0	0	0	-1	0	0
7	0	0	0	0	0	0	1	0	0	0	0	0	0	-2	-1	-1
8	0	0	0	0	0	0	0	1	0	0	0	0	0	-1	-1	0
9	0	0	0	0	0	0	0	0	1	0	0	0	0	0	-2	0
10	0	0	0	0	0	0	0	0	0	1	0	0	1	-3	-1	-1
11	0	0	0	0	0	0	0	0	0	0	1	0	1	0	0	0
12	0	0	0	0	0	0	0	0	0	0	0	1	1	-2	0	0

To solve this problem actually, we need only to write down two matrices: the dimensional matrix and the matrix containing the coefficients of the solution. The intervening steps, though useful for understanding, are not necessary for solution.

Finally, the dimensionless groups are written using as exponents successive rows of the matrix of solutions.

COMPLETE SET OF DIMENSIONLESS PRODUCTS

$$F\left(\frac{L}{D}, \frac{tv}{D}, \frac{vD\rho}{\mu}, \frac{p}{\rho v^2}, \frac{v^2}{Dg}, \frac{v}{c}, \frac{\rho v^2 D}{\sigma}, \frac{\mu}{\rho\mathscr{D}}, \frac{D^2}{k}, \frac{v^3 D\rho}{k_m T}, \beta T, \frac{k_m}{\hat{C}_p \mu}\right) = 0$$

We can operate on these groups by combining dimensionless products to give

IN TERMS OF COMMON GROUPS

$$F\left(\frac{L}{D}, \frac{tv}{D}, \frac{vD\rho}{\mu}, \frac{p}{\rho v^2}, \frac{D^3 p g^2 \rho \Delta C_0}{\mu^2}, \frac{v}{c},\right.$$

$$\left.\frac{\rho v^2 D}{\sigma}, \frac{\mu}{\rho\mathscr{D}}, \frac{k_m D}{\mathscr{D}}, \frac{hD}{k_m}, \frac{D^3 \rho^2 \beta T}{\mu^2}, \frac{k_m}{\hat{C}_p \mu}\right) = 0$$

F [dimensionless length, dimensionless time, Reynolds number, pressure coefficient, Grashof number (mass transfer), Mach number, Froude number, Schmidt number, Nusselt number (mass transfer), Nusselt number (heat transfer), Grashof number (heat transfer), Prandtl number] $= 0$

□ □ □

Fundamental units

One problem we have bypassed above in the dimensional analysis of variables is that of selecting a set of fundamental units. (This is associated with the rank of the dimensional matrix and the number of dimensionless products.)

The number that gives the distance from an origin on an axis to some point on the axis is called the coordinate of the point. The variable which gives the magnitude of the scalar associated with the particular coordinate is called the coordinate of magnitude. The dimension of an entity in the LtM system is $(L^a t^b M^c)$ and the coordinate of magnitude of the entity is x, where L, t, and M are in the "original units." If new units are introduced such that

1 original length unit $= A$ new length units
1 original time unit $= B$ new time units
1 original mass unit $= C$ new mass units

what is the coordinate of magnitude, $\bar{x}$, in terms of the new units? One can easily show that

$$\bar{x} = xA^a B^b C^c \qquad\qquad\qquad (3A\text{-}7)$$

where Eq. (3A-7) defines a coordinate transformation. If

$$y = f(x_1, x_2, \ldots, x_n) \qquad\qquad\qquad (3A\text{-}8)$$

and the basic system of measurement changes, then Eq. (3A-8) is dimensionally homogeneous if and only if

$$\bar{y} = f(\bar{x}_1, \bar{x}_2, \ldots, \bar{x}_n) \qquad\qquad\qquad (3A\text{-}9)$$

This condition is expressed by the statement that the equation is "invariant" under the coordinate transformation Eq. (3A-7). Since

$$\bar{y} = y\,A^a B^b C^c = yK$$
$$\bar{x}_1 = x_1 A^a B^b C^c = x_1 K_1$$

.
.
.

$$\bar{x}_n = x_n A^a B^b C^c = x_n K_n$$

(3A-10)

then

$$Kf(x_1, x_2, \ldots, x_n) = f(K_1 x_1, K_2 x_2, \ldots, K_n x_n) \qquad (3A\text{-}11)$$

Thus the function $f(x_1, x_2, \ldots, x_n)$ is dimensionally homogeneous if and only if Eq. (3A-11) is an identity in the variables and parameters $(x_1, x_2, \ldots, x_n; A, B, C)$.

Mathematically this is satisfactory but let us consider a problem which will lead to an apparent paradox.

Example 3A-2 Apparent paradox in dimensional analysis caused by selection of fundamental units[1] A ball of diameter D is fixed in a stream of liquid and kept at a temperature T above that of the liquid at a great distance from the ball. If the velocity of the stream is v and the liquid has heat capacity $\hat{C}_p$ per *unit volume* and thermal conductivity k, what is the rate r of heat transferred from the ball to the liquid?

Solution 1 In the *LtMT* system

DIMENSIONAL MATRIX

	r	$\hat{C}_p$	v	T	D	k
M	1	1	0	0	0	1
L	2	-1	1	0	1	1
t	-2	-2	-1	0	0	-3
T	-1	-1	0	1	0	-1

[1] L. Brand, *Arch. Rat. Mech. Anal.*, **1**:35 (1957).

$$\det = -1$$
$$r = 4$$
$$N = n - r = 2 \text{ dimensionless products}$$

SOLUTION MATRIX

	r	$\hat{C}_p$	v	T	D	k
1	1	0	0	-1	-1	-1
2	0	1	1	0	1	-1

and

$$\pi_1 = \frac{r}{DkT}$$
$$\pi_2 = \frac{Dv\,\hat{C}_p}{k}$$

or

$$r = DkTf(\pi_2)$$

Solution 2 In the *LtM* system

DIMENSIONAL MATRIX

	r	$\hat{C}_p$	v	T	D	k
M	1	0	0	1	0	0
L	2	-3	1	2	1	-1
t	-3	0	-1	-2	0	-1

$$\det = -1$$
$$r = 3$$
$$N = n - r = 6 - 3 = 3$$

and

SOLUTION MATRIX

	r	$\hat{C}_p$	v	T	D	k
1	1	0	0	-1	-1	-1
2	0	1	0	0	3	0
3	0	0	1	0	-2	-1

and

$$\pi_1 = \frac{r}{DkT}$$

$$\pi_2 = \hat{C}_p D^3$$

$$\pi_3 = \frac{v}{D^2 k}$$

or

$$r = DkTg(\pi_1, \pi_2)$$

Thus the knowledge that $T = L^2 M / t^2$ from the molecular theory of heat leads to an apparent paradox. In solution 1, the assumption that $LtMT$ are the fundamental units implies that

$$\bar{x} = xM^a L^b t^c T^d \tag{3A-12}$$

is dimensionless where $a = b = c = d = 0$. In solution 2, the assumption that LtM are the fundamental units implies that

$$\bar{x} = xM^a L^b t^c T^d \tag{3A-13}$$

is dimensionless when $a = -d$, $b = -2d$, $c = 2d$.

Thus, the conditions of Eq. (3A-13) are included in Eq. (3A-12) when $d = 0$, and they are less restrictive than the conditions of Eq. (3A-12). Only experiment can decide which is correct. In most problems in momentum, heat, and mass transfer, we add a dimensional constant such as g_c or the conversion for temperature above[1] to our list of variables when using four fundamental units as in solution 1, and we have the same number of dimensionless products as in solution 2. The addition of this *constant* to the list of *variables* is formalism which results from the physics of the situation, not from the mathematics.

□ □ □

Problems

3.1 Write the fundamental units of the following quantities using the system of units indicated: (*a*) Diameter (system: *FMt*), (*b*) density (*FLt*), (*c*) acceleration (*FML*), (*d*) viscosity (*MLt*), (*e*) coefficient of thermal expansion (increase in volume per unit volume per degree change in temperature) (*FMLtT*).

[1]D. Q. Kern, "Process Heat Transfer," McGraw-Hill Book Company, New York, 1950.

3.2   The discharge through a horizontal capillary tube depends upon the pressure drop per unit length, the diameter, and the viscosity.

Quantity	Symbol	Dimensions
Discharge	Q	$\dfrac{L^3}{t}$
Pressure drop/length	$\Delta p/L$	$\dfrac{M}{L^2 t^2}$
Diameter	D	L
Viscosity	μ	$\dfrac{M}{Lt}$

(a) How many dimensionless groups are there?
(b) Find the functional form of the equation.

3.3   Using dimensional analysis, arrange the following variables into several dimensionless groups:

pipe diameter D, ft

fluid velocity v, ft/sec

pipe length L, ft

fluid viscosity μ, lbm/sec ft

fluid density ρ, lbm/ft^3

Show by dimensional analysis how these groups are related to pressure drop in lbf/in.^{2}.

3.4   Using D (diameter), ρ (density), and g (acceleration of gravity) as $\tilde{Q}$'s, form a dimensionless group using as the additional variable μ (viscosity). Use the MLt system and solve the equations (do not do by inspection).

3.5   The calibration curve for a particular flowmeter has the form

$$Q = f(D, \rho, \mu, \Delta p) = \text{volumetric flow rate}$$

Using dimensional analysis, find the required dimensionless variables π_1 and π_2 where $F(\pi_1, \pi_2) = 0$. Let $\tilde{Q}_1 = D$, $\tilde{Q}_2 = \rho$, $\tilde{Q}_3 = \Delta p$.[1]

3.6   The pressure drop in a viscous incompressible fluid flowing in a length L may be represented functionally as

$$\Delta p = f(\mu, \rho, \langle v \rangle, D, L, k)$$

where $\mu =$ viscosity

$\rho =$ density

$\langle v \rangle =$ bulk velocity

$D =$ diameter

[1] H. Rouse, "Advanced Mechanics of Fluid," John Wiley & Sons, Inc., p. 13, New York, 1959.

$L = $ length

$k = $ roughness

Using dimensional analysis find the correct representation for the pressure drop in terms of dimensionless groups.

3.7 Dimensional analysis is useful in the design of centrifugal pumps. The pressure rise across a pump (proportional to the head developed by the pump) is affected by the fluid density ρ, the angular velocity ω, the impeller diameter D, the volumetric flow rate Q, and the fluid viscosity μ. Find the pertinent dimensionless groups, choosing them so that Δp, Q, and μ each appear in one group only. Find similar expressions replacing the pressure rise first by the power input to the pump, and then by the efficiency of the pump.[1]

3.8 The scaling ratios for centrifugal pumps are $Q/ND^3 = $ constant, $H/(ND)^2 = $ constant, where Q is the volumetric flow rate (ft³/sec), H is the pump head (ft), and N is rotational speed (rpm). Determine the head and discharge of a 1:4 model of a centrifugal pump that produces 20 ft³/sec at a head of 96 ft when turning 240 rpm. The model operates at 1,200 rpm.

3.9 You are trying to simulate dynamically the cooling system of a proposed sodium-cooled nuclear reactor by constructing a 1:4 model of the cooling system using water. These water experiments are to be carried out at 70°F and will simulate flowing sodium at 1000°F. Data on sodium reactor:

$$\text{Hydraulic diameter} = \frac{4(\text{flow cross-sectional area})}{\text{wetted perimeter}} = 1.768 \times 10^{-2}\text{ft}$$

(This replaces the ordinary tube diameter for flows in noncircular channels.) Bulk sodium velocity = 20 ft/sec. Pressure drop across the core = 30 psig. For the water model to be dynamically similar,

(a) What must the water velocity be?

(b) What is the expected pressure drop in the water model?

3.10 A rough method of scaling liquid mixing tanks and impellers is to keep the power input per unit volume constant. If it is desired to construct a prototype of an existing properly baffled mixer by a ratio of 2:1, by what ratio must the tank diameter D and the impeller speed N be changed? Consider the mixers to be geometrically similar and to be operating in the completely turbulent region.[2]

3.11 Frequently in heat transfer problems with flowing fluids the six variables D, ρ, h, v, μ, k may be combined into dimensionless groups which will characterize experimental data. Find these dimensionless groups by inspection with the restriction that h and v only appear in separate terms. D is the diameter, ρ is the density, v is the velocity, μ is the viscosity, h is the heat transfer coefficient and has dimensions Btu/hr ft² °F, and k is the thermal conductivity and has dimensions Btu/hr ft °F.

3.12 In a forced convection heat transfer process the quantities listed below are pertinent to the problem:

Tube diameter D

Thermal conductivity of the fluid k

[1]C. O. Bennett and J. E. Myers, "Momentum, Heat, and Mass Transfer," p. 163, McGraw-Hill Book Company, New York, 1962.

[2]R. B. Bird, W. E. Stewart, and E. N. Lightfoot, "Transport Phenomena," John Wiley & Sons, Inc., New York, 1960.

Velocity of the fluid v

Density of the fluid ρ

Viscosity of the fluid μ

Specific heat of the fluid $\widehat{C}_p$

Heat transfer coefficient h

Using dimensional analysis find the functional relationship among these parameters.

3.13 In a forced convection mass transfer process the quantities listed below are pertinent to the problem:

$$k_L = \frac{\text{lb moles}}{\text{ft}^2\ \text{hr}} \qquad c = \frac{\text{lb moles}}{\text{ft}^3}$$

Mass transfer coefficient	$\dfrac{(k_L)}{c}$	$\dfrac{L}{t}$
Characteristic length	D	L
Diffusivity	$\mathscr{D}_{AB}$	$\dfrac{L^2}{t}$
Fluid density	ρ	$\dfrac{M}{L^3}$
Velocity of fluid	v	$\dfrac{L}{t}$
Fluid viscosity	μ	$\dfrac{M}{Lt}$

Using dimensional analysis find the functional relationship among these parameters.

3.14 The energy equation for heat transfer from a flat plate of length H at temperature T_0 suspended in a large body of fluid at temperature T_1 is

$$\rho \widehat{C}_p \left[v_y \frac{\partial(T - T_1)}{\partial y} + v_z \frac{\partial(T - T_1)}{\partial z} \right] = k \frac{\partial^2(T - T_1)}{\partial y^2}$$

This equation is nonlinear and is difficult to solve. Usually dimensional analysis is applied to such equations. Given the following definitions

$$\Theta = \frac{T - T_1}{T_0 - T_1} \qquad \xi = \frac{z}{H} \qquad \eta = \left(\frac{B}{\mu \alpha H} \right)^{1/4} y$$

$$\phi_z = \left(\frac{\mu}{B \alpha H} \right)^{1/2} v_z \qquad \phi_y = \left(\frac{\mu H}{\alpha^3 B} \right)^{1/4} v_y$$

where $\alpha = k/\rho \widehat{C}_p$ and $B = \rho g \beta (T_0 - T_1)$, write the above equation in dimensionless form.

4
Application of the Macroscopic Balances to Flow Measurement

In this chapter, we illustrate the practical application of our macroscopic balances by considering a particular type of equipment: flow measuring devices and manometers. These devices indicate the volumetric or mass flow rate of fluids, and require primarily mass and mechanical energy balances for their mathematical description. The macroscopic energy balance and the macroscopic momentum balance could be illustrated in a parallel fashion here if it were common practice to have meters indicating momentum or heat flux; however, these quantities are usually calculated from combinations of other meter readings, for example, temperature and flow rate. (One could, of course, construct a heat or momentum fluxmeter by relating these fluxes to temperature, velocities, etc., as is done for flowmeters, but this is seldom economically desirable. The basic idea, however, would be exactly the same as that of the flowmeter, i.e., to relate some changing property of the flowing stream to the heat or momentum flux.)

There are many types of flowmeters used commercially. A common type is the head meter which operates by relating the pressure drop

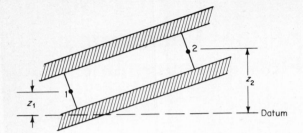

Fig. 4.1-1 Pressure difference in inclined conduit.

through a flow restriction to either the point velocity (pitot tube) or average velocity (orifice, venturi) via the mechanical energy balance. Another common type of meter is the area meter, which determines flow based on a variable flow area. Other meters, such as the gas meter in your house, operate on a positive displacement principle (much like a piston *pump* running as a *motor* using the fluid to drive the pump rather than vice versa) and measure the actual volume of fluid passing a given point. There are many special meters which, for example, measure current from an electric generator driven by the flowing stream (anemometer), relate dissipation of heat to flow (hot wire anemometer), or relate change in some property of the fluid as it flows to velocity (magnetic flowmeter).

We will discuss head meters in some detail since they are very common and they provide us with examples of the application of the mechanical energy balance. Rotameters, which operate on a different principle (variable area), are common and will also be discussed.

4.1 MANOMETERS[1]

Meters which operate on the principle of relating rate of flow to pressure drop across a restriction (head meters) require a pressure measurement. Most often a manometer, typically a U tube, is used to obtain this pressure drop.

If a conduit is filled with a fluid that is not moving, Bernoulli's equation, Eq. (2.4-48), may be applied to find the pressure difference between any two points, since there is no frictional loss. In Fig. 4.1-1 the pressure difference is found between points 1 and 2 with Bernoulli's equation (where the kinetic energy term drops out) and

$$\frac{\Delta p}{\rho} + g\,\Delta z = 0 \tag{4.1-1}$$

[1] See Figs. 4A-1, 4A-2, and 4A-3 for pictures of various types of manometers.

or, solving for the pressure difference between points 1 and 2,

$$\Delta p = p_2 - p_1 = - \rho g(z_2 - z_1) \tag{4.1-2}$$

A manometer can be used to measure this pressure difference by "balancing columns of fluid." Figure 4.1-2 is a sketch of a system for which we might wish to know the pressure difference between points 1 and 5. The tank on the left and the left leg of the manometer above point 2 are filled with fluid A; the manometer contains an appropriate fluid, fluid B, between points 2 and 4; the right leg of the manometer above point 4 and the tank on the right contain fluid C. Applying Bernoulli's equation between points 1 and 2 gives an expression for the pressure difference between points 1 and 2 in terms of the difference in height of points 1 and 2 above the datum:

$$p_1 - p_2 = - \rho_A g(z_1 - z_2) \tag{4.1-3}$$

The pressure difference desired is the difference between points 1 and 5. We may write this pressure difference as the sum of pressure differences:

$$p_1 - p_5 = (p_1 - p_2) + (p_2 - p_3) + (p_3 - p_4) + (p_4 - p_5) \tag{4.1-4}$$

and applying the Bernoulli equation to each Δp:

$$p_1 - p_5 = - [\rho_A g(z_1 - z_2) + \underbrace{\rho_B g(z_2 - z_3) + \rho_B g(z_3 - z_4)}_{\rho_B g(z_2 - z_4)} + \rho_C g(z_4 - z_5)] \tag{4.1-5}$$

In many actual process plants and in many research applications pressure devices which convert a pressure signal to an electrical signal (transducers) are used. These transducers normally sense pressure

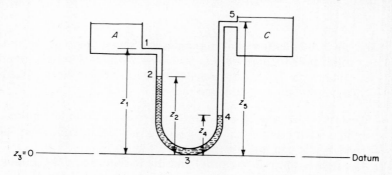

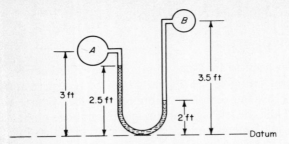

Fig. 4.1-3 System for Example 4.1-1.

change with a diaphragm or bellows and the subsequent displacement is converted to an electric signal. Most often this type of transducer measures a pressure difference between the two sides of the diaphragm.

Example 4.1-1 Simple manometer Consider in the system shown in Fig. 4.1-3 that we are interested in the pressure difference between two pipes A and B where both pipes contain *water* ($\rho = 62.4$ lbm/ft^3). The manometer fluid has a specific gravity of 1.1. The pressure difference using the procedure above is

$$p_A - p_B = -\left[1 \left(62.4 \frac{\text{lbm}}{\text{ft}^3} \right) \left(\frac{32.2 \text{ ft}}{\text{sec}^2} \right) \left(\frac{\text{lbf sec}^2}{32.2 \text{ lbm ft}} \right) (3 - 2.5)\text{ft} \right.$$

$$\left. + (1.1)\,(62.4)\,\frac{32.2}{32.2}\,(2.5 - 2.0) + (1)\,(62.4)\,\frac{32.2}{32.2}\,(2 - 3.5) \right]$$

$$= -62.4\,[\,(3 - 2.5) + 1.1\,(2.5 - 2.0) + (2 - 3.5)\,]$$

$$= (-62.4)\,(-0.45) = 28.2\frac{\text{lbf}}{\text{ft}^2} = 0.195\frac{\text{lbf}}{\text{in.}^2}$$

□ □ □

Example 4.1-2 Differential manometer Shown in Fig. 4.1-4 is an arrangement called a *differential manometer*, which is used to obtain large-scale differences in height from small pressure differences. The principle of this manometer rests on two facts:

1. Reservoirs A and B are large in cross section compared to the manometer tube, and so changes in the interface position give negligible change in z ($z_1 = z_4$).
2. The denser fluid (in the tube) and the less-dense fluid (in the tube and both reservoirs) differ in density by a small amount.

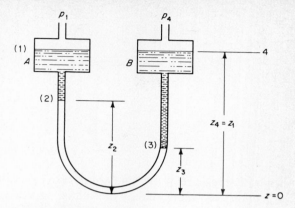

Fig. 4.1-4 Differential manometer.

Denote the density of the heavier fluid by ρ_h and the lighter fluid by ρ_l. The pressure difference may be written as

$$p_4 - p_1 = (p_4 - p_3) + (p_3 - p_2) + (p_2 - p_1)$$
$$= -\rho_l g (z_4 - z_3) - \rho_h g(z_3 - z_2) - \rho_l g(z_2 - z_1)$$

but since $z_4 = z_1$

$$p_4 - p_1 = -\rho_h g(z_3 - z_2) - \rho_l g(z_2 - z_3)$$
$$= g(z_2 - z_3)(\rho_h - \rho_l)$$

Solving:

$$z_2 - z_3 = \frac{p_4 - p_1}{g(\rho_h - \rho_l)}$$

The closer in magnitude ρ_h is to ρ_l, the more the displacement for a given pressure drop.

□ □ □

4.2   PITOT TUBE

Figure 4.2-1 shows a pitot tube. It is inserted into the flow stream with the opening, point 6, perpendicular to the flow. Fluid flows into the opening until the pressure builds to a level that withstands the velocity of the stream and the fluid directly in the front of the opening remains stationary. This point of stationary flow represented by point 6 is called the stagnation point. The difference in pressure between the stagnation point and

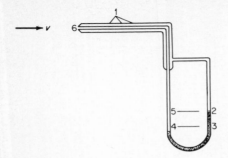

Fig. 4.2-1 Pitot tube.

the static pressure of the fluid at this cross section, which we may assume approximately independent of radius, represents the pressure rise associated with the deceleration of the fluid. Thus, this pressure rise is indicative of the local velocity of the fluid. The static pressure in the pipe is assumed to be the pressure measured at an opening parallel to the flow. Most often pitot tubes are constructed, as in Fig. 4.2-1, of a double-walled tube with static pressure measured through holes parallel to the flow on the outer wall, point 1. A manometer is connected to the pitot tube as shown in the sketch and the pressure difference between points 1 and 6 is determined. We assume that the static pressure measured at (1) is the same as would have been measured at (6) in the absence of the pitot tube; that is, we assume that at (1) the flow is essentially back to the undisturbed steady-state pattern.

   We may describe the rise in pressure associated with the deceleration of the fluid by applying the mechanical energy balance to a system swept out by the fluid, which decelerates to nearly zero velocity at the mouth of the pitot tube as shown in Fig. 4.2-2. The boundaries of the system are *path lines*, which are the lines a fluid "particle" traces out in space as time progresses. In steady flow, which we are considering, this is also a streamline (the locus of the instantaneous direction of the velocity field), and so there is no flow across the lateral boundary of our system. At the upstream boundary, fluid enters at a high velocity v_7, and at the downstream boundary it leaves at a low velocity v_6, which is so low that the kinetic energy is insignificant—v_6 approaches zero. Writing the mechanical energy balance for this system

$$\frac{\langle v_6 \rangle^2 - \langle v_7 \rangle^2}{2} + g\,\Delta z + \frac{p_6 - p_7}{\rho} = -\widehat{lw} - \widehat{W} \qquad (4.2\text{-}1)$$

   If we arbitrarily represent the lost work term as some fraction C of the kinetic energy change

$$C\,\frac{\langle v_6 \rangle^2 - \langle v_7 \rangle^2}{2} = -\widehat{lw} \qquad (4.2\text{-}2)$$

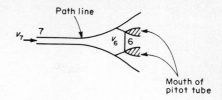

Fig. 4.2-2 Mouth of pitot tube.

we may write, assuming that $\langle v_6 \rangle^2$ is much less than $\langle v_7 \rangle^2$

$$(1 - C) \langle v_7 \rangle^2 = 2 \frac{p_6 - p_7}{\rho} \tag{4.2-3}$$

or

$$\langle v_7 \rangle = \sqrt{\frac{1}{1 - C}} \; \sqrt{\frac{2 \, \Delta p}{\rho}} \tag{4.2-4}$$

Since $\sqrt{1/1 - C}$ is simply another coefficient, let us call it C_p, the pitot tube coefficient, and write

$$\langle v_7 \rangle = C_p \sqrt{\frac{2 \, \Delta p}{\rho}} \tag{4.2-5}$$

We *must notice* three things about this relationship:

1. The coefficient C_p accounts for the lost work while decelerating and accelerating the fluid. This cannot in general be calculated and C_p is usually determined experimentally for the given instrument.
2. The $\langle v_7 \rangle$ determined is the average velocity *over the inlet area* to the system in Fig. 4.2-2, *not for the pipe*. In fact, as the area of the pitot tube is made smaller and smaller, $\langle v_7 \rangle$ approaches the *local velocity*.
3. The density in Eq. (4.2-5) is that of the *flowing* fluid, not the manometer fluid.

As mentioned above, we assume $p_7 = p_1$, and we obtain our Δp from the manometer as in the preceding section.

Example 4.2-1 Calculation of flow rate from pitot traverse A cylindrical duct 2 ft in diameter carries air at atmospheric pressure. A pitot tube traverse is made of the duct using a water-filled manometer with the following result:

Radial distance from wall, ft	Height difference in manometer, in.
0.1	1
0.3	2
0.5	4
0.7	5
0.9	6

Calculate the approximate velocity profile and volumetric flow rate:

Data: $\rho_{air} = 0.0808 \dfrac{lbm}{ft^3}$

$$\rho_{HOH} = 62.4 \frac{lbm}{ft^3}$$

Coefficient $= C_p = 0.99$

Solution The height differences may be converted to pressures and the pressures to velocities as follows: Observe that the density of air is much less than that of water. If the pitot tube is as shown in Fig. 4.2-1, we may write

$$p_6 - p_1 = (p_6 - p_4) + (p_4 - p_2) + (p_2 - p_1)$$

$$= -\rho_{air} g(z_6 - z_4) - \rho_{HOH} g(z_4 - z_2)$$

$$- \rho_{air} g(z_2 - z_1)$$

but since $z_5 = z_2$

$$p_6 - p_1 = g(z_2 - z_4)(\rho_{HOH} - \rho_{air})$$

but since $\rho_{HOH} >> \rho_{air}$

$$p_6 - p_1 = \rho_{HOH} g(z_2 - z_4)$$

The expression relating the pressure drop to velocity is Eq. (4.2-5). Substituting:

$$\langle v \rangle = C_p \sqrt{\frac{2 \, \Delta p}{\rho}} = C_p \sqrt{\frac{2\rho_{HOH}(z_2 - z_4)g}{\rho_{air}}}$$

$$\langle v \rangle = 0.99 \sqrt{\frac{(2)(62.4)}{0.0808} \left(\frac{lbm}{ft^3}\right)\left(\frac{ft^3}{lbm}\right)(z_2 - z_4)\left(in. \, \frac{ft}{12 \, in.}\right)\left(\frac{32 \, ft}{sec^2}\right)}$$

$$= 11.2 \sqrt{z_2 - z_4} \qquad z_2 - z_4 \text{ in inches, } \langle v \rangle \text{ in ft/sec}$$

From this relation we can calculate, for each height difference, a velocity:

Radial distance from wall, ft	$\langle v \rangle$, ft/sec
0.1	11.2
0.3	15.8
0.5	22.4
0.7	25
0.9	26.4

This profile is plotted in Fig. 4.2-3. Note that, as usually happens, our experimental data scatter somewhat. We would have to decide in a real world situation whether we had some good reason for drawing the smooth curve shown through the data or if in fact the curve should go through all the points. Here, however, we will assume that the smoothing of the data as shown will give us sufficient accuracy.

We now wish to get the volumetric flow rate by

$$Q = \int_A v \, dA$$

This relation may be written as

$$Q = \int_0^r v \, 2\pi r \, dr$$

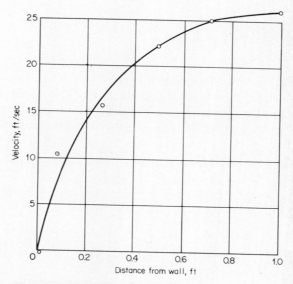

Fig. 4.2-3 Velocity profile.

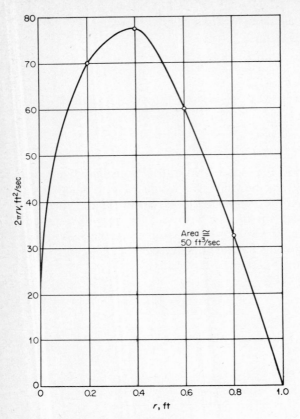

Fig. 4.2-4 Graphical integration.

if we assume no variation in the θ direction. Since we do not have an explicit expression for v as a function of r, but have only Fig. 4.2-3, we will integrate graphically by plotting $2\pi r v$ versus r, as shown in Fig. 4.2-4. The area under this curve is the required volumetric flow rate.

Distance from wall, ft	r, ft	\<v\> (from Fig. 4.2-3)	$2\pi r v$
0	1	0	0
0.2	0.8	14	70.4
0.4	0.6	20.8	78.4
0.6	0.4	24	60.3
0.8	0.2	25.7	32.3
1.0	0.0	26.5	0

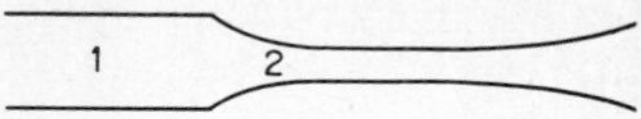

Fig. 4.3-1 Sketch of venturi meter.

Note that the integration and smoothing of velocity profile have been done fairly crudely—this is all that is justified based on the original data. We could, of course, use some sort of numerical integration on a computer but we would do little or nothing to improve the accuracy of the calculated 50 ft^3/sec flow rate.

□ □ □

4.3 VENTURI METER

Figure 4.3-1 shows a sketch of a venturi meter. The flow of a fluid through the constriction of a venturi meter causes a pressure drop which may be related to the average flow velocity with the mechanical energy balance and mass balance.

From the mass balance the volumetric flow rate at points 1 and 2 is the same if $\rho_1 = \rho_2$:

$$\rho_1 A_1 \langle v_1 \rangle = \rho_2 A_2 \langle v_2 \rangle \tag{4.3-1}$$

$$Q_1 = Q_2 \tag{4.3-2}$$

The mechanical energy balance yields

$$\frac{\langle v_2 \rangle^2 - \langle v_1 \rangle^2}{2} + g(z_2 - z_1) + \frac{p_2 - p_1}{\rho} = -\widehat{lw} - \widehat{W} \tag{4.3-3}$$

As with the pitot tube before, let us represent the lost work term as some fraction of the kinetic energy change (we will use the same symbol even though the number will not be the same; we will replace it later anyhow):

$$C \frac{\langle v_2 \rangle^2 - \langle v_1 \rangle^2}{2} = -\widehat{lw} \tag{4.3-4}$$

Then

$$(1 - C) \frac{\langle v_2 \rangle^2 - \langle v_1 \rangle^2}{2} = -\frac{p_2 - p_1}{\rho} \tag{4.3-5}$$

But the mass balance in Eq. (4.3-1) permits us to replace $\langle v_1 \rangle$ in terms of $\langle v_2 \rangle$:

$$\langle v_1 \rangle = \langle v_2 \rangle \frac{A_2}{A_1} \tag{4.3-6}$$

Substituting and solving for $\langle v_2 \rangle^2$, we have

$$\langle v_2 \rangle^2 = \frac{-2}{1-C} \frac{p_2 - p_1}{\rho(1 - A_2{}^2/A_1{}^2)} \tag{4.3-7}$$

If we define the venturi coefficient C_v as

$$C_v = \sqrt{\frac{1}{1-C}} \tag{4.3-8}$$

then

$$\langle v_2 \rangle = C_v \sqrt{\frac{-2(p_2 - p_1)}{\rho(1 - A_2{}^2/A_1{}^2)}} \tag{4.3-9}$$

It is customary to denote D_2/D_1 as β, and so since

$$\frac{A_2{}^2}{A_1{}^2} = \left(\frac{\pi D_2{}^2}{4}\right)^2 \left(\frac{4}{\pi D_1{}^2}\right)^2 = \frac{D_2{}^4}{D_1{}^4} = \beta^4 \tag{4.3-10}$$

then

$$\langle v_2 \rangle = C_v \sqrt{\frac{-(p_2 - p_1)}{\rho(1 - \beta^4)}} \tag{4.3-11}$$

The name "velocity of approach factor" is given to $1/\sqrt{1 - \beta^4}$.

Strictly speaking, C_v must be obtained experimentally, but most venturi meters have so little lost work that a value of 0.975 is a good approximation for C_v in turbulent flow. The coefficient falls off severely at low Reynolds numbers, though, as shown in Fig. 4.3-2.[1,2] We are not usually interested in the velocity at the point of minimum area, but rather in the mass flow rate, the volumetric flow rate, or the average velocity of the fluid in the pipe. The mass flow rate is

$$w = \langle v_2 \rangle A_2 \rho \tag{4.3-12}$$

[1] ASME Power Test Code, part 5, "Measurement of Quantity of Materials," The American Society of Mechanical Engineers, New York, 1959.
[2] "Fluid Meters: Their Theory and Correlations," 4th ed., The American Society of Mechanical Engineers, New York, 1937.

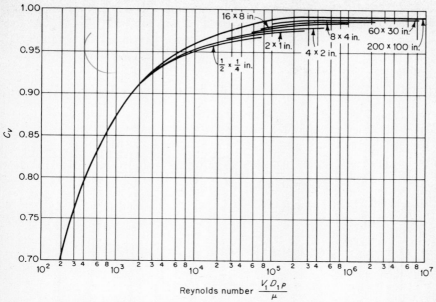

Fig. 4.3-2 Coefficient C_v for venturi meter. (*"Fluid Meters: Their Theory and Application,"* 4th ed., *ASME, 1937.*)

The volumetric flow rate is

$$Q = \frac{w}{\rho} = \frac{C_v A_2}{\sqrt{1 - \beta^4}} \sqrt{\frac{2(p_1 - p_2)}{\rho}} \qquad (4.3\text{-}13)$$

The *average velocity in the pipe* can be obtained from Eq. (4.3-11) and the mass balance:

$$\langle v_1 \rangle = \frac{A_2 \langle v_2 \rangle \rho}{A_1 \rho} = \frac{A_2 \langle v_2 \rangle}{A_1} \qquad (4.3\text{-}14)$$

    A venturi meter is the most accurate of the head meters. The *permanent pressure loss* in a venturi meter is much lower than with other meters of this type (such as the orifice meter). The disadvantage of the venturi meter is its cost and, in some applications, its size. The venturi meter, like flow nozzles and orifices, should not be installed close to fittings or obstructions.[1]

[1]See R. H. Perry, C. H. Chilton, and S. D. Kirkpatrick (eds.), "Chemical Engineers Handbook," 4th ed., p. 5-12, table 5-2, McGraw-Hill Book Company, New York, 1963, to locate venturi and orifice meters in relation to fittings.

Example 4.3-1 Calculation of flow from venturi data A venturi meter with throat diameter of 1 1/2 in. is installed in a schedule 40 4-in. line to measure the flow of water. The pressure drop occurring in the meter is 50 in. Hg. What is the flow rate of water? (Remember the water above the mercury.)

Solution   Assume $\rho_{H_2O} = 62.4$ lb$m$/ft^3, $C_v = 0.975$. From Eq. (4.3-13)

$$Q = \frac{C_v A_2}{\sqrt{1 - \beta^4}} \sqrt{\frac{2(p_1 - p_2)}{\rho}}$$

$$Q = \frac{0.975 \, [1.5/(12 \times 2)]^2 \pi}{\sqrt{1 - (1.5/4.026)^4}} \sqrt{\frac{(2)(32.2)(50/12)(13.6 - 1.0)(62.4)}{62.4}}$$

$$Q = \frac{0.0119}{\sqrt{0.981}} \sqrt{3,380}$$

$$Q = \frac{0.0119}{0.99} 58.2 = 0.70 \text{ ft}^3/\text{sec}$$

$$= 0.70 \left(60 \frac{\text{sec}}{\text{min}}\right) \left(7.48 \frac{\text{gal}}{\text{ft}^3}\right)$$

$$= 315 \text{ gal/min} \qquad \text{(This is turbulent.)}$$

□ □ □

4.4 ORIFICE METER

An orifice (basically, a hole in a plate) placed normal to flow in a pipeline is a restriction and causes pressure drop. This pressure drop is related to the average velocity of the fluid via the mechanical energy balance in the same manner as for the venturi meter. Figure 4.4-1 shows a sketch of a pipe containing an orifice.[1] Upstream of the orifice, the average velocity of the fluid is the bulk velocity in the pipe $\langle v_1 \rangle$, and the area is A_1. Down-

---

[1] See Figs. 4A-4 and 4A-5 for a picture of an orifice and installation.

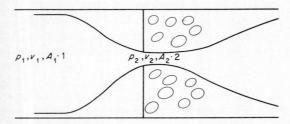

$p_1, v_1, A_1 \cdot 1$ $p_2, v_2, A_2 \cdot 2$

Fig. 4.4-1 Orifice meter.

stream of the orifice there is a point where the cross-sectional area of the flowing stream is a minimum (*vena contracta*) and the average velocity at this point of minimum area is $\langle v_2 \rangle$, and the area is A_2. The stream continues to contract after passing through the orifice because for a fluid particle near the pipe wall to pass through the orifice it must acquire a radial velocity in the inward direction. At the same time, the fluid particle is moving down the pipe. By the time the drag of the surrounding fluid on the fluid particle overcomes the radial velocity, the particle has already passed through the orifice and is some distance downstream. The pipe still remains full of fluid; that is, we are not saying that the fluid coming directly from the orifice is surrounded by gas—it is surrounded by circulating fluid.

The flow through the orifice is not unlike the flow through a venturi, in that the flowing stream contracts and then reexpands, but in the orifice there is an area of turbulence just behind the orifice plate which dissipates a great deal of mechanical energy into internal energy; therefore, the orifice has much more lost work associated with it, and therefore a higher permanent pressure loss. Sketched in Fig. 4.4-2 is a qualitative comparison of the pressure profiles in each of the meters.

We *could* apply the mechanical energy balance, replacing the lost work in terms of the kinetic energy as before, and obtain as before [Eq. (4.3-7)]

$$\langle v_2 \rangle^2 = \frac{-2}{1 - C} \frac{p_2 - p_1}{\rho(1 - A_2{}^2/A_1{}^2)} \tag{4.4-1}$$

The problem with this relation is that the position of the vena contracta changes with the velocity. Our pressure taps are at a fixed point,

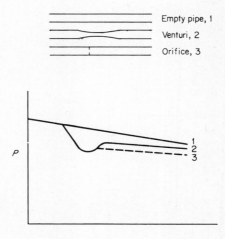

Empty pipe, 1

Venturi, 2

Orifice, 3

P

1
2
3

Fig. 4.4-2 Comparison of pressure profiles for venturi and orifice.

since one can hardly drill a new pressure tap each time the velocity changes. We therefore do not know exactly what the flow area is at the point where we measure pressure, nor do we always measure the downstream pressure at the vena contracta. We therefore usually write Eq. (4.4-1) with the area of the orifice, A_0, in place of A_2, and define an orifice coefficient which in effect is

$$C_0 = \sqrt{\frac{1}{1-C} \frac{1-A_0^2/A_1^2}{1-A_2^2/A_1^2}} \qquad (4.4\text{-}2)$$

so that

$$\langle v_2 \rangle = C_0 \sqrt{\frac{-2(p_2 - p_1)}{\rho(1-\beta^4)}} \qquad (4.4\text{-}3)$$

where

$$\beta = \frac{D_0}{D_1} \qquad (4.4\text{-}4)$$

for an orifice.

Since the point of minimum area does change with velocity, *the orifice coefficient also changes with velocity.* Values of the orifice coefficient have been determined experimentally for various values of β and are plotted versus the Reynolds number in Fig. 4.4-3. More extensive sets of these values are tabulated in reference texts.[1]

[1] J. H. Perry (ed.), "Chemical Engineers Handbook," 3rd ed., p. 405, McGraw-Hill Book Company, New York, 1950.

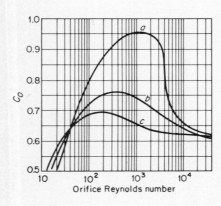

Fig. 4.4-3 Orifice coefficient. (a) $D_0/D_1 = 0.80$; (b) $D_0/D_1 = 0.60$; (c) $D_0/D_1 = 0.20$. [*Tuve and Sprenkle, Instruments*, **6**: 201 (*1933*).]

A natural question at this point is: Where do you put the pressure taps for an orifice? An orifice with *flange taps* has taps 1 in. on either side of the orifice which are part of the flange. An orifice with pipe taps has taps 2 1/2 pipe diameters upstream and 8 pipe diameters downstream. With this arrangement only the *permanent* pressure loss caused by the orifice is used. This pressure drop is smaller than for either flange taps or taps at the point of minimum area. An orifice with *vena contracta taps* has upstream taps 1/2 to 2 pipe diameters from the orifice and the downstream tap is located varying distances from the orifice depending on pipe size and orifice size.[1] *Flow disturbances in the line near the orifice may change the value of the coefficient. Fittings and disturbances should be no closer than 5 pipe diameters from the orifice on the downstream side, and no closer than 20 pipe diameters on the upstream side.*

Example 4.4-1 Calculation of flow from orifice data   A 2-in.-diameter orifice with flange taps is to be used to measure the flow rate of 100°F crude oil flowing in a 4-in. schedule 40 pipe. The manometer reads 30 in. Hg with a 1.1 specific gravity fluid filling the manometer taps. If the viscosity of the oil is 5 cp at 100°F and the specific gravity of the oil is 0.9, what is the flow rate?

Solution   Assume $C_0 = 0.61$.

$$\langle v_2 \rangle = C_0 \sqrt{\frac{-2(p_2 - p_1)}{\rho(1 - \beta^4)}}$$

$$\beta = \frac{2}{4.026} = 0.497 \qquad \beta^4 = 0.061$$

$$\langle v_2 \rangle = 0.61 \sqrt{\frac{(2)(32.2)(30/12)(13.6 - 1.1)}{1 - 0.061}} = 0.61 \sqrt{\frac{2,010}{0.939}}$$

$$\langle v_2 \rangle = 29 \text{ ft/sec}$$

□ □ □

It is not always possible to *calculate* the performance of an orifice to within the tolerance one wishes to determine the flow rate. In such instances, the orifice must be calibrated after it is in place. There are a variety of ways of doing this depending on the types of tankage, scales, etc., that are available. One typical way would be to let the fluid flow into a tank at a constant flow rate and measure the increase of level of the tank with time.

[1] See R. H. Perry, C. H. Chilton, and S. D. Kirkpatrick (eds.), "Chemical Engineers Handbook," 4th ed., p. 5–12, table 5-2, McGraw-Hill Book Company, New York, 1963, to locate venturi and orifice taps in relation to fittings.

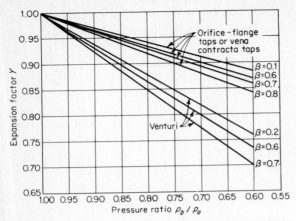

Fig. 4.4-4 Expansion factors for air flowmeters. (*Fluid Meters—Report, ASME, 1951.*)

When orifice and venturi meters are used with compressible fluids, a correction is applied to Eqs. (4.3-11) and (4.4-3) such that

$$\langle v_2 \rangle = C \sqrt{\frac{-2(p_2 - p_1)}{\rho(1 - \beta^4)}} \; Y \sqrt{\tilde{\rho}} \tag{4.4-5}$$

When C is the venturi or orifice coefficient, Y is a dimensionless correction factor given in Fig. 4.4-4[1] and $\tilde{\rho}$ is the upstream density of the fluid.

4.5 ROTAMETER

An area meter, such as a rotameter, operates under a constant pressure difference and varies the free flow area rather than vice versa. Figure 4.5-1 is a sketch of a rotameter.[2] A rotameter is usually constructed from a tapered glass tube and a float of some sort. The float is inside the glass tube, and the fluid to be metered is directed through the tube. A "float," in fact, does not float. This is simply a term which has been used traditionally in discussing these devices. The name rotameter came from the fact that early floats rotated for stability; most tubes now have integral ribs to guide the float. The float will rise in the tube to such a height, depending on the flow rate of the fluid, that the flow area between the tube and the float will allow the required flow of fluid to pass under the pressure difference across the float. At different flow rates, since the tube is tapered, the float will rise to different heights. The position of the float

[1] V. L. Streeter (ed.), "Handbook of Fluid Dynamics," p. 14–14, McGraw-Hill Book Company, New York, 1961.
[2] See Fig. 4A-6 for a picture of a rotameter.

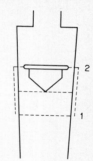

Fig. 4.5-1 Sketch of a rotameter.

can be calibrated in terms of flow rate, giving a means of measuring flow. The basic equation for flow in an *orifice* can be adapted to use for a *rotameter*. The area of the *orifice* corresponds to the area between the float and the tube. The area of the pipe corresponds to the area of the tube at the float position in question. A force balance across the float gives, neglecting frictional drag forces from fluid or tube guides on the edge of the float and from the tube wall on the fluid,

$$\underbrace{\rho_f V_f g}_{\text{Gravity}} = \underbrace{\rho V_f g}_{\text{Buoyancy}} + \underbrace{A_f(p_1 - p_2)}_{\text{Pressure}} \tag{4.5-1}$$

Gravity   Buoyancy Pressure

where V_f is the volume of the float, ρ_f is the density of the float, and A_f is the maximum cross-sectional area of the float. Solving for the pressure difference across the float gives

$$p_1 - p_2 = \frac{V_f(\rho_f - \rho)g}{A_f} \tag{4.5-2}$$

Substituting this pressure difference in Eq. (4.4-3) gives the equation for the rotameter:

$$\langle v_r \rangle = C'_r \sqrt{\frac{2gV_f(\rho_f - \rho)}{A_f\rho(1 - \beta^4)}} \tag{4.5-3}$$

where

$$\beta = \frac{A_2 - A_f}{A_2}$$

The velocity of approach factor is usually included in the rotameter coefficient, so that the final equation is

$$\langle v_r \rangle = C_r \sqrt{\frac{2gV_f(\rho_f - \rho)}{A_f\rho}} \tag{4.5-4}$$

As with the orifice coefficient the rotameter coefficient varies with the properties of the fluid and also with the float shape. Values of the ro-

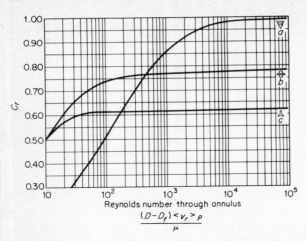

Fig. 4.5-2 Rotameter discharge coefficients. (*G. G. Brown and Associates,* "*Unit Operations,*" *John Wiley & Sons, Inc., New York, 1950.*)

tameter coefficient with different float shapes for various values of the Reynolds number are plotted in Fig. 4.5-2.

As with the orifice and venturi, we again are interested not so much in the velocity as in the volumetric flow rate. This may be obtained, as before, by multiplying the average velocity by the free cross-sectional area available for flow. The reader is cautioned, however, to make sure that he uses the area *over which the average was taken* (here the annular cross section) rather than a different area. Formulas can equally well be developed which give, for example, the average velocity based on the total cross-sectional area of the tube.

Example 4.5-1 Calculation of flow from rotameter data A rotameter with a diameter of 2.5 in. at the top and 1.5 in. at the bottom of a 100-mm tube is used to measure the flow rate of alcohol, specific gravity of 0.8 and viscosity of 1.5 cp. The 1-lbm float (Fig. 4.5-2b) is stainless steel, with specific gravity 8.0 and a diameter of 1.5 in. If the reading is 50 mm, what is the flow rate in gallons per minute?

Solution Assume $C_r = 0.78$.

$$\langle v_r \rangle = C_r \sqrt{\frac{2gV_f(\rho_f - \rho)}{A_f \rho}}$$

$$\langle v_r \rangle = 0.78 \sqrt{\frac{(2)(32.2)(0.002)(8 - 0.8)(62.4)}{[1.25/(12 \times 2)]^2 \pi (62.4)(0.8)}}$$

$$\langle v_r \rangle = 0.78 \sqrt{\frac{(64.4)(0.002)(7.2)}{(0.0122)(0.8)}} = 0.78 \sqrt{94.5}$$

$$\langle v_r \rangle = (0.78)(9.75) = 7.6 \text{ ft/sec}$$

$$\text{Re} = \frac{(D - D_f)\langle v_r \rangle \rho}{\mu} = \frac{(2 - 1.5)(7.6)(62.4)(0.8)}{1.5} = 126$$

$$C_r = 0.76 \qquad \langle v_r \rangle = 7.4 \text{ ft/sec}$$

$$w = \langle v_r \rangle \, \rho A_r = (7.4)(62.4)(0.8)\,\frac{\pi}{144}\,(1^2 - 0.75^2)$$

$$w = 3.5 \text{ lb}m/\text{sec}$$

$$Q = \frac{(3.5)(60)}{(8.33)(0.8)} = 31.6 \text{ gal/min}$$

☐ ☐ ☐

Rotameters are relatively expensive pieces of equipment, and usually are salvaged from one application to the next. For this reason, it is not uncommon for an engineer to find it necessary to use a rotameter for a fluid different from the one for which it had been calibrated. In this next example, we consider the effect of such a change in fluid on the rotameter calibration.

Example 4.5-2 Change of fluid with existing rotameter After the rotameter of Example 4.5-1 had been installed, the equipment was changed to measure the flow rate of water at 60°F. What is the flow rate of water in gallons per minute when the reading is at 50 mm?

Solution If the condition is quite different from the one which was used to calibrate, then recalibration is necessary. However, when the condition is not very different, then an approximate calibration can be used; that is, assume that the same rotameter coefficient holds.

Assume C_r is the same.

$$\frac{\langle v_r \rangle_{\text{H}_2\text{O}}}{\langle v_r \rangle_{\text{Al}}} = \sqrt{\frac{(\rho_f - \rho_{\text{H}_2\text{O}})/\rho_{\text{H}_2\text{O}}}{(\rho_f - \rho_{\text{H}_2\text{O}})/\rho_{\text{Al}}}}$$

$$\frac{\langle v_r \rangle_{\text{H}_2\text{O}}}{\langle v_r \rangle_{\text{Al}}} = \sqrt{0.8\,\frac{8 - 1}{8 - 0.8}} = \sqrt{\frac{(0.8)(7.0)}{7.2}}$$

$$\frac{\langle v_r \rangle_{\text{H}_2\text{O}}}{\langle v_r \rangle_{\text{Al}}} = 0.88$$

$$\langle v_r \rangle_{\text{H}_2\text{O}} = 0.88\,\langle v_r \rangle_{\text{Al}}$$

$$Q = 0.88\,Q_{\text{Al}} = 27.8 \text{ gal/min}$$

☐ ☐ ☐

In this chapter, we have considered a variety of flowmeters which illustrate applications of our macroscopic balances. There are, in fact, many many varieties of flowmeters, most of which we do not discuss.

Here, we have restricted our discussion to those types of flowmeters which present particular illustrations of the application of the macroscopic balances. The selection of flowmeters, like the selection of most items of processing equipment, depends strongly on the type of fluid used—whether or not it is corrosive, contains suspended solids, the temperature level, etc. Most flowmeters are made by companies which specialize in this type of device, and in practical selection of flowmeters the engineer frequently gets data for the specific device from a company which holds a patent on the specific device.

APPENDIX 4A

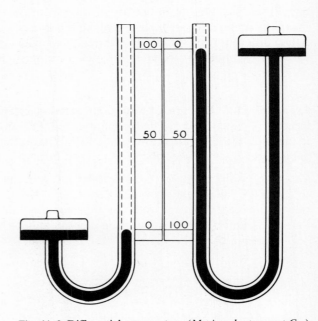

Fig. 4A-2 Differential manometer. (*Meriam Instrument Co.*)

Fig. 4A-1 Manometer. (*Meriam Instrument Co.*)

Fig. 4A-3 Inclined manometer. (*Meriam Instrument Co.*)

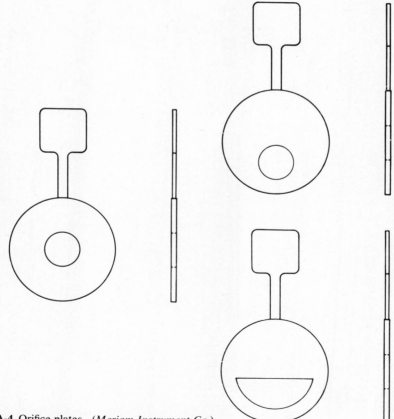

Fig. 4A-4 Orifice plates. (*Meriam Instrument Co.*)

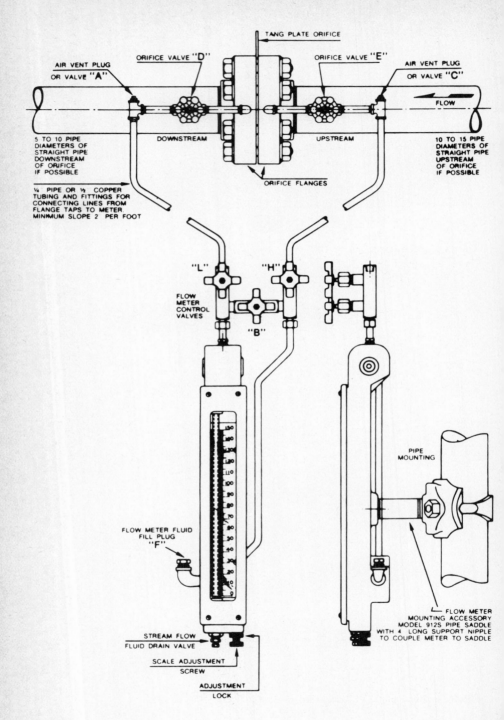

AIR VENT PLUG
OR VALVE "A"

ORIFICE VALVE "D"

TANG PLATE ORIFICE

ORIFICE VALVE "E"

AIR VENT PLUG
OR VALVE "C"

FLOW

DOWNSTREAM

UPSTREAM

5 TO 10 PIPE
DIAMETERS OF
STRAIGHT PIPE
DOWNSTREAM
OF ORIFICE
IF POSSIBLE

ORIFICE FLANGES

10 TO 15 PIPE
DIAMETERS OF
STRAIGHT PIPE
UPSTREAM
OF ORIFICE
IF POSSIBLE

¼ PIPE OR ½ COPPER
TUBING AND FITTINGS FOR
CONNECTING LINES FROM
FLANGE TAPS TO METER.
MINIMUM SLOPE 2 PER FOOT

"L"

"H"

FLOW
METER
CONTROL
VALVES

"B"

PIPE
MOUNTING

FLOW METER FLUID
FILL PLUG
"F"

STREAM FLOW
FLUID DRAIN VALVE

SCALE ADJUSTMENT
SCREW

ADJUSTMENT
LOCK

FLOW METER
MOUNTING ACCESSORY
MODEL 912S PIPE SADDLE
WITH 4 LONG SUPPORT NIPPLE
TO COUPLE METER TO SADDLE

162

FLOW METER INSTRUCTIONS
INSTALLATION AND OPERATION
FOR WATER, AIR OR GAS
FLOW MEASUREMENT

INSTALLATION:

1. Mount the flow meter vertically at a convenient reading height.

2. Connect the orifice flange pressure taps through valves to the flow meter with ¼" pipe or ½" copper tubing and fittings as shown. Piping should slope upward at least 2" per foot. Upstream orifice tap connects to the instrument well of the flow meter at the "H" connection point. Downstream orifice tap connects to the glass tube side of the flow meter at the "L" connection point.

WATER FLOW METER OPERATION:

a. Close both orifice valves "D" and "E"

b. Open flow meter control valves "L", "B", & "H"

c. Remove air vent plugs "A" and "C" or open air venting valves if used. (Similar to Model 944V50)

d. Slowly open orifice valves "D" and "E" to fill the flow meter and piping with water until it flows from the vent openings. All the air in the system must be eliminated.

e. Close orifice valves "D" and "E".

f. Close air vent valves or replace vent plugs "A" and "C".

g. Remove fluid fill plug "F" and pour approximately 1½ pounds of mercury in the flow meter well. The mercury will displace the water and help eliminate air entrapment in the system.

h. Replace fluid fill plug "F".

i. Set zero scale graduation to the mercury meniscus by turning the scale adjustment screw on the flow meter. Lock the adjustment screw.

j. Open orifice valves "D" and "E" to put flow meter under pressure.

k. Close flow meter bypass valve "B".

l. Read water flow rate as indicated by the mercury indicating fluid column.

AIR AND GAS FLOW METER OPERATION:

a. Close orifice valves "D" and "E".

b. Open flow meter control valves "B", "L", and "H".

c. Fill flow meter at "F" with approximately 3 fluid ounces of oil base indicating fluid or 1½ pounds of mercury as required per scale legend.

d. Set zero scale graduation to fluid meniscus with scale adjustment. Secure with locknut.

e. Open orifice valves "D" and "E".

f. Close flow meter bypass valve "B".

g. Read air or gas flow rate as indicated by oil or mercury indicating fluid column.

To check flow meter for zero reference scale setting while in service, close flow meter valves "L" and "H" and open bypass valve "B". After scale is adjusted put flow meter in service by opening valves "L" and "H" and slowly closing bypass valve "B".

In freezing temperatures, water flow meters must be removed from service. Open all flow meter valves, drain water and mercury indicating fluid from flow meter.

Fig. 4A-5 Orifice installation.
(*Meriam Instrument Co.*)

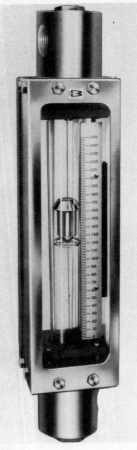

Fig. 4A-6 Rotameter. (*Brooks Instrument Co.*)

PROBLEMS

4.1 An orifice is being used to measure the flow of water through a line. A mercury-water manometer is connected across the orifice as shown below, the lines being filled with water up to the mercury-water interface. If the manometer reading is 5 in., what is the pressure difference between the two locations in the pipe? Suppose the pipe had been vertical with water flowing upward, and the distance between the two manometer taps was 2 ft. What would the pressure difference between the two taps have been when the manometer reading was 5 in.?

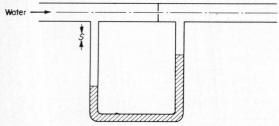

In the vertical case, what would the pressure difference have been if ethyl acetate were substituted for the water?

4.2

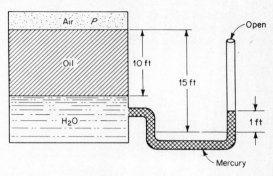

What is the pressure p?   The specific gravity of oil $= 0.8$.   The specific gravity of mercury $= 13.6$.

4.3

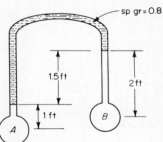

The liquid at A and B is water.[1]

    (*a*) Find $p_A - p_B$.
    (*b*) If the atmospheric pressure is 14.7 psia, find the gage pressure at A.

4.4

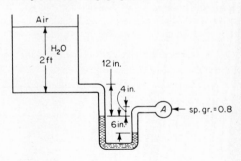

Find the pressure in the drum A at position A.[2]   The manometer fluid is Hg.

[1]V. L. Streeter, "Fluid Mechanics," 4th ed., p. 37, McGraw-Hill Book Company, New York, 1966.
[2]I. H. Shames, "Mechanics of Fluids," p. 59, McGraw-Hill Book Company, New York, 1962.

4.5 Shown is a 12-in. circular duct in which a gas with $\rho = 0.01$ lb$m$ ft^3 and $\mu = 0.1$ cp is flowing. A pitot tube attached to a differential manometer is inserted in the duct as shown. Flow is laminar, and the pitot tube coefficient may be taken as 1.0.

(a) Calculate the velocity (if any) in ft/sec using the pitot tube equation. If no flow is present, explain.

(b) Calculate the mass flow rate in lb$m$/sec if possible—if not, explain why in detail.

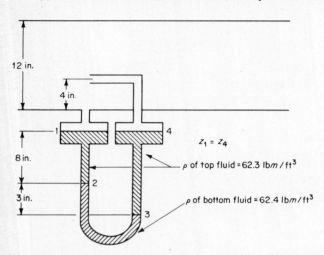

12 in.

4 in.

8 in.

3 in.

$z_1 = z_4$

ρ of top fluid $= 62.3$ lb$m$/ft^3

ρ of bottom fluid $= 62.4$ lb$m$/ft^3

4.6 A pitot tube traverse was made in a 3-in. schedule 40 steel pipe in which water at 58°F was flowing. By integrating

$$\int_A v\, dA = \int_0^R v 2\pi r\, dr = \pi \int_0^{R^2} v\, dr^2 = \pi \int_0^{R^2} C \sqrt{\frac{2g(\rho_m - \rho)}{\rho}}\ \Delta h\ dr^2$$

graphically, we get a flow rate of 0.1336 ft^3/sec (C is the coefficient). At the same time as the traverse the discharge into a weigh tank was $1,400$ lb$m$ of water in 194.4 sec. Find the pitot tube coefficient. Explain the integration above.

4.7 You are calibrating a pitot tube in a 4-in. schedule 40 pipe in which water is flowing at 60°F. The following data are recorded:

$\dfrac{y}{R}$	h (inches of manometer fluid, sp.gr. = 1.2)
0.1	13.60
0.2	16.59
0.3	18.49
0.5	21.55
0.7	23.70
0.9	25.55
1.0	26.30

where y is the distance measured from the wall. At the same time, a sample of the flow is taken by discharging the entire stream into a weighing tank. The discharge rate was found to be 1,410 lbm in 60 sec. Assuming that the velocity profile is symmetrical, find the pitot tube coefficient.

4.8   Fuel oil is flowing into an open-hearth-furnace complex and is entering through a venturi meter as shown below.

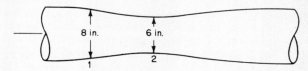

The lost work between points 1 and 2 is estimated at 5 percent of $-\Delta p/\rho$ between 1 and 2. If the oil sp. gr. is 1.4 and the measured pressure drop is $p_2 - p_1 = -8$ psig, (a) show a control volume and label all control surfaces and forces, and (b) calculate the oil flow rate in gallons per minute.

4.9   A venturi meter is used to determine rate of flow in a pipe. The diameter at section 1 is 6 in. and at section 2 is 4 in. (See sketch.) Find the discharge through the pipe when $p_1 - p_2 = 3$ psi, for oil, sp. gr. 0.6.

4.10   A venturi meter is placed in a 4-in.-ID pipe on an angle of 30° with the horizontal. Water at room temperature flows through the venturi which has a throat of 2 in. If the pressure drop in the meter is measured by a mercury manometer, find the velocity of water in the 4-in. pipe.

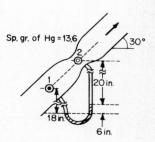

Sp. gr. of Hg = 13.6

30°

20 in.

18 in.

6 in.

4.11   Water at 60°F is flowing in a 3-in.-ID pipeline at the rate of 50 lbm/sec. The flow is to be metered either by a venturi meter or an orifice meter which is attached to a manometer which will give the *same* differential pressure. A pump operating at an efficiency of 70 percent supplies flow in the system.

   (a) Which meter gives the greater permanent pressure loss?

   (b) How much will the *daily cost* of the electrical energy to operate the pump be reduced if a venturi meter is used in the line rather than an orifice meter? Electricity costs 3 cents per kilowatthour. For the orifice

   $C_0 = 0.61$

For the venturi

$$C_v = 0.985 \qquad \beta = 0.5$$

and 1 ft lbf $= 1.356 \times 10^{-4}$ kwh.

4.12 Water at 70°F is flowing in a 3-in.-ID pipeline and is being metered by an orifice meter where $\beta = 0.5$. Calculate the flow rate in gallons per minute if the Δp across the orifice is 28 in. of manometer fluid where the manometer fluid has a specific gravity of 1.7. Justify your assumption on C_0, the orifice discharge coefficient.

4.13 Water at 70°F is flowing in a 4-in.-ID pipeline and is being metered by an orifice meter where $\beta = 0.6$. Calculate the flow rate in gallons per minute if the Δp across the orifice is 26 in. of manometer fluid where the manometer fluid has a sp. gr. $= 1.6$. Justify your assumption on C_0, the discharge coefficient.

4.14 Water is being metered by an orifice as shown below. Calculate the water flow rate in gallons per minute.

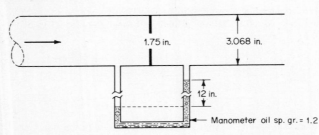

1.75 in.   3.068 in.

12 in.

Manometer oil sp. gr. = 1.2

4.15 Water is metered by an orifice. (See sketch.) Calculate the flow rate in gallons per minute.

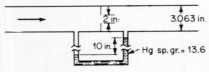

2 in.   3.063 in.

10 in.   Hg sp. gr. = 13.6

4.16 Calculate the flow rate of ethylene glycol at 70°F which is being metered by a rotameter which has no scale. Data:

Ethylene glycol at 70°F

Sp. gr. $= 1.09$
$\mu = 20$ cp

Rotameter

Float type b as per Fig. 4.5-2
$D_f = 1$ in.
$A_f = 0.787$ in.^{2}
$\rho_f = 500$ lbm/ft^3
$V_f = 1.325$ in.^{3}
$D =$ diameter of tube at level of float $= 1.500$ in.

4.17 A rotameter originally purchased to meter acetone (sp. gr. = 0.792) had a stainless-steel float (sp. gr. = 7.92) and a flow range from 0 to 200 lb$m$/min of acetone. The instrument is to be converted to measure the flow of water from 0 to 100 lb$m$/min by changing the float. If the new float is to be the same size and shape as the original, what should its density be?

4.18 A rotameter was designed to measure 0 to 100 gal/min of water using a steel float (sp. gr. = 7.92). It is now desired to utilize the same rotameter to measure 0 to 70 gal/min of ethylene glycol (sp. gr. = 1.1) by using a float with the same size and shape as the original. Find the required density of the float and assume that the rotameter coefficient C_r is the same in both cases.

5
Momentum Transfer in Fluid Flow

In this chapter, we discuss the manner in which fluids behave while flowing. The transport coefficient, viscosity, is introduced through Newton's law of viscosity. Since many fluids do not follow this simple law, a first discussion of "non-Newtonian" or complex fluids is also introduced. A viscous fluid has a velocity profile other than being in plug flow as we have sometimes assumed. The derivation of the Hagen-Poiseuille expression demonstrates the velocity profile in laminar pipe flow.

For most practical situations, the drag and therefore the velocity gradient is largest next to a boundary, while in the mainstream it is often possible to assume an ideal fluid (one with constant density and zero viscosity). The practical approach to describing fluid flow in a real system is to model flow next to a boundary (the boundary layer) and in the mainstream (ideal fluid) and join these two at their common boundary. To understand this procedure, we introduce the concepts of streamlines and stream functions. The boundary layer concept is introduced by considering flow past a flat plate. This concept is then used to discuss flow in the entrance to a pipe. Turbulent flow is introduced and an expression for turbulent flow in a pipe, the universal velocity profile, is discussed.

This chapter provides physical understanding for the practical topics taken up in Chap. 6, where we use design equations based on physical laws, dimensional analysis, and experimental results to determine flow in many practical situations.

5.1 NEWTONIAN FLUIDS

A Newtonian fluid is one that follows Newton's law of viscosity where the shear stress τ in the z direction on a fluid surface of constant y by the fluid in the region where y is smaller is

$$\tau_{yz} = -\mu \frac{dv_z}{dy} \tag{5.1-1}$$

The proportionality constant μ in Eq. (5.1-1) is the coefficient of viscosity. The above equation can be found in some texts defined without the minus sign. The reason for including the minus sign is that the shear stress can also be interpreted as a momentum flux. Units on shear stress (pounds force per square foot) are the same as the units on flux of momentum (pounds mass feet per second squared foot squared), since

$$\frac{lbf}{ft^2} = \left(\frac{lbm\ ft}{sec^2\ ft^2}\right)\left(\frac{lbf\ sec^2}{32.2\ lbm\ ft}\right)$$

We know that momentum tends to go down the velocity gradient; that is, momentum is transferred from regions of higher velocity to regions of lower velocity (this is a consequence of the second law of thermodynamics). The viscosity as defined in the equation above will then be a positive number.

For a typical example of how the shear stress varies with distance from a surface, refer to Fig. 5.1-1, which illustrates shear stress between two parallel planes, separated by fluid flowing in the z direction because of steady motion of the top plane. The reason the profile is shown as

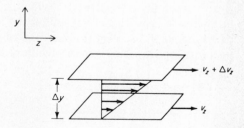

Fig. 5.1-1 Shear between two parallel plates.

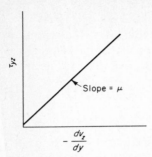

Fig. **5.1-2** Shear stress versus velocity gradient—Newtonian fluid.

linear is that all the momentum transferred from the top plane to the fluid is ultimately transferred to the bottom plane; that is, none of the *momentum* is "dissipated" within the fluid itself. This means that the momentum flux (since the area does not change) is a constant. As a consequence of Eq. (5.1-1), since the momentum flux is constant and the viscosity is constant, the velocity gradient is a constant, as shown in Fig. 5.1-2.

The shear stress τ is a momentum flux. The shear stress or momentum flux is a force per unit area and the coefficient of viscosity has dimensions, in the cgs system, of $gm\,m/(cm\,sec)$. This unit is called the poise. (This is a large unit, and so viscosity is usually expressed in centipoise, $cp = 1/100$ poise.) In the $LtMF$ system, the unit of viscosity is lbm/ft sec. To convert viscosity measured in centipoise to viscosity in pounds mass per foot second, one multiples by 6.72×10^{-4}.

We frequently refer to the action of viscosity as being "fluid friction." In fact, viscosity is not exhibited in this manner. Rather than friction from layers of fluid sliding over one another (much as consecutive cards in a deck of cards will slide over each other), the momentum is transferred by molecules in a slower-moving layer of fluid jumping to a faster-moving region of fluid as shown in Fig. 5.1-3. An analogy might be as follows: Suppose that you are standing one day on roller skates at the roller skating rink, and a friend of yours skates by at 10 mph and tosses you an anvil which he happens to be holding in his arms. Needless to say, the jump of this anvil from a region of higher velocity (him) to a region of lower velocity (you) will induce a velocity in the lower-velocity region, namely, you. This is much like the situation as it happens in a fluid.

It is often convenient to divide the fluid viscosity by the fluid density to give $\nu = \mu/\rho$, where ν is called the kinematic coefficient of viscosity. The kinematic viscosity is a measure of the inherent resistance of the fluid to transition from laminar to turbulent flow as will be seen later when we

$v_1 > v_2$ Fig. **5.1-3** Mechanistic picture of momentum transfer.

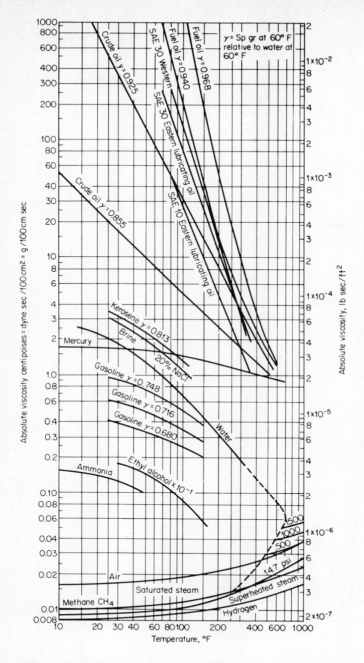

Fig. 5.1-4 Viscosity of fluids. (S. Whitaker, *"Introduction to Fluid Mechanics,"* Prentice-Hall, Inc., Englewood Cliffs, N.J., 1968, reprinted by permission of the publishers.)

discuss the Reynolds number. It measures the intrinsic resistance of the fluid itself as opposed to effects of flow variables such as velocity. The kinematic viscosity indicates the rate of diffusion of momentum and has units cm^2/sec (1 stoke $= 1$ cm^2/sec—this is a large unit and centistokes, cs, are most often used). Fluids with a low kinematic viscosity support turbulence more readily than those fluids with a high kinematic viscosity. For a Newtonian fluid, the viscosity is a state property; that is, it is a function of pressure, temperature, and composition.[1] Figures 5.1-4 and 5.1-5 give values of viscosity for some common fluids.

Example 5.1-1 Calculation of shear stress For flow of a fluid between two fixed parallel plates of length L, the velocity distribution in the z direction is a function of the plate separation as follows:

$$v_z = \frac{(p_1 - p_2) B^2}{2\mu L} \left[1 - \left(\frac{y}{B} \right)^2 \right]$$

where $p_1 - p_2$ is the pressure drop over the length of the plate, $2B$ is the plate separation, and y is the distance from the center line. Determine the shear stress at the plate walls caused by water flowing between the plates under a pressure drop $(p_1 - p_2) = 1$ lbf/in.^{2}, if B is 0.1 ft and L is 10 ft.

$$(a) \qquad \tau_{yz} = -\mu \frac{dv_z}{dy}$$

From the given velocity profile:

$$\frac{dv_z}{dy} = -\frac{(p_1 - p_2) y}{\mu L}$$

Substituting in equation (a):

$$\therefore \tau_{yz} \bigg|_{y=B} = \frac{(p_1 - p_2) B}{L} = \frac{(1 \text{ lbf/in.}^2)(0.1 \text{ ft})}{10 \text{ ft}} = 0.01 \text{ lbf/in.}^2$$

where $\tau_{yz} \bigg|_{y=B}$ is the shear stress at the wall.

□ □ □

---

[1] Normally the viscosity of a gas increases with temperature; the viscosity of a liquid decreases with temperature.

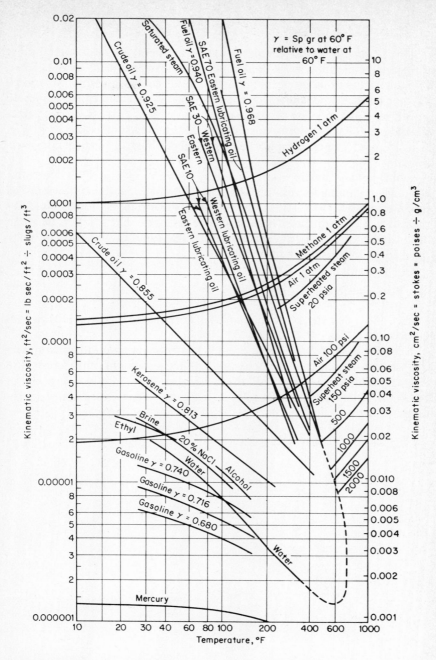

Fig. 5.1-5 Kinematic viscosity of fluids. (*S. Whitaker, "Introduction to Fluid Mechanics," Prentice-Hall, Inc., Englewood Cliffs, N. J., 1968, reprinted by permission of the publishers.*)

5.2 COMPLEX FLUIDS

A plot of shear stress versus velocity gradient for gases and most homogeneous nonpolymeric liquids yields a straight line as in Fig. 5.2-1. A large number of industrially important fluids such as polymer solutions, polymer melts, suspensions, starches, emulsions, and pastes, however, do not follow Newton's law of viscosity; a plot of shear stress versus velocity gradient for these fluids does not yield the Newtonian line as in Fig. 5.2-1. A plot for such complex fluids may give different types of curves as shown.

A Bingham plastic, represented by curve c on Fig. 5.2-1, does not flow until the applied stress reaches a yield point τ_0, and then for stresses higher than the yield stress the plot of shear stress versus velocity gradient is a straight line (Newtonian behavior). The relation between shear stress and velocity gradient for this type of material may be represented by the Bingham model:

$$\tau_{yz} = \tau_0 - \mu_p \frac{dv_z}{dy} \tag{5.2-1}$$

where μ_p is called the plastic viscosity. Toothpaste is a typical Bingham plastic.

A pseudoplastic fluid, Fig. 5.2-1b, has behavior such that the apparent viscosity decreases as velocity gradient increases. In other words, the harder these fluids are "pushed" the easier they flow. A power-law model may be used to represent the relation between shear stress and velocity gradient for these fluids:

$$\tau_{yz} = - m \left| \frac{dv_z}{dy} \right|^{n-1} \frac{dv_z}{dy} \tag{5.2-2}$$

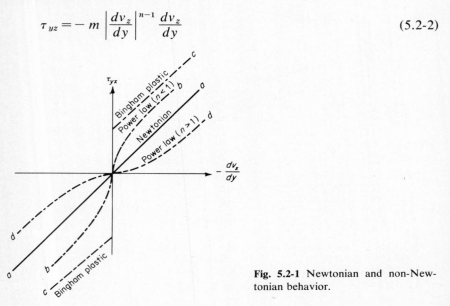

Fig. 5.2-1 Newtonian and non-Newtonian behavior.

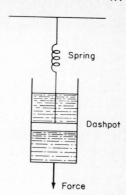

Spring

Dashpot

Fig. 5.2-2 Simple mechanical analogy for viscoelasticity.

Force

where $n < 1$. When $n = 1$, Eq. (5.2-2) reduces to Newton's law of viscosity and m is equivalent to μ. Polymer melts and polymer solutions are typical examples of pseudoplastic fluids.

A dilatant fluid, Fig. 5.2-1d, is a fluid where the apparent viscosity increases as velocity gradient increases. In this case, the harder the fluids are "pushed" the harder it is to "push" them. The power-law model, Eq. (5.2-2), may be used to represent behavior of dilatant fluids using $n > 1$. Suspensions are typically dilatant.

Viscoelastic fluids comprise another important class of fluids. Viscoelastic fluids have a strain recovery or recoil when stress is released. (In a purely viscous fluid, stress and strain relax immediately.) A simple mechanical analog of a viscoelastic fluid is that of a dashpot and a spring connected in series as in Fig. 5.2-2. The dashpot considered alone is analogous to a viscous fluid; the spring is analogous to the elastic behavior. When a force is applied, the spring stretches and the dashpot moves. When the force is released, the dashpot remains in place but the spring recoils and may oscillate. Viscoelastic fluids include a "spring constant" and time-dependent behavior:

$$\tau_{yz} + \lambda \frac{d\tau_{yz}}{dt} = -\mu \frac{dv_y}{dz} \qquad (5.2-3)$$

where λ is the elastic term. Equation (5.2-3) is a simple analog of the mechanical system of the sketch. Rubber cement is viscoelastic.

Some other types of non-Newtonian fluid behavior are time-dependent. These fluids are often said to exhibit memory. This is because the fluid appears to remember what happened to it in the immediate past; that is, how long it has been sheared or for how long it has been stretched. Fluids which show a gradual decrease in apparent viscosity when a constant stress is applied are called thixotropic. Fluids which show an increase in apparent viscosity with time are called rheopectic. The study of complex fluids is a large and important area of fluid mechanics.

A typical example of complex fluid is the toy called "Silly Putty." This material, if an attempt is made to shear it abruptly, acts like a solid—that is, the viscosity goes almost to infinity and it exhibits almost purely elastic behavior. It will, if thrown against a table, bounce. However, if left undisturbed on a table under the force of gravity, at these low shear rates it will exhibit a much lower apparent viscosity and will tend to flow like a fluid, as anyone knows who has ever left a piece of Silly Putty on top of his expensive hi-fi set. Another example for the culinary-minded reader of complex behavior is the so-called Weissenberg effect, which is commonly exhibited by many doughs when mixed with mechanical agitators. Anyone who has mixed much dough in a kitchen realizes that the dough has a most annoying habit of climbing up the beaters rather than going to the outside of the bowl as an ordinary Newtonian fluid would do. This effect comes from the so-called normal stresses. We refer the interested reader to texts such as Fredrickson.[1]

Even though we have used somewhat superficial examples to introduce complex fluid behavior, the reader should realize that we frequently must concern ourselves with complex fluids in practice. For example, many textile fibers are made by extrusion, which involves the flow of highly non-Newtonian fluid, a molten polymer. In many instances of pipeline flow we can effect drag reduction by the addition of a polymer which induces non-Newtonian behavior into the fluid. Many mixing processes in the synthetic rubber industry involve mixing of fluids which behave in a non-Newtonian manner. Production of such simple cosmetic articles as toothpaste, face cream, etc., involves flow of complex fluids—frequently with the accompanying problems of heat transfer and sometimes mass transfer. This very interesting area is covered in great detail in a number of advanced textbooks.

5.3 LAMINAR FLOW

The equations we use to describe the movement of fluids are formulated for a continuum. We know that all materials are composed of molecules; however, for most problems with which we deal the distance between molecules is so small that the continuum approach is valid.[2] Thus we

[1]A. G. Fredrickson, "Principles and Applications of Rheology," Prentice-Hall, Inc., Englewood Cliffs, N.J., 1964.
[2]We may resort to the molecular approach, as we did above to explain how momentum may be transferred between "layers" of fluid. There are also, of course, certain problems of practical interest where the continuum approach may not apply, such as motion where the size of the system approaches the mean free path of molecules. For example, "slip flow" problems are of this type: the motion of gas in the pores of a catalyst where pore size is of the order of the mean free path or motion in the upper atmosphere where the mean free path of molecules becomes very large.

postulate that velocity, density, pressure, etc., behave as continuous functions. (This is a continuum assumption just like the assumption we made in Chap. 2 and justified in some detail in that chapter.) In the type of flow which we call laminar flow, layers of fluids slide over one another smoothly—much as when one pushes on the top of a deck of cards, the top card slides over the next card which transfers momentum to the succeeding card, etc., and on down through the deck. We know because of our discussion earlier that momentum is, in fact, not transferred this way in a fluid; however, it is a useful analogy for laminar flow. One can also conceive of momentum being transferred by clumps of fluid moving from one region to another within the fluid. At the wall, momentum is transferred much as a basketball player transfers momentum to the floor by his bouncing a basketball. (This is a somewhat imperfect analogy because the basketball contains elastic energy, which is not present to any great extent in a normal fluid, but the idea is essentially the same.)

In laminar flow, the fluid particles move in smooth paths or in layers analogous to the deck of cards. It is important that the reader realize that by fluid particle we do not mean a molecule of fluid. Rather, by fluid particle we mean some identifiable region in the fluid that is moving as a lump at the moment. The confusion results from our application of a continuum model to a physical situation which is made up of molecules. Momentum is converted at the containing walls, slowing down the layers closer to the wall. Newton's law of viscosity, Eq. (5.1-1), describes this situation.

If the velocity of the fluid becomes high enough, the fluid no longer moves in layers but "tumbles" and momentum is converted (and kinetic energy is dissipated) by "clumps" of fluid bumping into each other. This type of flow is not easily described and we must resort to statistical theories (time averaging) and experiments to describe such flow. This type of flow is called turbulent.

The result of momentum conversion at the containing walls and in the fluid in laminar flow is that a velocity profile develops such that fluid near the walls travels more slowly than the fluid away from the wall. If, for example, we have two flat plates separated by a distance y, and at time $t = 0$ we set the upper plate in motion with velocity v, the fluid gains momentum and a velocity profile is established. (With the deck of cards the top card will move with the velocity of your hand; the next card will have a lower velocity, etc., until the bottom card next to the table will have zero velocity.)

One of the most useful laminar velocity profiles is for flow in a circular conduit (for example, a pipe). Let us consider this case. A fluid is flowing in laminar flow in a pipe of radius R and length L under the influence of a pressure drop Δp across the pipe. Let us determine the velocity profile for steady flow by writing a force balance on an imaginary circular column of fluid as indicated in Fig. 5.3-1.

Fig. 5.3-1 Forces on fluid flowing in a pipe.

For steady flow, the force caused by the pressure drop must be bal-anced by the viscous force at the boundary of the circular column. If the pressure drop is such that the net pressure force is to the right, the force caused by this pressure drop is the pressure drop times the area on which it is effective

$$F_p = (p|_{z=0} - p|_{z=L})\pi r^2 \tag{5.3-1}$$

The viscous force retarding the motion is the shear stress τ_{rz} at the radius r of the column times the area of shear $2\pi rL$:

$$F_v = \tau_{rz}(2\pi rL) \tag{5.3-2}$$

At steady state $F_p + F_v = 0$ and

$$\tau_{rz} = \frac{(p|_{z=0} - p|_{z=L})r}{2L} \tag{5.3-3}$$

Substituting Newton's law of viscosity

$$-\mu \frac{dv_z}{dr} = \frac{(p|_{z=0} - p|_{z=L})r}{2L} \tag{5.3-4}$$

Integrating Eq. (5.3-4) from 0 to r, and applying the boundary condition that at $r = R$, $v_z = 0$

$$v_z = -\frac{\Delta p \, \bar{R}^2}{4\mu L}\left[1 - \left(\frac{r}{R}\right)^2\right] \tag{5.3-5}$$

Equation (5.3-5) gives the velocity in the tube as a function of the radius. Figure 5.3-2 shows the velocity profile predicted by Eq. (5.3-5). The maximum velocity occurs at $r = 0$, and differentiation will show that

$$v_{z,\text{max}} = \frac{-\Delta p \, R^2}{4\mu L} \tag{5.3-6}$$

The area average velocity for a cross section is found by integration:

$$\langle v_z \rangle = \frac{\int_0^R v_z \, 2\pi r \, dr}{\int_0^R 2\pi r \, dr} \tag{5.3-7}$$

to give

$$\langle v_z \rangle = \frac{-\Delta p \, R^2}{8\mu L} = \frac{1}{2} v_{z,\text{max}} \tag{5.3-8}$$

The volumetric flow rate Q in volume per unit time is the cross-sectional area of the tube multiplied by the average velocity:

$$Q = \pi R^2 \langle v_z \rangle = \frac{-\pi R^4 \, \Delta p}{8\mu L} \tag{5.3-9}$$

Equation (5.3-9) is called the Hagen-Poiseuille law for laminar flow in tubes.

The total drag on the fluid resulting from the shear force transmitted from the fluid to the tube wall is the product of the wetted area of the wall and the shear stress at the wall.

$$\begin{aligned} F_{\text{drag}} &= (2\pi RL)\tau_{rz} \big|_{r=R} = 2\pi RL \left(-\mu \frac{dv_z}{dr} \right)_{r=R} \\ &= -\pi \, \Delta p \, R^2 = 8\pi\mu L \, \langle v_z \rangle \end{aligned} \tag{5.3-10}$$

which shows that the drag force is proportional to the average velocity.

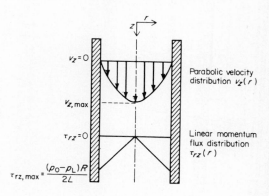

$v_z = 0$

$v_{z,\text{max}}$

$\tau_{rz} = 0$

$\tau_{rz,\text{max}} = \dfrac{(p_0 - p_L)R}{2L}$

Parabolic velocity distribution $v_z(r)$

Linear momentum flux distribution $\tau_{rz}(r)$

Fig. 5.3-2 Velocity profile and momentum flux distribution for laminar flow in a pipe. (*R. B. Bird, W. E. Stewart, and E. N. Lightfoot, "Transport Phenomena," John Wiley & Sons, Inc., New York, 1960.*)

5.4 FLOW OF IDEAL FLUIDS

Many laminar flow problems, especially flow at a point removed from a boundary, can be solved assuming that the fluid is an ideal fluid. An ideal fluid is defined as one with constant density and zero viscosity. Before we can discuss the flow of ideal fluids, we must define certain concepts which permit us to describe the flow field of some real fluids as well as ideal fluids. A *streamline* is an imaginary line in the fluid which has the direction of the local velocity at all points. (In *steady* flow, a streamline is, in fact, the line along which fluid moves.) Since the line has the direction of the fluid velocity at all points, there is, therefore, no flow across a streamline—the flow is *along* a streamline. If one finds the streamlines for an ideal fluid, the velocities and pressures are related by Bernoulli's equation.

A fluid "particle" follows a so-called *path line*. In steady flow the "particle" path is streamline. In unsteady flow a streamline shifts as the direction of velocity changes, and a fluid "particle" can *shift* from one streamline to another, so that the particle path is not a streamline.

One can also define another type of line which is called a *streak line*. A streak line is the instantaneous path exhibited by a tracer which is emitted at a point in the fluid; for example, if one places in the flowing fluid a small hypodermic needle emitting dye in a thin stream, a streak line is the position of the dye line.

A *stream tube* is a tube whose surface is made from all streamlines passing through a closed curve. There is, therefore, no flow through the imaginary walls of the tube, and for steady flow a stream tube is fixed in space. Figure 5.4-1 is a stream tube (in two dimensions) bounded by two streamlines. The *stream function* ψ is a quantity defined in such a way that it is constant along a streamline. The stream tube in the figure is made of two streamlines which may be represented by the functions $\psi = \text{constant}_1 = \psi_1$ and $\psi = \text{constant}_2 = \psi_2$.

The mass rate of flow in the stream tube across a line C in the sketch

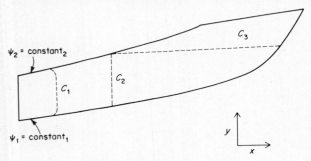

Fig. 5.4-1 2-D stream tube.

must be constant, since there is no flow across the streamlines. Let us develop a relationship between values of the stream function and velocity in a stream tube. Since the flow rate within a stream tube is constant, we may integrate across any path to evaluate the mass flow rate in a stream tube. Let us first choose the path indicated by C_1. If C_1 has length ds then

$$w = \int_{C_1} \rho(\mathbf{v} \cdot \mathbf{n}) \, ds = \int_{C_1} \rho v \cos \alpha \, ds \tag{5.4-1}$$

where α is the angle between velocity vector and outward-directed normal and we define the value of the stream function so that

$$w = (\psi_1 - \psi_2)\rho \tag{5.4-2}$$

Since we have not restricted the distance between the streamlines considered, we can let this difference shrink to a differential difference in the stream function. The equation

$$\int \rho \, d\psi = \text{constant} \tag{5.4-3}$$

is true for any path C_1, C_2, C_3, etc., which joins the two streamlines. For path C_2, $\delta s = \delta y$ and $v_x = v \cos \alpha_2$; for path C_3, $\delta s = \delta x$ and $v_y = v \cos \alpha_3$; therefore

$$\delta\psi = -v_x \mid_x \delta y \tag{5.4-4}$$

$$\delta\psi = v_y \mid_y \delta x \tag{5.4-5}$$

Therefore in the limit

$$v_x = -\frac{\partial\psi}{\partial y} \tag{5.4-6}$$

$$v_y = \frac{\partial\psi}{\partial x} \tag{5.4-7}$$

These are partial derivatives because ψ is a function of both x and y. The potential causing flow along a streamline (such as pressure or gravity) is related to the velocity along a streamline and this potential is orthogonal to the streamline so that

$$v_x = -\frac{\partial\phi}{\partial x} \tag{5.4-8}$$

$$v_y = -\frac{\partial \phi}{\partial y} \tag{5.4-9}$$

From Eqs. (5.4-6) to (5.4-9)

$$\frac{\partial \phi}{\partial x} = \frac{\partial \psi}{\partial y} \tag{5.4-10}$$

$$\frac{\partial \phi}{\partial y} = -\frac{\partial \psi}{\partial x} \tag{5.4-11}$$

Equations (5.4-10) and (5.4-11) are the Cauchy-Riemann equations which must be satisfied by the real and imaginary parts of an analytic function

$$w(z) = \phi(x, y) + i\psi(x, y) \tag{5.4-12}$$

where z and w are complex numbers.[1] An analytic function is a well-behaved function, that is, one which possesses a derivative at every point. Any analytic function yields a pair of functions ϕ and ψ which are the velocity potential and stream function for *some* ideal flow problem. The curves $\phi = $ constant and $\psi = $ constant are the equipotentials and streamlines of the problem.

Rather than become involved in an extended mathematical discussion, we will present a couple of simple examples which will illustrate the utility of the stream function and the velocity potential by utilizing their identification as the real and imaginary parts of a complex function.[2]

Example 5.4-1 Flow of an ideal fluid around a cylinder Show that the complex potential w describes ideal flow around a circular cylinder of radius R with approach velocity v_∞ if

$$(a) \qquad w(z) = v_\infty\left(z + \frac{R^2}{z}\right)$$

Solution To see what flow situation is described, examine the streamlines—since there is no flow across a streamline, any streamline can be interpreted as the boundary of a solid object. Since streamlines are located

---

[1] R. V. Churchill, "Complex Variables and Applications," McGraw-Hill Book Company, New York, 1960.
[2] The reader is referred to such texts as Churchill, *ibid.*, or Milne-Thomson, "Theoretical Hydrodynamics," The Macmillan Company, New York, 1955.

as constant values of the stream function, we first obtain the stream function using the fact that

(b) $w(z) = \phi + i\psi$

To separate the given complex potential $w(z)$ into its real and imaginary parts, we substitute in equation (a):

$z = x + iy$

Thus

(c) $w = v_\infty \left(x + iy + \dfrac{R^2}{x + iy} \right)$

Multiplying and dividing the fractions in equation (c) by $x - iy$ and rearranging

$$w = \underbrace{v_\infty x \left(1 + \frac{R^2}{x^2 + y^2} \right)}_{\phi} + \underbrace{iv_\infty y \left(1 - \frac{R^2}{x^2 + y^2} \right)}_{\psi}$$

where we can compare equation (b) and immediately identify ϕ and ψ as shown.

If we examine the streamline represented by $\psi = 0$ we have

$$v_\infty y \left(1 - \frac{R^2}{x^2 + y^2} \right) = 0$$

so that either $y = 0$ or $1 - R^2/(x^2 + y^2) = 0$ since $v_\infty \neq 0$ in general. The line $y = 0$ is the x axis and from the second condition

$$1 - \frac{R^2}{x^2 + y^2} = 0$$

or

(d) $x^2 + y^2 = R^2$

Equation (d) is the locus of a circle with radius R. Therefore the streamline $\psi = 0$ consists of the x axis and a circle of radius R about the origin, and so w may be taken to describe 2-D potential flow around an in-

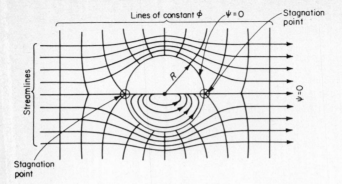

Fig. 5.4-2 Potential flow outside and inside a cylindrical shape. (*L. Prandtl, and O. G. Tietjens, "Fundamentals of Hydro- and Aerodynamics," Dover Publications, Inc., New York, reprinted by permission of the publishers.*)

finite (in the direction normal to the xy plane) cylinder.

This answers the question posed; however, by substituting arbitrary values of x and y in ϕ and ψ we can plot the entire flow network as shown in Fig. 5.4-2. We can also examine the magnitude of the components and the velocity vector by using the fact that

$$(e) \qquad v^2 = v_x^2 + v_y^2$$

and either calculating v_x and v_y with Eqs. (5.4-8) and (5.4-9) or alternatively by observing that since w is an analytic function of z its derivative is independent of the way dz approaches zero. From this it may be shown that[1]

$$\frac{dw}{dz} = \frac{\partial \phi}{\partial x} + i\frac{\partial \psi}{\partial x} = -i\frac{\partial \phi}{\partial y} + \frac{\partial \psi}{\partial y}$$

that is,

$$(f) \qquad \frac{dw}{dz} = v_x + iv_y$$

This permits us to obtain v_x and v_y simultaneously. Here

$$(g) \qquad \frac{dw}{dz} = \frac{d}{dz}\left[v_\infty\left(z + \frac{R^2}{z}\right)\right] = v_\infty\left(1 - \frac{R^2}{z^2}\right)$$

[1] R. V. Churchill, "Complex Variables and Applications," McGraw-Hill Book Company, New York, 1960.

To separate real and imaginary parts as in equation (f) we use the polar form

$$z = re^{i\theta} = r(\cos \theta + i \sin \theta)$$

so that in (g):

$$z^{-2} = (re^{i\theta})^{-2} = r^{-2} [\cos (- 2\theta) + i \sin (-2\theta)]$$

Substituting in equation (g) using the facts that

$$\cos (-2\theta) = \cos (2\theta)$$
$$\sin (-2\theta) = -\sin (2\theta)$$

we get

(h) $\qquad \dfrac{dw}{dz} = v_\infty \left[1 - \dfrac{R^2}{r^2} (\cos 2\theta - i \sin 2\theta) \right]$

Comparing equations (f) and (h)

(i) $\qquad v_x = v_\infty \dfrac{R^2}{r^2} (\cos 2\theta) - 1$

(j) $\qquad v_y = v_\infty \dfrac{R^2}{r^2} \sin 2\theta$

If we examine the total velocity at $r = R$ using equation (e)

$$v^2 = 4v_\infty^2 \sin^2 \theta$$

The minima of this function are at $n\pi$ ($0°$, $180°$, $360°$, etc.) and the maxima at $(n + \frac{1}{2})\pi$ ($90°$, $270°$, $450°$, etc.). This is obvious from the flow network if one remembers that closely spaced network lines imply high velocity and conversely. The points at $0°$ and $180°$ for $r = R$ have $v = 0$ and are accordingly called *stagnation points*.

Using equations (i) and (j) we can obtain the velocity anywhere in the flow field. If we are given the pressure at one point we know the pressure field as well since we can calculate the pressure at any other point by calculating the velocities at the two points and using the Bernoulli equation:

$$\frac{v^2 - v_0^2}{2} + \frac{p - p_0}{\rho} = 0$$

□ □ □

Fig. 5.4-3 Flow in a corner.

Example 5.4-2   Flow of an ideal fluid in a corner   Show that the complex potential $w = -v_\infty z^2$ can describe the flow in a corner. What are the velocity components $v_x(x,y)$ and $v_y(x,y)$? What is the physical significance of v_∞?

Solution

$$w = -v_\infty z^2 = -v_\infty (x + iy)^2$$

$$= -v_\infty \underset{\phi}{[(x^2 - y^2)} + i \underset{\psi}{(2xy)]}$$

Note that the streamline $\psi = 0$ consists of the x and y axes—these form the corner.

Therefore

$$v_x = -\frac{\partial \phi}{\partial x} = -\frac{\partial \psi}{\partial y} = 2xv_\infty$$

$$v_y = -\frac{\partial \phi}{\partial y} = \frac{\partial \psi}{\partial x} = -2yv_\infty$$

$$v^2 = v_x{}^2 + v_y{}^2 = 4v_\infty{}^2 (x^2 + y^2) = 4v_\infty{}^2 R^2$$

$$v_\infty = \frac{1}{2}\frac{v}{R} = \frac{1}{2} \text{ (angular velocity)}$$

The solution is shown in Fig. 5.4-3. This solution can be generalized to non-90° corners.

□ □ □

5.5 TURBULENT FLOW

There are two dominant mechanisms by which momentum is transferred in a flowing fluid. Both laminar and turbulent flow are characterized by a pressure loss due to momentum transfer from the fluid to the containing walls or to objects in the flow stream. The dominating force in laminar flow is the viscous force. In turbulent flow the mechanism of momentum transfer causing pressure loss is mainly the "tumbling" of the fluid. This pressure loss is due to the formation of *eddies*. The dominant force in turbulent flow is the inertial force. The inertial forces arise from the rapid acceleration and deceleration of the fluid from velocity fluctuations. The Reynolds number is the dimensionless product that represents the ratio of inertial to viscous forces. As the fluid moves faster in a given flow situation, such as the flow of water in a pipe, the inertial forces become larger relative to the viscous forces and eventually flow shifts from laminar (viscous-dominated) to turbulent (inertia-dominated) flow. The transition occurs at about $Re = 2,100$ for flow in a pipe. For a flat plate the transition occurs at about $Re_x = 3 \times 10^6$. Note that Re_x is based on the length from the nose of the plate and the free-stream velocity. It is not surprising that transition occurs at a different numerical value than in a pipe—one is an external flow where the profile continues to develop and the other an internal flow which theoretically reaches a fully developed state.

In turbulent flow, velocity at a point varies stochastically in magnitude and direction as time progresses. If we measure velocity at a point in a fluid in turbulent flow we obtain a plot of velocity similar to Fig. 5.5-1. Turbulence is a random process. If we took the frequency spectrum shown in Fig. 5.5-1 and fed it to a loudspeaker we would hear not a musical note, but rather a roar. This is the reason that people who survive tornadoes frequently describe the sound as like the roar of a freight train; in fact, the roar of a freight train is sound which contains almost all frequencies in the spectrum in some sort of random fashion. The turbulence in a tornado gives much the same tone when translated to

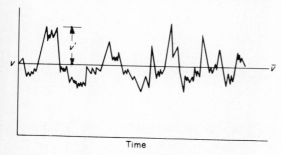

Fig. 5.5-1 Point velocity for turbulent flow.

acoustical waves. We describe turbulence by a time-averaged velocity:

$$\bar{v} = \frac{\int_0^t v\, dt}{\int_0^t dt} \tag{5.5-1}$$

where v is the instantaneous velocity and is a function of time. The average velocity is represented in Fig. 5.5-1 by the straight line. At any time, the instantaneous velocity is the sum of the average velocity and the velocity fluctuation v':

$$v = \bar{v} + v' \tag{5.5-2}$$

In studying turbulence we usually refer to components of the velocity in the various coordinate directions. For the x direction we use v_x for the component, $\bar{v}_x$ for the time-averaged component, and v'_x for the fluctuation in the component. For the y direction we use v_y, $\bar{v}_y$, and v'_y, and for the z direction v_z, $\bar{v}_z$, and v'_z.

For a detailed treatment of turbulence the reader is referred to texts such as Hinze.[1] Here we will emphasize application of principles from turbulent flow rather than derivations.

The velocity distribution and average velocity for *laminar* flow in a tube are, as we have already seen,

$$v_z = v_{z,\max}\left[1 - \left(\frac{r}{R}\right)^2\right] \tag{5.5-3}$$

$$\langle v_z \rangle = \frac{1}{2} v_{z,\max} \tag{5.5-4}$$

and the pressure drop is directly proportional to volume rate of flow. For *turbulent* flow, experiments show that the *time-averaged* velocity is approximately

$$\bar{v}_z \cong \bar{v}_{z,\max}\left(1 - \frac{r}{R}\right)^{1/7} \tag{5.5-5}$$

$$\bar{v}_z \cong \frac{4}{5} v_{z,\max} \tag{5.5-6}$$

if $10^4 < \mathrm{Re} < 10^5$. Figure 5.5-2 is a qualitative comparison of laminar and turbulent velocity profiles in a pipe.

[1] J. O. Hinze, "Turbulence," McGraw-Hill Book Company, New York, 1959.

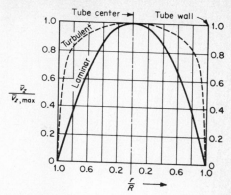

Fig. 5.5-2 Comparison of laminar and turbulent velocity profiles.

We can arbitrarily divide turbulent flow in a pipe into three regions: (1) the turbulent core, (2) a buffer region which lies between this core and the domain which is influenced by the presence of the pipe wall, and (3) this domain.

Example 5.5-1 Average velocity for turbulent flow Use the one-seventh-power law and determine the average velocity for turbulent flow in a pipe in terms of the maximum velocity.

Solution

$$\langle \bar{v}_z \rangle = \frac{\int_0^{2\pi} \int_0^R \bar{v}_z \, r \, dr \, d\theta}{\int_0^{2\pi} \int_0^R r \, dr \, d\theta}$$

$$\langle \bar{v}_z \rangle = \frac{2}{R^2} \int_0^1 \bar{v}_{z,\max} R^2 \left(1 - \frac{r}{R} \right)^{1/7} \frac{r}{R} \, d\left(\frac{r}{R} \right)$$

This expression may be integrated by substitution of

$$1 - \frac{r}{R} = \tau$$

$$\frac{r}{R} = 1 - \tau$$

$$d\left(\frac{r}{R} \right) = - \, d\tau$$

yielding

$$\frac{2\bar{v}_{z,\max} R^2}{R^2} \int_1^0 \tau^{1/7} (1 - \tau) \, (-d\tau)$$

or $\quad -2\bar{v}_{z,max}\left(\dfrac{7}{8}\tau^{8/7} - \dfrac{7}{15}\tau^{15/7}\right)\bigg|_1^0$

$$\langle\bar{v}_z\rangle = -2\bar{v}_{z,max}\left(\frac{7}{15}-\frac{7}{8}\right) = -2\bar{v}_{z,max}\left(\frac{56-105}{120}\right)$$

$$\langle\bar{v}_z\rangle = \frac{49}{60}\bar{v}_{z,max}$$

Experimentally

$$\langle\bar{v}_z\rangle \cong \frac{4}{5}\bar{v}_{z,max}$$

This is much the same problem as is covered in Appendix 2A.

□ □ □

A velocity distribution equation for turbulent flow in a tube can be derived by assuming that the shear stresses in turbulent flow are proportional to the fluctuating velocities v'_y and v'_z. This model was formulated by Prandtl who assumed a "particle" of fluid to be displaced a distance ℓ (called the mixing length)[1] before its momentum is changed by the new environment as shown in Fig. 5.5-3. He related the velocity fluctuations to the time-averaged velocity gradient as

$$v'_z = \ell\,\frac{d\bar{v}_z}{dy} \tag{5.5-7}$$

and assumed that $v'_y \cong v'_z$. If the shear stress in the fluid is assumed to retain the form of Newton's law of viscosity then

$$\bar{\tau}_{yz} = -(\mu+\eta)\frac{d\bar{v}_z}{dy} = \bar{\tau}^l_{yz} + \bar{\tau}^t_{yz} \tag{5.5-8}$$

[1] This is not a very satisfying model physically since no fluid has identifiable "particles" on a macroscopic scale.

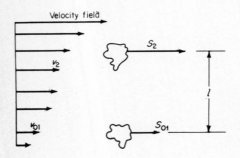

Fig. 5.5-3 Prandtl model. "Particle" of fluid moving at v_0 moves length l before changing velocity via momentum interchange to the local velocity at the new point.

where μ is the fluid viscosity, η is called the eddy viscosity (momentum transport coefficient due to turbulence), and

$$\bar{\tau}^t_{yz} = \overline{\rho v'_y \, v'_z} \tag{5.5-9}$$

This quantity is sometimes called the Reynolds *stress* even though it is the product of two velocities and a density. Examining the units one can see it has units of stress, which, again, are the same units as momentum flux.

Therefore

$$\bar{\tau}^t_{yz} = \rho \ell^2 \left| \frac{d\bar{v}_z}{dy} \right| \frac{d\bar{v}_z}{dy} \tag{5.5-10}$$

where ℓ is the mixing length and

$$\eta = \rho \ell^2 \left| \frac{d\bar{v}_z}{dy} \right| \tag{5.5-11}$$

Note: η is not a fluid property; it depends on flow field. It is *defined* in such a way that it is related to shear stress in the form of Eq. (5.5-8). As turbulence decreases, Eq. (5.5-8) approaches Newton's law of viscosity because $\eta \rightarrow 0$. For turbulent flow past a boundary, $\eta = 0$ at the surface while μ is negligible at distances from the surface large when compared to δ, the thickness of the viscous sublayer.

The concept of mixing length, together with the time-averaged velocity, may be used to obtain an expression for velocity during steady-state turbulent flow in a pipe. Close to the pipe wall, where η is negligible, the shear stress is assumed given by Eq. (5.5-8), where the thickness over which μ dominates is $y = \delta$ (the viscous sublayer). The shear stress at the wall[1] is

$$\tau_0 = - \mu \frac{dv_z}{dy} \bigg|_{y=0} \tag{5.5-12}$$

Dividing both sides of Eq. (5.5-12) by the density ρ gives

$$\frac{\tau_0}{\rho} = - \nu \frac{dv_z}{dy} \qquad y \leqslant \delta \tag{5.5-13}$$

where $\nu = \mu/\rho$ is the kinematic viscosity. We assume that $\tau = \tau_0$

---

[1] We drop the bar ($\bar{}$) for convenience of writing.

throughout the viscous sublayer and that $\delta \ll R$, and so we can apply the rectangular coordinate form.

We define a "shear" velocity

$$v^* = \sqrt{\frac{\tau_0}{\rho}} \tag{5.5-14}$$

Then Eq. (5.5-14) may be written

$$\frac{v_z}{v^*} = \frac{v^* y}{\nu} \qquad y \leq \delta \tag{5.5-15}$$

For $y > \delta$, μ is assumed to be much less than η and

$$\tau = \rho \ell^2 \left| \frac{dv_z}{dy} \right| \frac{dv_z}{dy} \tag{5.5-16}$$

Since ℓ has the dimensions of length we may assume for a simple model

$$\ell = Ky \tag{5.5-17}$$

Substituting Eq. (5.5-17) into Eq. (5.5-16) and rearranging

$$\frac{dv_z}{v^*} = \frac{1}{K} \frac{dy}{y} \tag{5.5-18}$$

We now assume the shear stress to be constant at τ_0 throughout the turbulent core. Integrating Eq. (5.5-18) yields

$$\frac{v_z}{v^*} = \frac{1}{K} \ln y + \text{constant} \tag{5.5-19}$$

The constant of integration may be found by assuming that v_z is zero at some value of y, say y_0 (but not at $y = 0$ since $\ln 0 = - \infty$)

$$\frac{v_z}{v^*} = \frac{1}{K} \ln \frac{y}{y_0} \tag{5.5-20}$$

It is customary to write Eq. (5.5-20) by multiplying the numerator and denominator of the argument of the logarithm by v^*/ν to give

$$\frac{v_z}{v^*} = \frac{1}{K} \left(\ln \frac{y v^*}{\nu} - \ln \frac{y_0 v^*}{\nu} \right) \tag{5.5-21}$$

or $\quad \dfrac{v_z}{v^*} = \dfrac{1}{K} \ln \dfrac{yv^*}{\nu} + C$ $\hspace{3cm}$ (5.5-22)

where C is zero in the viscous sublayer and is determined from experiment for the buffer zone and the turbulent core. Equation (5.5-22) is called the universal velocity distribution since v_z/v^* is a universal function of yv^*/ν. Data on turbulent flow in pipes with $4{,}000 \leqq \text{Re} \leqq 3.2 \times 10^6$ may be fit by the following forms.

$$\dfrac{v_z}{v^*} = \dfrac{yv^*}{\nu} \qquad 0 < \dfrac{yv^*}{\nu} < 5 \qquad \text{(viscous sublayer)} \qquad (5.5\text{-}23)$$

$$\dfrac{v_z}{v^*} = 5 \ln \dfrac{yv^*}{\nu} - 3.05 \qquad 5 < \dfrac{yv^*}{\nu} < 30 \qquad \text{(buffer zone)}$$
$$\hspace{9cm}(5.5\text{-}24)$$

$$\dfrac{v_z}{v^*} = 2.5 \ln \dfrac{yv^*}{\nu} + 5.5 \qquad 30 < \dfrac{yv^*}{\nu} \qquad \text{(turbulent core)} \quad (5.5\text{-}25)$$

Figure 5.5-4 is a plot of these three equations, where $v^+ = v/v^*$ and $y^+ = yv^*/\nu$.

Example 5.5-2 Size of the sublayer and buffer zone in turbulent flow
Figure 5.5-4 is a correlation of the data for turbulent flow in a smooth

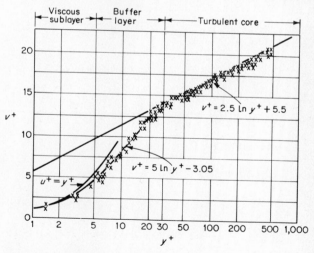

Fig. 5.5-4 Universal velocity profile. (*C. O. Bennett, and J. E. Myers, "Momentum, Heat, and Mass Transfer," McGraw-Hill Book Company, New York, 1962.*)

circular pipe. The lines representing the equation for the viscous sublayer, the buffer zone, and the turbulent core are shown. If water is flowing in a 1-in. schedule 40 steel pipe at a bulk velocity of 5 ft/sec, and if the pressure drop over 1,000 ft of pipe is 30 psig, what is the thickness of the viscous sublayer? Of the buffer zone? At 60°F: $\mu = 1.13$ cp, $\rho = 62.4$ lb/ft^3; ID $= 1.049$ in.

Solution

$$\mathrm{Re} = \frac{1.049}{12}\,\frac{5}{1.13}\,\frac{62.4}{0.000672} = 35{,}900$$

Sublayer

$$\frac{v_z}{v^*} = \frac{yv^*}{v} = 5$$

$$y = \frac{5v}{v^*} = \frac{5\mu}{\rho}\sqrt{\frac{\rho}{\tau_0}} = \frac{5\mu}{\sqrt{\rho\tau_0}}$$

A force balance on the fluid in the pipe yields

$$\tau_0 \pi D L = -\,\Delta p\,\frac{\pi D^2}{4}$$

Substituting

$$y = 5\mu\sqrt{\frac{4L}{\rho D(-\Delta p)}} = 10\mu\sqrt{\frac{L}{\rho D(-\Delta p)}}$$

$$y = [(10)(1.13)\mathrm{cp}]\left(\frac{6.72\times 10^{-4}\,\mathrm{lbm}}{\mathrm{cp\ ft\ sec}}\right)$$

$$\left\{\frac{1000\ \mathrm{ft}}{(62.4\ \mathrm{lbm/ft^3})(1.049\ \mathrm{ft/12})(32.2\ \mathrm{lbm\ ft/lbf\ sec^2})[(30)(144)\,\mathrm{lbf/ft^2}]}\right\}^{1/2}$$

$$y = 0.000268\ \mathrm{ft} = 0.00322\ \mathrm{in.}$$

Edge of buffer zone

$$y = \frac{30v}{v^*} = (6)(0.000268\ \mathrm{ft}) = 0.00161\ \mathrm{ft} = 0.0193\ \mathrm{in.}$$

Buffer zone $= 0.00268 - 0.00161 = 0.00134\ \mathrm{ft} = 0.0161\ \mathrm{in.}$

□ □ □

In practice, pipes are not the idealized smooth cylindrical objects that one might like for one's theory. Real pipes have protuberances on the wall which stick out into the flow field. The protuberances are caused by the way the pipes are manufactured.

For rough pipes, a given protuberance is characterized by an effective roughness height k and the relative roughness k/D, where D is the pipe diameter. Wall roughness does not affect laminar flow; therefore, in turbulent flow roughness will not have an effect if the roughness is smaller than the thickness of the viscous sublayer. Pipes where roughness is less than the thickness of the viscous sublayer are called hydraulically smooth. When the roughness height is greater than the thickness of the viscous sublayer it will affect the "tumbling" of the fluid, increasing turbulence and changing the velocity profile. This is from form drag, explained in Chap. 6. The universal velocity distribution can be used for rough pipes by assuming $y_0 = k$ (in other words, assuming that the true inside diameter of the pipe is the diameter out to the viscous sublayer rather than to the pipe wall). Using this assumption

$$\frac{v_z}{v^*} = \frac{1}{K} \ln \frac{y}{k} + C_2 \tag{5.5-26}$$

This equation is valid for the completely rough region $yv^*/\nu > 70$ with $K = 0.4$ and $C_2 = 8.5$. For $yv^*/\nu < 5$ the pipe appears hydraulically smooth. In between, there is no simple equation to express the velocity.

5.6 THE BOUNDARY LAYER

If we consider flow over a flat plate from a uniform free-stream velocity profile, we will see as shown in Fig. 5.6-1 that the flow will be retarded at the surface of the plate. The fluid particles which are retarded near the plate transfer momentum from the fluid particles which are farther from the plate and so on, and thus drag is transmitted into the mainstream fluid. If we plot the position where the velocity reaches 99 percent of the free-stream velocity we will find that we build up a gradually thickening "boundary" layer on the plate as shown in Fig. 5.6-1. This boundary layer is not a physical entity but rather an arbitrary boundary which we have chosen. It is, however, a useful concept because it is within the

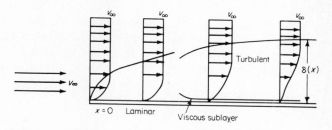

Fig. 5.6-1 Boundary layer development.

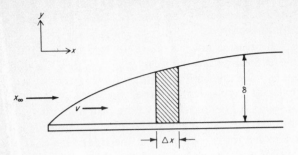

Fig. 5.6-2 Boundary layer.

boundary layer that most of the momentum is transferred. We can see that this is true as a consequence of the velocity profile. The momentum transfer is proportional to the velocity gradient. Since the velocity gradient is almost zero outside the boundary layer, very little momentum is transferred outside the boundary layer. We have really divided the flow field into a region in which momentum transfer occurs and a region where very little momentum transfer occurs.

As was mentioned during the discussion of ideal fluids, when describing viscous flows it is possible to describe the flow by finding the velocity profile near the boundary taking viscosity into account, and joining this velocity profile with the ideal flow solution which is valid far from the boundary. The velocity in the boundary layer may be found from the simplified general equations for flow of viscous fluids, which we call the boundary layer equations. Application of the momentum equation and appropriate assumptions permits the development of equations for boundary layer growth and the drag associated with the boundary.

Consider Fig. 5.6-2 where fluid is flowing in the x direction past the top half of a flat plate. The approach, or free-stream, velocity is assumed to have a flat profile with velocity v_∞. As the fluid passes along the plate, the drag on the fluid causes a velocity reduction such that the velocity is zero at the plate, and then the velocity increases as a function of y until it reaches the approach velocity v_∞. The boundary layer thickness δ is usually defined as the normal distance from the surface to the point where the fluid velocity has a value of $0.99\, v_\infty$. For smooth upstream boundaries, the boundary layer is initially laminar, and the fluid moves in smooth layers. The thickness of the boundary layer increases from the leading edge ($x = 0$) until the boundary layer becomes unstable, and at a characteristic Reynolds number

$$\mathrm{Re}_x = \frac{x v_\infty \rho}{\mu} \tag{5.6-1}$$

the flow becomes turbulent ($\mathrm{Re}_x = 3 \times 10^6$ for a flat plate).

We have earlier pointed out that the Reynolds number is the ratio of the inertial forces to the viscous forces. We might expect the viscous forces to be related, not to the distance from the front of the plate, but more logically to the thickness of the boundary layer. In fact, we *could* define a Reynolds number based on boundary layer thickness, but this thickness is not usually at our disposal, and since the boundary layer thickness is proportional to the distance from the front of the plate (which we do know) we use the more convenient distance. For $2 \times 10^5 <$ $Re_x < 3 \times 10^6$ the boundary layer flow is transitional. For $Re_x < 2 \times 10^5$ flow in the boundary layer is laminar. As one method of developing a relationship for a drag force on a flat plate, we can consider the so-called Von Karman integral momentum technique. This technique is very widely applicable and we apply it here only in a very simple context.

A momentum balance can be applied to the boundary layer in steady flow. In Fig. 5.6-2 consider an element as shown fixed in space. The net force in the x direction is equal to the net momentum transfer to the element in a unit time.

External force on element: $-\tau_0 \, \Delta x - \dfrac{dp}{dx} \, \Delta x \, \delta$

From a steady-state mass balance:

Mass flow out top of element = mass flow in left side − mass flow out right side $= \displaystyle\int_0^\delta \rho v_x \, dy \bigg|_{x_0} - \int_0^\delta \rho v_x \, dy \bigg|_{x_0+\Delta x}$

(Output − input) rate of momentum = output at top + output on right − input on left $= \left[\displaystyle\int_0^\delta \rho v_x v_\infty \, dy \bigg|_{x_0} - \int_0^\delta \rho v_x v_\infty \, dy \bigg|_{x_0+\Delta x} \right] -$

$\displaystyle\int_0^\delta \rho v_x v_x \, dy \bigg|_{x_0} + \int_0^\delta \rho v_x v_x \, dy \bigg|_{x_0+\Delta x}$

Substituting these terms in a momentum balance, dividing by Δx, and taking the limit as $\Delta x \to 0$ yields

$$-\tau_0 - \frac{dp}{dx} \, \delta = \frac{d}{dx} \left(\int_0^\delta \rho v_x{}^2 \, dy - \int_0^\delta \rho v_x v_\infty \, dy \right) \tag{5.6-2}$$

For the flat plate, $dp/dx = 0$, v_∞ is constant, and

$$\tau_0 = \frac{d}{dx} \int_0^\delta \rho(v_\infty - v_x) v_x \, dy \tag{5.6-3}$$

If we assume a velocity distribution of the form

$$\frac{v_x}{v_\infty} = f\left(\frac{y}{\delta}\right) \tag{5.6-4}$$

where δ is the boundary layer thickness, expressions for the thickness of the boundary layer and the shear stress on the flat plate can be found.

For *laminar* flow a possible form of Eq. (5.6-4) is

$$\frac{v_x}{v_\infty} = \frac{3}{2}\frac{y}{\delta} - \frac{1}{2}\left(\frac{y}{\delta}\right)^3 \qquad 0 \le y \le \delta \tag{5.6-5}$$

$$\frac{v_x}{v_\infty} = 1 \qquad y > \delta$$

Substituting Eq. (5.6-5) into Eq. (5.6-3) and using the fact that $\tau_0 = \mu(dv_x/dy)\big|_{y=0}$ it may be shown that

$$\frac{\delta}{x} = \frac{4.64}{\sqrt{\mathrm{Re}_x}} \qquad \text{or} \qquad \delta = 4.64\sqrt{\frac{\nu x}{v_\infty}} \tag{5.6-6}$$

and

$$\tau_0 = 0.332\sqrt{\frac{\mu\rho v_\infty^3}{x}} \tag{5.6-7}$$

For *turbulent* flow we arbitrarily apply either the universal velocity distribution or the one-seventh-power law for *pipe* flow to our flat-plate problem. The latter is

$$\frac{v_x}{v_\infty} = \left(\frac{y}{\delta}\right)^{1/7} \tag{5.6-8}$$

From Eq. (5.6-8) we obtain for the boundary layer thickness and the shear stress[1]

$$\frac{\delta}{x} = \frac{0.38}{\mathrm{Re}_x^{1/5}} \tag{5.6-9}$$

$$\tau_0 = 0.023\,\rho v_\infty^2 \left(\frac{1}{\mathrm{Re}_\delta}\right)^{1/4} \tag{5.6-10}$$

[1]See, for example, V. L. Streeter, "Fluid Mechanics," 4th ed., McGraw-Hill Book Company, 1966, or W. M. Rosenhow and H. Choi, "Heat, Mass, and Momentum Transfer," Prentice-Hall, Inc., Englewood Cliffs, N.J., 1961.

Note that the first Reynolds number is based on a characteristic length of distance from the nose of the plate; the second on the boundary layer thickness. Equation (5.6-9) can obviously be solved for δ and substituted in Eq. (5.6-10).

The boundary layer concept may be used to interpret entrance and developed flow in a pipe. Figure 5.6-3 shows the entrance to a pipe for the two situations of laminar and turbulent flow past the entrance region. In Fig. 5.6-3(a), as the fluid enters the pipe the laminar boundary layer grows and intersects itself at point 1. Downstream of point 1 the flow is laminar, the velocity profile is parabolic, and the Reynolds number is less than 2,100. In Fig. 5.6-3(b), the fluid enters the pipe and before merging the boundary layer grows, becomes turbulent, and then intersects itself. Downstream of point 1 the flow is turbulent, the velocity profile is blunt, and the Reynolds number is greater than 2,100. We usually assume a viscous sublayer to remain the entire length of the pipe.

PROBLEMS

5.1 Two horizontal square parallel plates measuring 10 by 10 in. are separated by a film of water at 80°F. The bottom plate is bolted down, and the top plate, in contact with the water, is free to move. A spring gage is attached to the top plate and a horizontal force of 1 lbf is applied to the top plate. If the distance between the plates were 0.1 in., how fast would the top plate move at steady state? How fast would the top plate move if the distance were $\frac{1}{8}$ in.? If the horizontal force were increased to 1.6 lbf and the distance between the plates were $\frac{1}{6}$ in., how fast would the top plate move?

5.2 Compare the plots of τ_{yz} versus (dv_z/dy) for a Newtonian fluid in laminar flow in a pipe and for a pseudoplastic fluid with $n = 0.8$. Assume the viscosity of the Newtonian fluid is $\mu = 2$ cp and the apparent parameter of the non-Newtonian fluid $m = 2$ cp. (Use the veloci-

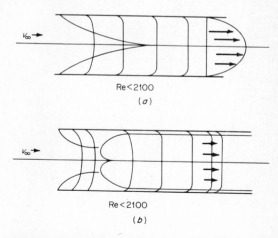

Re<2100

(a)

Re<2100

(b)

Fig. 5.6-3 Developing laminar and turbulent flow in a pipe.

ty gradient from the Newtonian fluid in the power-law expression.)

5.3 Glycerine at 26.5°C is flowing through a horizontal tube 1 ft in length and 0.1-in. ID. For a pressure drop of 40 psi the flow rate is 0.00398 ft³/min. The density of glycerine at 26.5°C is 1.261 gm/cm³. What is the viscosity of the glycerine? Glycerine may be regarded as a Newtonian fluid.

5.4 Find the velocity profile, average velocity, and free surface velocity for a Newtonian fluid flowing down a vertical wall of length L. Use a control volume (CV) of thickness Δx inside the fluid and measure x from the free surface. Remember that because the liquid surface is "free" (unbounded physically) at $x = 0$, the pressure at all points, vertical as well as horizontal, is atmospheric.

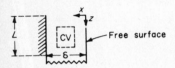

5.5 Derive the equations for the velocity distribution and for the pressure drop in laminar, Newtonian flow through a slit of height y_0 and infinite width.

5.6 An incompressible fluid is in steady-state laminar flow in the annular space between two horizontal, concentric pipes of length L. To find the expressions for the velocity distribution and pressure drop, the momentum balance may be applied to a control volume which is an annular shell of thickness dr. Show the differential equation obtained

$$\frac{d}{dr} r\tau_{rz} = \frac{p_0 - p_L}{L} r$$

where r is radius and τ_{rz} is shear stress in the z direction on the face of the shell perpendicular to the r direction.

5.7 Consider the flow of air over an airfoil as shown below:

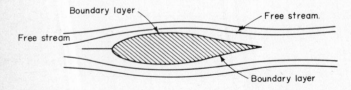

(*a*) What forces are acting in the free stream (ideal flow)?
(*b*) What forces are acting in the boundary layer?

5.8 Consider the two-dimensional ideal flow described by the potential function

$$\phi = -\frac{\Gamma}{2\pi}\theta$$

and stream function

$$\psi = -\frac{\Gamma}{2\pi}\ln r$$

where r and θ are the usual polar coordinates. Sketch the flow field. This flow field represents a *vortex*—similar to the vortex seen in a draining bathtub. What does the ideal solution yield as r approaches zero? The ideal solution does not describe the real world situation for small r because of large viscous forces in this region.

5.9 Sketch the flow field for the potential function given by

$$\phi = \frac{K}{2\pi} \ln r$$

and stream function

$$\psi = \frac{K}{2\pi} \theta$$

where K is a constant. Note that these are the interchanged functions for the vortex above. Depending on the sign of K, the flow represents a source or a sink. The point at $r = 0$ is again a singularity. This type of mathematical description is sometimes used for flow of oil into a well bore from an underground formation. The solution is not good close to the well.

5.10 Shown are data taken at short equal time intervals from a hot wire anemometer mounted in a turbulent air stream. Plot the data, and determine the mean velocity and the mean square fluctuation.

Time	Velocity, ft/sec
1	76
2	103
3	118
4	102
5	112
6	100
7	97
8	87
9	96
10	105
11	85
12	80
13	92
14	78
15	100
16	101
17	92
18	115
19	112
20	81

5.11 A smooth pipe carries water at a bulk velocity of 15 ft/sec. The inside diameter of the pipe is 1 in. Use the universal velocity profile to calculate the shear stress at the wall.

5.12 Water at 60°F flows through a straight section of 100 ft of 1-in. schedule 40 pipe. At a bulk velocity of 5 ft/sec the total pressure drop over the 100 ft is 3 psi. Calculate the thickness of the viscous sublayer inside the pipe. What is the equivalent roughness height for this pipe? Would the pipe be considered completely rough or hydraulically smooth under the given conditions?

5.13 For steady flow: Briefly discuss (a) how ideal flow differs from laminar flow, and (b) how laminar flow differs from turbulent flow. (c) On a sketch such as below, draw in a laminar boundary layer forming at the tip of the flat plate (upper side only). At some point downstream, show on the sketch how transition to a turbulent boundary layer occurs. Label where the laminar sublayer, the buffer layer, and the turbulent core should be on the diagram. (d) Describe what would happen if we increased the approach velocity.

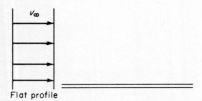

Flat profile

5.14 The boundary layer which forms at the entrance to a circular pipe may be considered to behave like the boundary layer on a flat plate. Estimate the length needed to reach fully developed flow (i.e., for the edge of the boundary layer to reach the center line) for water at 68°F entering a pipe of 2-in. ID at a uniform velocity of (a) 50 ft/sec, (b) 5 ft/sec, (c) 0.05 ft/sec.[1]

5.15 Assume that the velocity distribution for laminar flow over a flat plate is given by $v_x/v_\infty = \frac{3}{2}(y/\delta) - \frac{1}{2}(y/\delta)^3$ (Eq. 5.6-5). Obtain the expression for the boundary layer thickness relative to distance along the plate, Eq. (5.6-6), and the expression for shear stress at the plate, Eq. (5.6-7).

5.16 For turbulent flow using Eq. (5.6-8), $v_x/v_\infty = (y/\delta)^{1/7}$, obtain the expression for boundary layer thickness, Eq. (5.6-9), and shear stress at the plate, Eq. (5.6-10).

[1]C. O. Bennett and J. E. Myers, "Momentum, Heat, and Mass Transfer," p. 146, prob. 13.10, McGraw-Hill Book Company, New York, 1962.

6
Momentum Transfer Coefficients

This is the first of a series of three parallel chapters: Chaps. 6, 9, and 12. In each of these three chapters, we introduce a *coefficient* via which we calculate momentum transfer, heat transfer, or mass transfer. The reason for introducing such coefficients is that for most real situations we are involved with turbulent flow, for which we have no exact mathematical solution. Therefore, we correlate based on reasoning derived from dimensional analysis and the writing of differential equations (which in general we cannot solve). The coefficients are used for calculations in turbulent flow regimes. This is not to say that similar coefficients are not valid for laminar flow—they are, but frequently for laminar flow conditions we can obtain exact solutions.

In this chapter, we consider predicting mechanical energy losses in fluid flow using coefficient correlations such as the friction factor chart. The friction factor for flow in a pipe is discussed and is used to estimate the lost work. We also consider calculation of flow in porous media. This topic is very important since many operations utilize porous media, for example, fixed bed reactors, packed absorption columns, and filters.

We consider general examples of the use of the friction factor chart and various design equations to determine pressure losses in flow systems and to calculate flow rates. A brief discussion of flow in pipe networks is included and the hydraulic radius concept for flow in noncircular conduits is also introduced. We also calculate pressure losses for flow in packed beds and in filters.

6.1 DRAG COEFFICIENTS

We now introduce the friction factor or drag coefficient, which is the empirical coefficient we use to calculate drag (momentum transfer). These coefficients are applicable to two general classes of flow problems: (1) external flows and (2) internal flows.

By external flows we mean flows over the outside surface of something, for example, flow over a flat plate or flow over the outer surface of a sphere or cylinder. By internal flow, we mean flow inside conduits, for example, flow inside pipes. There is a distinct difference in these two types of flow. As we go farther and farther from the surface of an object with external flow, the velocity is less and less affected by the boundary of the object; eventually we reach the so-called free-stream velocity, which is unaffected by the boundary. With developed internal flow, however, we do *not* approach such a constant velocity, but instead, the stream is everywhere affected by the boundary. Although the basic concept of the drag coefficient or friction factor is precisely the same for the two types of flow, it is based on a different velocity in each case: the free-stream velocity for external flows and the bulk velocity for internal flows.

In our discussion below, we will find that bodies with external flows experience two types of drag. "Skin friction" arises as a result of momentum transfer from a fluid to the surface or vice versa, and "form drag" arises as a result of an adverse pressure gradient existing from front to back of the object (a pressure gradient whose net force opposes the motion of the body through the fluid). For example, consider the case of a discus thrower who decided to throw the discus "flat" rather than "edge on." Why didn't this work as well? The basic reason is the difference between form drag and skin friction. When the discus is thrown "edge on," as in the normal field event, the primary resistance to its passage through the air is the skin friction and the pressure gradient from front to back of the discus has very little effect. However, if one attempts to throw the discus "flat" into the airstream, a large pressure differential develops between the front and back of the discus and most of the momentum is transferred by this process. (There is, of course, lift involved when the discus is thrown "edge on," since it acts somewhat like an airfoil. We neglect this effect for purposes of this illustration.)

Skin friction versus form drag can also be illustrated by contrasting divers in the aquatic sense and the air sense. A person jumps from a diving board into a swimming pool with as little of his body cross section normal to his motion as possible, because there is a distinct increase in drag as this cross section increases, as anyone knows who has performed a "belly flop" into a swimming pool. The sky diver, by contrast, falls with his body axis normal to his motion, because this is the way of maximizing the form drag to prolong free fall.

Another real example of form drag is the practice of "drafting" with racing machines. As students of racing well know, if a pursuing car can place its nose only a foot or so from the rear of the car preceding, it will be "sucked along" by a favorable pressure gradient which exists in the wake of the car ahead. When practiced on city streets, this is called "tailgating" and leads to fatalities; furthermore, it is ineffective except at racing speeds.

Consider the case of flow over a curved surface as shown in Fig. 6.1-1 (such as an airplane wing). At very low speeds, flow is essentially laminar and the air follows the contour of the wing as is shown in Fig. 6.1-1a. The majority of the drag in this case results from skin friction, which is the transfer of momentum from the wing, which is moving, to the surrounding air, which is stationary. Under certain circumstances, however, as flow is increased the boundary layer can separate from the wing as shown in Fig. 6.1-1b. To understand why the boundary layer separates, imagine a small particle of air which is going to pass over the wing. The mechanical energy balance developed in Chap. 2 can be applied to this particle of air as a system. To simplify things, let us assume that the particle does not change position significantly in a gravitational field, and therefore the change in potential energy is negligible in passing across the wing. The air, therefore, has two types of mechanical energy: kinetic energy and pressure energy. The mechanical energy does not stay constant because the air undergoes momentum interchange processes with the surrounding pieces of air as it goes across the wing, and, therefore, the important terms in the mechanical energy balance for the air particle are the kinetic energy, the pressure energy, and the lost work due to friction. (There is no shaft work.) As you know from basic physics, the lift on the

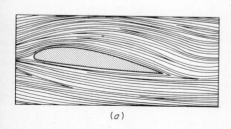   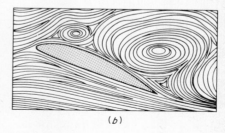

(a)   (b)

Fig. 6.1-1 Flow around an airfoil (a) without and (b) with separation. (*L. Prandtl, and O.G. Tietjens, "Fundamentals of Hydro- and Aerodynamics," Dover Publications, Inc., New York, reprinted by permission of the publishers.*)

airfoil results from the kinetic energy increase in passing across the upper surface of the airfoil; since this form of mechanical energy can come only from the pressure energy, the pressure decreases, giving lift. As the air comes off the back edge of the airfoil, it once again slows down and its kinetic energy is reconverted to pressure energy. There is, however, a net loss of mechanical energy because of the frictional processes as it traverses the airfoil. This means that the kinetic energy, when transformed to pressure energy, is insufficient to regain the original level of pressure. The pressure gradient, in fact, becomes negative. Since flow tends to go down the pressure gradient this means that flow may actually reverse next to the airfoil, the boundary layer separate, and the wing "stall." This is of great concern to a pilot, because this means that he has lost the lift on the latter part of the wing.

The region downstream from the separation point is called the wake. The effect of separation and the subsequent wake is to decrease the amount of kinetic energy transfer back to pressure energy and to increase transfer to internal energy (loss of mechanical energy).

Bodies are often streamlined to cause the separation point to be as far downstream as possible. If there is no separation, the primary loss is caused by the shear in the boundary layer, that is, skin friction. Notice that even though we call this term skin friction, it is not really due to the layer of air molecules adjacent to the surface of the body rubbing on the surface. We know that these molecules are almost always attached to the surface. Skin friction is simply a term we use to describe the momentum transfer from boundary layer formation. An extreme example of skin friction as contrasted with pressure or form drag is flow past an infinitesimally thin flat plate. If we move the flat plate "edge on," all the drag is due to skin friction. If we move the flat plate normal to the flow, almost all the drag is due to the pressure difference.

Let us now introduce the definition of the coefficient that we use to correlate drag behavior either for internal or external flow. (Refer to Example 3.2-2.) Drag behavior is correlated in terms of a drag coefficient. The drag coefficient is *defined* by the equation

$$F_{\text{drag}} = \frac{1}{2} C_D \rho v_\infty^2 A \tag{6.1-1}$$

where

$$C_D = C_D' + C_D'' \tag{6.1-2}$$

with C_D' and C_D'' the skin friction and form drag coefficients respectively. Notice that this definition makes the drag coefficient a dimensionless proportionality factor which relates the drag force to a characteristic kinetic energy of the flowing stream and a characteristic area of the body. If we did not separate the kinetic energy or area terms from the coefficient,

we would require a different coefficient each time we changed fluid velocity or size of the body. By defining the coefficient in this manner, the coefficient is always the same for *similar* flow situations, that is, situations (for
totally immersed objects) where Reynolds number and body shape are the
same.

The form drag coefficient C_D'' is usually estimated from physical
model studies since there is no theory sufficient to calculate it. The skin
friction coefficient C_D' is a function of the Reynolds number. This coefficient may be estimated by writing

$$F_{\text{drag from skin friction}} = C_D' \frac{1}{2} \rho v_\infty^2 A = \int_A \tau_0 \, dA \qquad (6.1\text{-}3)$$

and substituting for τ_0 in terms of velocity gradient and fluid properties.

In order to use the drag coefficient for a particular object, it is necessary to know the appropriate energy to use and the appropriate area term.
We distinguish two cases.

For internal flows, the velocity in the kinetic energy term is normally
the bulk velocity. This velocity can be any velocity one wishes to choose
but conventionally the bulk velocity is used. The area that is used for internal flows is the inside surface area of the conduit, that is, the circumfer-
ence multiplied by the length.

For external flows, two completely different definitions are used.
The characteristic velocity is usually taken to be the free-stream velocity.
Notice that there is equivalence between flowing air past an object and
moving object through air. Therefore, when we say "free-stream velocity" for the case of an object moving through quiescent air, we mean the
velocity of air as seen from the object (the velocity of the object). The
characteristic area usually used is the cross-sectional area of the object
normal to flow. We could use any other particular area, for example, the
external surface area of the object; however, in most of our problems we
will use the cross-sectional area normal to flow. (For objects of similar
shape, the various areas change in the same proportion as size changes.)

In Fig. 6.1-2, flow past a cylinder is considered for (*a*) flow of an
ideal fluid and (*b*) flow of a real fluid with separation. Figure 6.1-3 is a
plot of the pressure distribution versus angle for each of the situations in
Fig. 6.1-2. Notice that the distributions are different.

In general, form drag is dominant for flow past blunt objects while
skin friction dominates with streamlined objects. Figure 6.1-4*a* presents
the drag coefficient for flow past an infinite cylinder.

We can interpret the shape of the curve presented in terms of the
flow field. In Fig. 6.1-5 the flow field around a cylinder is shown for increasing Reynolds numbers. Up to Reynolds numbers of about 1.0, the
flow is laminar with no boundary layer separation. As the Reynolds
number is increased, separation begins at the rear stagnation point, and

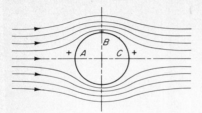

Fig. 6.1-2a Frictionless flow about a circular cylinder. (*H. Schlicting, "Boundary Layer Theory," McGraw-Hill Book Company, New York, 1968.*)

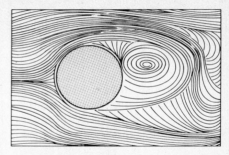

Fig. 6.1-2b Instantaneous photograph of flow with complete boundary layer separation in the wake of a circular cylinder, after Prandtl-Tietjens. (*L. Prandtl, and O. G. Tietjens, "Fundamentals of Hydro- and Aerodynamics," Dover Publications, Inc., New York, reprinted by permission of the publishers.*)

the curve flattens. The separation point moves forward on the object until flow is separated over about half the object. The drop in the curve (Fig. 6.1-4a) above Re = 10^5 is caused by transition of the boundary layer flow from laminar to turbulent. Simultaneously with this transition the separation point moves sharply rearward. The pressure gradient increases but now acts over a smaller area, so that drag is reduced. This is the principle used in the turbulence promoters on the wing of a jet aircraft. These devices trigger turbulent flow, which gives more losses, but keeps the boundary layer attached to the wing and prevents stalling. This is also the reason a golf ball is dimpled—the dimpling reduces form drag by tripping turbulent boundary layer flow.[1]

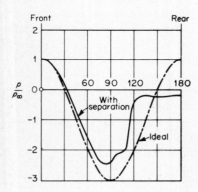

Fig. 6.1-3 Pressure distribution around a circular cylinder. (*H. Schlicting, "Boundary Layer Theory," McGraw-Hill Book Company, New York, 1968.*)

[1] Bluff bodies also shed vortices. This causes an erratic path, for example, a "knuckleball" in baseball. The dimpling of a golf ball also helps minimize the vortex shedding effect.

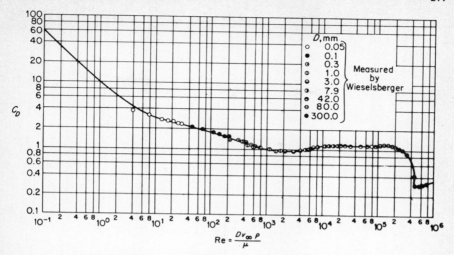

Fig. 6.1-4a Drag coefficient for circular cylinders as a function of the Reynolds number. (*H. Schlicting, "Boundary Layer Theory," McGraw-Hill Book Company, New York, 1968.*)

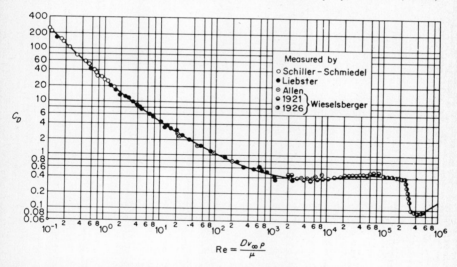

Fig. 6.1-4b Drag coefficient for spheres as a function of the Reynolds number. (*H. Schlicting, "Boundary Layer Theory," McGraw-Hill Book Company, New York, 1968.*)

Example 6.1-1   Drag on a cylinder   Calculate the drag force on a distillation column 50 ft high and 8 ft in diameter in a wind of 40 mph. Use the properties of air given.

$$\nu = \frac{\mu}{\rho} = 16.88 \times 10^{-5}\,\text{ft}^2/\text{sec}$$

$$\rho = 0.0735 \quad \text{lb}m/\text{ft}^3$$

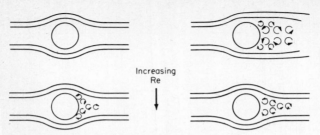

Fig. 6.1-5 Flow around cylinder.

Solution

$$v_\infty = \left(\frac{40 \text{ mi}}{\text{hr}}\right)\left(\frac{\text{hr}}{3{,}600 \text{ sec}}\right)\left(\frac{5{,}280 \text{ ft}}{\text{mi}}\right) = 58.7 \text{ ft/sec}$$

$$\text{Re} = \frac{Dv_\infty\rho}{\mu} = \frac{(8)\,(58.7)}{(16.88)\,(10^{-5})} = 2.78 \times 10^6$$

From Fig. 6.1-4a $C_D = 0.4$

$$F_d = \frac{1}{2}\, C_D\rho v_\infty^2 DL$$

$$F_D = \frac{(0.4)(0.0735)(58.7)^2\,(8)\,(50)}{(2)(32.2)}$$

$$F_D = 630 \quad \text{lbf}$$

Note: The drag coefficient is relatively constant, and so if we have a hurricane wind of 200 mph the drag force would go up by 25 times. (Actually one must integrate the force over the column, taking into account the velocity profile caused by the ground effect.)

□ □ □

Figure 6.1-4b shows the drag coefficient for flow past a sphere. We can interpret the shape of the curve presented in a similar manner as the cylinder.

For the sphere it is common to use the following equations to estimate drag coefficients. These equations fit various portions of the drag coefficient curve. For the Stokes law or creeping flow region, $\text{Re} < 0.1$, the drag coefficient is

$$C_D = \frac{24}{\text{Re}} \tag{6.1-4}$$

For the intermediate region, $2 < \text{Re} < 5 \times 10^2$, an empirical equation for the drag coefficient is

$$C_D = \frac{18.5}{\text{Re}^{3/5}} \tag{6.1-5}$$

and for the following region up to $\mathrm{Re} = 10^5$ the drag coefficient is given by (Newton's law region)

$$C_D = 0.44 \tag{6.1-6}$$

Example 6.1-2 Drag on a submerged object Calculate the drag force and terminal velocity for a glass sphere of 1 mm diameter and specific gravity of 2.5 falling through water at room temperature.

Solution

$$\mathrm{Re} = (0.1 \text{ cm}) \left(2.54 \, \frac{\text{cm}}{\text{in.}} \right) \left(\frac{\text{ft}}{12 \text{ in.}} \right) v_\infty \left(\frac{\text{ft}}{\text{sec}} \right)$$

$$\left(\frac{62.4 \text{ lb}m \text{ cp ft sec}}{\text{ft}^3 \, 1 \text{ cp } (6.72 \times 10^{-4}) \text{ lb}m} \right) = 1{,}960 \, v_\infty$$

$(a) \qquad \Sigma F = 0 = F_\text{drag} + F_\text{grav} + F_\text{buoy} = F_\text{drag} - \frac{4}{3} \pi R^3 (\rho_s - \rho)$

but

$$F_\text{drag} = \frac{1}{2} C_D \rho v_\infty^2 A = \frac{1}{2} C_D \rho v_\infty^2 \pi R^2$$

Substituting in equation (a) and solving for C_D

$$C_D = \frac{4}{3} \frac{D}{v_\infty^2} \frac{\rho_s - \rho}{\rho}$$

Assume

$$C_D = 0.44 = \frac{4}{3} \frac{(32.2)(2.54 \times 10^{-1})}{12 \, v_\infty^2} \left(\frac{2.5 - 1}{1} \right)$$

$$v_\infty = (3.08)^{1/2} = 1.76 \text{ ft/sec}$$

Check drag coefficient assumption

$$\mathrm{Re} = (1{,}980)(1.76) = 3{,}460 \qquad \text{(Newton's law region)}$$

$$F_\text{drag} = \frac{1}{2} C_D \rho v_\infty^2 A$$

$$= \frac{1}{2} (0.44)(62.4) \left(\frac{\text{lb}m}{\text{ft}^3} \right) \left(\frac{\text{lb}f \text{ sec}^2}{32.2 \text{ lb}m \text{ ft}} \right) (1.76)^2 \left(\frac{\text{ft}^2}{\text{sec}^2} \right)$$

$$\frac{\pi}{4} \left(\frac{2.54 \times 10^{-1}}{12} \right)^2 \text{ ft}^2$$

$$= 0.000462 \text{ lb}f$$

□ □ □

6.2 FRICTION FACTORS

One of the most important applications of flow inside conduits is that of
flow inside circular conduits (pipes and tubes). There are literally millions
of miles of process piping used in industrial applications. Table 6.2-1
gives sizes of commercial pipe. Because of the overriding preponderance
of this type of conduit flow, the calculations associated with determining
pressure drop, etc., tend to be discussed in nomenclature peculiar to this
case. For example, instead of talking about drag coefficients we refer to
friction factors. Calculations are done applying the identical basic principles, and so the equations are essentially the same as those we have just
discussed but different in nomenclature.

The skin friction or pressure loss in pipes caused by viscous drag is
usually obtained from empirically determined friction factor charts and
the mechanical energy balance. In Chap. 3 the friction factor chart was
introduced via a dimensional analysis argument. The variables associated
with flow in a pipe were analyzed to show that the pressure loss during
flow of a fluid in a pipe is a function of the Reynolds number and the relative roughness of the pipe. The meaning of drag coefficients (which
include the friction factor) was discussed in Sec. 6.1. The flow behavior
in both laminar and turbulent flow in conduits is summarized with the use
of the friction factor chart.

The drag coefficient is arbitrarily defined as relating the drag force
proportional to a characteristic area A and a characteristic kinetic energy
per unit volume KE. The dimensionless proportionality coefficient is
called the friction factor f in discussion of pipe flow.

$$F_{\text{drag}} = A \ (KE) f \tag{6.2-1}$$

The characteristic area used for fluid flowing through a pipe is the wetted
area $2\pi RL$, where R is the pipe radius and L is the pipe length, because
this is an area normal to momentum transfer. The kinetic energy per unit
volume is $\frac{1}{2} \rho \langle v \rangle^2$ and the drag force is

$$F_{\text{drag}} = (2\pi RL) \left(\frac{1}{2} \rho \langle v \rangle^2 \right) f \tag{6.2-2}$$

Unfortunately, there are at least *two* friction factors commonly used
to describe flow in pipes. The friction factor in the equation above is
known as the Fanning friction factor. Many books contain tabulations of
a friction factor known as the Darcy (sometimes Blasius) friction factor
which is equal to four times the Fanning friction factor. This means that
the two friction factors are easily interconverted; however, it is necessary
to be extremely careful not to use a *chart* for Darcy friction factor with

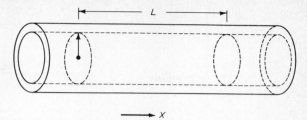

Fig. 6.2-1 System for steady flow in horizontal pipe.

equations containing Fanning friction factors and vice versa.

Let us consider the application of the momentum balance and the friction factor to calculate pressure drop or flow rate at steady state in horizontal pipe of uniform diameter. If we make a momentum balance on a system consisting of the cylinder of space within the pipe as shown in Fig. 6.2-1, we see that in the x direction the drag force at the wall balances the difference in pressure force on each end of the cylinder, since there is no acceleration and the input of momentum equals the output of momentum (the velocity profile is assumed to be unchanging as we go down the pipe).

$$F_{\text{pres}} + F_{\text{drag}} = 0 \qquad (6.2\text{-}3)$$

$$p_1 \, \pi \, R^2 - p_2 \, \pi R^2 - (2\pi RL)\left(\frac{1}{2}\, \rho \langle v \rangle^2\right) f = 0 \qquad (6.2\text{-}4)$$

Solving:

$$p_2 - p_1 = \Delta p = - \frac{2 f L \rho \langle v \rangle^2}{D} \qquad (6.2\text{-}5)$$

The mechanical energy balance for this case reduces to

$$\frac{\Delta p}{\rho} = - \widehat{lw} \qquad (6.2\text{-}6)$$

Thus, substituting for Δp in terms of the friction factor,

$$\widehat{lw} = \frac{2 f L \langle v \rangle^2}{D} \qquad (6.2\text{-}7)$$

and since f is determined in horizontal, uniform-size pipe, f is really a vehicle to express $\widehat{lw}$. If the pipe is not horizontal we have an additional force involved: the body (gravity) force on the fluid in the system. Letting α be the angle a vector in the direction of flow makes with the gravity

Table 6.2-1 Properties of ferrous pipe[a]

Nominal pipe size, in.	Outside diameter, in.	Schedule	Wall thickness, in.	Inside diameter, in.	Cross-sectional area		Circumference, ft, or surface, ft²/ft of length		Capacity at 1 ft/sec velocity		Weight of plain-end pipe, lb/ft
					Metal, in.²	Flow, ft²	Outside	Inside	U.S. gal/min	lb/hr water	
1/8	0.405	10S	0.049	0.307	0.055	0.00051	0.106	0.0804	0.231	115.5	0.19
		40ST, 40S	0.068	0.269	0.072	0.00040	0.106	0.0705	0.179	89.5	0.24
		80XS, 80S	0.095	0.215	0.093	0.00025	0.106	0.0563	0.113	56.5	0.31
1/4	0.540	10S	0.065	0.410	0.097	0.00092	0.141	0.107	0.412	206.5	0.33
		40ST, 40S	0.088	0.364	0.125	0.00072	0.141	0.095	0.323	161.5	0.42
		80XS, 80S	0.119	0.302	0.157	0.00050	0.141	0.079	0.224	112.0	0.54
3/8	0.675	10S	0.065	0.545	0.125	0.00162	0.177	0.143	0.727	363.5	0.42
		40ST, 40S	0.091	0.493	0.167	0.00133	0.177	0.129	0.596	298.0	0.57
		80XS, 80S	0.126	0.423	0.217	0.00098	0.177	0.111	0.440	220.0	0.74
1/2	0.840	5S	0.065	0.710	0.158	0.00275	0.220	0.186	1.234	617.0	0.54
		10S	0.083	0.674	0.197	0.00248	0.220	0.176	1.112	556.0	0.67
		40ST, 40S	0.109	0.622	0.250	0.00211	0.220	0.163	0.945	472.0	0.85
		80XS, 80S	0.147	0.546	0.320	0.00163	0.220	0.143	0.730	365.0	1.09
		160	0.188	0.464	0.385	0.00117	0.220	0.122	0.527	263.5	1.31
		XX	0.294	0.252	0.504	0.00035	0.220	0.066	0.155	77.5	1.71
3/4	1.050	5S	0.065	0.920	0.201	0.00461	0.275	0.241	2.072	1036.0	0.69
		10S	0.083	0.884	0.252	0.00426	0.275	0.231	1.903	951.5	0.86
		40ST, 40S	0.113	0.824	0.333	0.00371	0.275	0.216	1.665	832.5	1.13
		80XS, 80S	0.154	0.742	0.433	0.00300	0.275	0.194	1.345	672.5	1.47
		160	0.219	0.612	0.572	0.00204	0.275	0.160	0.917	458.5	1.94
		XX	0.308	0.434	0.718	0.00103	0.275	0.114	0.461	230.5	2.44

Nominal	OD	Schedule									
1	1.315	5S	0.065	1.185	0.255	0.00768	0.344	0.310	3.449	1725	0.87
		10S	0.109	1.097	0.413	0.00656	0.344	0.287	2.946	1473	1.40
		40ST, 40S	0.133	1.049	0.494	0.00600	0.344	0.275	2.690	1345	1.68
		80XS, 80S	0.179	0.957	0.639	0.00499	0.344	0.250	2.240	1120	2.17
		160	0.250	0.815	0.836	0.00362	0.344	0.213	1.625	812.5	2.84
		XX	0.358	0.599	1.076	0.00196	0.344	0.157	0.878	439.0	3.66
1 1/4	1.660	5S	0.065	1.530	0.326	0.01277	0.435	0.401	5.73	2865	1.11
		10S	0.109	1.442	0.531	0.01134	0.435	0.378	5.09	2545	1.81
		40ST, 40S	0.140	1.380	0.668	0.01040	0.435	0.361	4.57	2285	2.27
		80XS, 80S	0.191	1.278	0.881	0.00891	0.435	0.335	3.99	1995	3.00
		160	0.250	1.160	1.107	0.00734	0.435	0.304	3.29	1645	3.76
		XX	0.382	0.896	1.534	0.00438	0.435	0.235	1.97	985	5.21
1 1/2	1.900	5S	0.065	1.770	0.375	0.01709	0.497	0.463	7.67	3835	1.28
		10S	0.109	1.682	0.614	0.01543	0.497	0.440	6.94	3465	2.09
		40ST, 40S	0.145	1.610	0.800	0.01414	0.497	0.421	6.34	3170	2.72
		80XS, 80S	0.200	1.500	1.069	0.01225	0.497	0.393	5.49	2745	3.63
		160	0.281	1.338	1.429	0.00976	0.497	0.350	4.38	2190	4.86
		XX	0.400	1.100	1.885	0.00660	0.497	0.288	2.96	1480	6.41
2	2.375	5S	0.065	2.245	0.472	0.02749	0.622	0.588	12.34	6170	1.61
		10S	0.109	2.157	0.776	0.02538	0.622	0.565	11.39	5695	2.64
		40ST, 40S	0.154	2.067	1.075	0.02330	0.622	0.541	10.45	5225	3.65
		80ST, 80S	0.218	1.939	1.477	0.02050	0.622	0.508	9.20	4600	5.02
		160	0.344	1.687	2.195	0.01552	0.622	0.436	6.97	3485	7.46
		XX	0.436	1.503	2.656	0.01232	0.622	0.393	5.53	2765	9.03
2 1/2	2.875	5S	0.083	2.709	0.728	0.04003	0.753	0.709	17.97	8985	2.48
		10S	0.120	2.635	1.039	0.03787	0.753	0.690	17.00	8500	3.53
		40ST, 40S	0.203	2.469	1.704	0.03322	0.753	0.647	14.92	7460	5.79
		80XS,80S	0.276	2.323	2.254	0.02942	0.753	0.608	13.20	6600	7.66
		160	0.375	2.125	2.945	0.2463	0.753	0.556	11.07	5535	10.01
		XX	0.552	1.771	4.028	0.01711	0.753	0.464	7.68	3840	13.70

Table 6.2-1 (Continued)

Nominal pipe size, in.	Outside diameter, in.	Schedule	Wall thickness, in.	Inside diameter, in.	Cross-sectional area Metal, in.²	Cross-sectional area Flow, ft²	Circumference, ft, or surface, ft²/ft of length Outside	Circumference, ft, or surface, ft²/ft of length Inside	Capacity at 1 ft/sec velocity U.S. gall/min	Capacity at 1 ft/sec velocity lb/hr water	Weight of plain-end pipe, lb/ft
3	3.500	5S	0.083	3.334	0.891	0.06063	0.916	0.873	27.21	13,605	3.03
		10S	0.120	3.260	1.274	0.05796	0.916	0.853	26.02	13,010	4.33
		40ST, 40S	0.216	3.068	2.228	0.05130	0.916	0.803	23.00	11,500	7.58
		80XS, 80S	0.300	2.900	3.016	0.04587	0.916	0.759	20.55	10,275	10.25
		160	0.438	2.624	4.213	0.03755	0.916	0.687	16.86	8430	14.31
		XX	0.600	2.300	5.466	0.02885	0.916	0.602	12.95	6475	18.58
3½	4.0	5S	0.083	3.834	1.021	0.08017	1.047	1.004	35.98	17,990	3.48
		10S	0.120	3.760	1.463	0.07711	1.047	0.984	34.61	17,305	4.97
		40ST, 40S	0.226	3.548	2.680	0.06870	1.047	0.929	30.80	15,400	9.11
		80XS, 80S	0.318	3.364	3.678	0.06170	1.047	0.881	27.70	13,850	12.51
4	4.5	5S	0.083	4.334	1.152	0.10245	1.178	1.135	46.0	23,000	3.92
		10S	0.120	4.260	1.651	0.09898	1.178	1.115	44.4	22,200	5.61
		40ST, 40S	0.237	4.026	3.17	0.08840	1.178	1.054	39.6	19,800	10.79
		80XS, 80S	0.337	3.826	4.41	0.07986	1.178	1.002	35.8	17,900	14.98
		120	0.438	3.624	5.58	0.07170	1.178	0.949	32.2	16,100	19.01
		160	0.531	3.438	6.62	0.06647	1.178	0.900	28.9	14,450	22.52
		XX	0.674	3.152	8.10	0.05419	1.178	0.825	24.3	12,150	27.54
5	5.563	5S	0.109	5.345	1.87	0.1558	1.456	1.399	69.9	34,950	6.36
		10S	0.134	5.295	2.29	0.1529	1.456	1.386	68.6	34,300	7.77
		40ST, 40S	0.258	5.047	4.30	0.1390	1.456	1.321	62.3	31,150	14.62
		80XS, 80S	0.375	4.813	6.11	0.1263	1.456	1.260	57.7	28,850	20.78
		120	0.500	4.563	7.95	0.1136	1.456	1.195	51.0	25,500	27.04
		160	0.625	4.313	9.70	0.1015	1.456	1.129	45.5	22,750	32.96
		XX	0.750	4.063	11.34	0.0900	1.456	1.064	40.4	20,200	38.55

6	6.25	5S	0.109	6.407	2.23	0.2239	1.734	1.667	100.5	50,250	7.60
		10S	0.134	6.357	2.73	0.2204	1.734	1.664	98.9	49,450	9.29
		40ST, 40S	0.280	6.065	5.58	0.2006	1.734	1.588	90.0	45,000	18.97
		80XS, 80S	0.432	5.761	8.40	0.1810	1.734	1.508	81.1	40,550	28.57
		120	0.562	5.501	10.70	0.1650	1.734	1.440	73.9	36,950	36.39
		160	0.719	5.187	13.34	0.1467	1.734	1.358	65.9	32,950	45.35
		XX	0.864	4.897	15.64	0.1308	1.734	1.282	58.7	29,350	53.16
8	8.625	5S	0.109	8.407	2.915	0.3855	2.258	2.201	173.0	86,500	9.93
		10S	148	8.329	3.941	0.3784	2.258	2.180	169.8	84,900	13.40
		20	0.250	8.125	6.578	0.3601	2.258	2.127	161.5	80,750	22.36
		30	0.277	8.071	7.260	0.3553	2.258	2.113	159.4	79,700	24.70
		40ST, 40S	0.322	7.981	8.396	0.3474	2.258	2.089	155.7	77,850	28.55
		60	0.406	7.813	10.48	0.3329	2.258	2.045	149.4	74,700	35.64
		80XS, 80S	0.500	7.625	12.76	0.3171	2.258	1.996	142.3	71,150	43.39
		100	0.594	7.437	14.99	0.3017	2.258	1.947	135.4	67,700	50.95
		120	0.719	7.187	17.86	0.2817	2.258	1.882	126.4	63,200	60.71
		140	0.812	7.001	19.93	0.2673	2.258	1.833	120.0	60,000	67.76
		XX	0.875	6.875	21.30	0.2578	2.258	1.800	115.7	57,850	72.42
		160	0.906	6.813	21.97	0.2532	2.258	1.784	113.5	56,750	74.69
10	10.75	5S	0.134	10.842	4.47	0.5993	2.814	2.744	269.0	134,500	15.23
		10S	0.165	10.420	5.49	0.5922	2.814	2.728	265.8	132,900	18.70
		20	0.250	10.250	8.25	0.5731	2.814	2.685	257.0	128,500	28.04
		30	0.307	10.136	10.07	0.5603	2.814	2.655	252.0	126,000	34.24
		40ST, 40S	0.365	10.020	11.91	0.5475	2.814	2.620	246.0	123,000	40.48
		80S, 60XS	0.500	9.750	16.10	0.5185	2.814	2.550	233.0	116,500	54.74
		80	0.594	9.562	18.95	0.4987	2.814	2.503	223.4	111,700	64.43
		100	0.719	9.312	22.66	0.4729	2.814	2.438	212.3	106,150	77.03
		120	0.844	9.062	26.27	0.4479	2.814	2.372	201.0	100,500	89.29
		140, XX	1.000	8.750	30.63	0.4176	2.814	2.291	188.0	94,000	104.13
		160	1.125	8.500	34.02	0.3941	2.814	2.225	177.0	88,500	115.64

Table 6.2-1 (Continued)

Nominal pipe size, in.	Outside diameter, in.	Schedule	Wall thickness, in.	Inside diameter, in.	Cross-sectional area Metal, in.²	Cross-sectional area Flow, ft²	Circumference, ft, or surface, ft²/ft of length Outside	Circumference, ft, or surface, ft²/ft of length Inside	Capacity at 1 ft/sec velocity U.S. gall/min	Capacity at 1 ft/sec velocity lb/hr water	Weight of plain-end pipe, lb/ft
12	12.75	5S	0.156	12.438	6.17	0.8438	3.338	3.26	378.7	189,350	22.22
		10S	0.180	12.390	7.11	0.8373	3.338	3.24	375.8	187,900	24.20
		20	0.250	12.250	9.82	0.8185	3.338	3.21	367.0	183,500	33.38
		30	0.330	12.090	12.88	0.7972	3.338	3.17	358.0	179,000	43.77
		ST, 40S	0.375	12.000	14.58	0.7854	3.338	3.14	352.5	176,250	49.56
		40	0.406	11.938	15.74	0.7773	3.338	3.13	349.0	174,500	53.52
		XS, 80S	0.500	11.750	19.24	0.7530	3.338	3.08	338.0	169,000	65.42
		60	0.562	11.626	21.52	0.7372	3.338	3.04	331.0	165,500	73.15
		80	0.688	11.374	26.07	0.7056	3.338	2.98	316.7	158,350	88.63
		100	0.844	11.062	31.57	0.6674	3.338	2.90	299.6	149,800	107.32
		120, XX	1.000	10.750	36.91	0.6303	3.338	2.81	283.0	141,500	125.49
		140	1.125	10.500	41.09	0.6013	3.338	2.75	270.0	135,000	139.67
		160	1.312	10.126	47.14	0.5592	3.338	2.65	251.0	125,500	160.27
14	14	5S	0.156	13.688	6.78	1.0219	3.665	3.58	459	229,500	22.76
		10S	0.188	13.624	8.16	1.0125	3.665	3.57	454	227,000	27.70
		10	0.250	13.500	10.80	0.9940	3.665	3.53	446	223,000	36.71
		20	0.312	13.376	13.42	0.9750	3.665	3.50	438	219,000	45.61
		30, ST	0.375	13.250	16.05	0.9575	3.665	3.47	430	215,000	54.57
		40	0.438	13.124	18.66	0.9397	3.665	3.44	422	211,000	63.44
		XS	0.500	13.000	21.21	0.9218	3.665	3.40	414	207,000	72.09
		60	0.594	12.812	25.02	0.8957	3.665	3.35	402	201,000	85.05
		80	0.750	12.500	31.22	0.8522	3.665	3.27	382	191,000	106.13
		100	0.938	12.124	38.49	0.8017	3.665	3.17	360	180,000	130.85
		120	1.094	11.812	44.36	0.7610	3.665	3.09	342	171,000	150.79
		140	1.250	11.500	50.07	0.7213	3.665	3.01	324	162,000	170.21
		160	1.406	11.188	55.63	0.6827	3.665	2.93	306	153,000	181.11

Size	Schedule									
16	5S	0.165	15.670	8.18	1.3393	4.189	4.10	601	300,500	27.87
	10S	0.188	15.624	9.34	1.3314	4.189	4.09	598	299,000	31.62
	10	0.250	15.500	12.37	1.3104	4.189	4.06	587	293,500	42.05
	20	0.312	15.376	15.38	1.2985	4.189	4.03	578	289,000	52.27
	30, ST	0.375	15.250	18.41	1.2680	4.189	3.99	568	284,000	62.58
	40, XS	0.500	15.000	24.35	1.2272	4.189	3.93	550	275,000	82.77
	60	0.656	14.688	31.62	1.1766	4.189	3.85	528	264,000	107.50
	80	0.844	14.312	40.19	1.1171	4.189	3.75	501	250,500	136.51
	100	1.031	13.938	48.48	1.0596	4.189	3.65	474	237,000	164.82
	120	1.219	13.562	56.61	1.0032	4.189	3.55	450	225,000	192.43
	140	1.438	13.124	65.79	0.9394	4.189	3.44	422	211,000	223.57
	160	1.594	12.812	72.14	0.8953	4.189	3.35	402	201,000	245.22
18	5S	0.165	17.670	9.25	1.7029	4.712	4.63	764	382,000	31.32
	10S	0.188	17.624	10.52	1.6941	4.712	4.61	760	379,400	35.48
	10	0.250	17.500	13.94	1.6703	4.712	4.58	750	375,000	47.39
	20	0.312	17.376	17.34	1.6468	4.712	4.55	739	369,500	58.94
	ST	0.375	17.250	20.76	1.6230	4.712	4.52	728	364,000	70.59
	30	0.438	17.124	24.16	1.5993	4.712	4.48	718	359,000	82.15
	XS	0.500	17.000	27.49	1.5763	4.712	4.45	707	353,500	93.45
	40	0.562	16.876	30.79	1.5533	4.712	4.42	697	348,500	104.67
	60	0.750	16.500	40.64	1.4849	4.712	4.32	666	333,000	138.17
	80	0.938	16.124	50.28	1.4180	4.712	4.22	636	318,000	170.92
	100	1.156	15.688	61.17	1.3423	4.712	4.11	602	301,000	207.96
	120	1.375	15.250	71.82	1.2684	4.712	3.99	569	284,500	244.14
	140	1.562	14.876	80.66	1.2070	4.712	3.89	540	270,000	274.22
	160	1.781	14.438	90.75	1.1370	4.712	3.78	510	255,000	308.50

Table 6.2-1 (Continued)

Nominal pipe size, in.	Outside diameter, in.,	Schedule	Wall thickness, in.	Inside diameter, in.	Cross-sectional area		Circumference, ft, or surface, ft²/ft of length		Capacity at 1 ft/sec velocity		Weight of plain-end pipe, lb/ft
					Metal, in.²	Flow, ft²	Outside	Inside	U.S. gal/min	lb/hr water	
20	20	5S	0.188	19.624	11.70	2.1004	5.236	5.14	943	471,500	39.76
		10S	0.218	19.564	13.55	2.0878	5.236	5.12	937	467,500	45.98
		10	0.250	19.500	15.51	2.0740	5.236	5.11	930	465,000	52.73
		20, ST	0.375	19.250	23.12	2.0211	5.236	5.04	902	451,000	78.60
		30, XS	0.500	19.000	30.63	1.9689	5.236	4.97	883	441,500	104.13
		40	0.594	18.812	36.21	1.9302	5.236	4.92	866	433,000	123.11
		60	0.812	18.376	48.95	1.8417	5.236	4.81	826	413,000	166.40
		80	1.031	17.938	61.44	1.7550	5.236	4.70	787	393,500	208.87
		100	1.281	17.438	75.33	1.6585	5.236	4.57	744	372,000	256.10
		120	1.500	17.000	87.18	1.5763	5.236	4.45	707	353,500	296.37
		140	1.750	16.500	100.3	1.4849	5.236	4.32	665	332,500	341.09
		160	1.969	16.062	111.5	1.4071	5.236	4.21	632	316,000	379.17
24	24	5S	0.218	23.564	16.29	3.0285	6.283	6.17	1359	679,500	55.08
		10,10S	0.250	23.500	18.65	3.012	6.283	6.15	1350	675,000	63.41
		20, ST	0.375	23.250	27.83	2.948	6.283	6.09	1325	662,500	94.62
		XS	0.500	23.000	36.90	2.885	6.283	6.02	1295	642,500	125.49
		30	0.562	22.876	41.39	2.854	6.283	5.99	1281	640,500	140.68
		40	0.688	22.624	50.39	2.792	6.283	5.92	1253	626,500	171.29
		60	0.969	22.062	70.11	2.655	6.283	5.78	1192	596,000	238.35
		80	1.219	21.562	87.24	2.536	6.283	5.64	1138	569,000	296.58
		100	1.531	20.938	108.1	2.391	6.283	5.48	1073	536,500	367.39
		120	1.812	20.376	126.3	2.264	6.283	5.33	1016	508,000	429.39
		140	2.062	19.876	142.1	2.155	6.283	5.20	965	482,500	483.12
		160	2.344	19.312	159.5	2.034	6.283	5.06	913	456,500	542.13

30	30									
	5S	0.250	29.500	23.37	4.746	7.854	7.72	2130	1,065,000	79.43
	10, 10S	0.312	29.376	29.10	4.707	7.854	7.69	2110	1,055,000	98.93
	ST	0.375	29.250	34.90	4.666	7.854	7.66	2094	1,084,000	118.65
	20,XS	0.500	29.000	46.34	4.587	7.854	7.59	2055	1,027,500	157.53
	30	0.625	28.750	57.68	4.508	7.854	7.53	2020	1,010,000	196.08

[a] Extracted from American National Standard Wrought Steel and Wrought Iron Pipe, ANSI B36.10-1970, and Stainless Steel Pipe, ANSI B36.19-1965, with permission of the publishers, the American Society of Mechanical Engineers, 345 East 47th St., New York, N.Y. 10017.

ST = standard wall, XS = extra strong wall, XX = double extra strong wall. Wrought iron pipe has slightly thicker walls, approximately 3 percent, but the same weight per foot as wrought steel pipe, because of lower density. Schedules 10, 20, 30, 40, 60, 80, 100, 120, 140, and 160 apply to steel pipe only. Decimal thicknesses for respective pipe sizes represent their nominal or average wall dimensions. Mill tolerances as high as 12½ percent are permitted.

Plain-end pipe is produced by a square cut. Pipe is also shipped from the mills threaded, with a threaded coupling on one end, or with the ends beveled for welding, or grooved or sized for patented couplings. Weights per foot for threaded and coupled pipe are slightly greater because of the weight of the coupling, but it is not available larger than 12 in., or lighter than schedule 30 sizes 8 through 12 in., or schedule 40 6 in. and smaller.

vector, and choosing the x direction along the pipe axis as shown in Fig. 6.2-2, a momentum balance yields

$$\Sigma F_x = 0 = F_{\text{pres}} + F_{\text{drag}} + (F_{\text{grav}})_x \qquad (6.2\text{-}8)$$

The friction factor gives us the drag force or lost work in *horizontal* pipe. We usually assume (an excellent assumption) that this is unaltered by pipe orientation, so that at the same Reynolds number in the same pipe the drag and lost work will be the same whether, for example, the pipe is vertical or horizontal.

$$0 = (p_2 - p_1)\pi R^2 + (2\pi RL)\left(\frac{1}{2}\rho\langle v\rangle^2\right)f - g\rho(\pi R^2 L)\cos\alpha \qquad (6.2\text{-}9)$$

Solving:

$$p_2 - p_1 = -\frac{2fL\rho\langle v\rangle^2}{D} + g\rho\,\overbrace{L\cos\alpha}^{-\Delta z} \qquad (6.2\text{-}10)$$

If we rearrange the above equation

$$\frac{\Delta p}{\rho} + g\,\Delta z = -\frac{2fL\langle v\rangle^2}{D} \qquad (6.2\text{-}11)$$

and compare with the mechanical energy balance for a uniform-size pipe with no shaft work

$$\frac{\Delta p}{\rho} + g\,\Delta z = -\widehat{lw} \qquad (6.2\text{-}12)$$

we see that

$$\widehat{lw} = \frac{2fL\langle v\rangle^2}{D} \qquad (6.2\text{-}13)$$

Thus, *the friction factor gives us a way to evaluate lost work.*

We now must have a way to determine f. If we make actual experiments (usually in horizontal, uniform-diameter pipe) we find

$$f = f\left(\text{Re}, \frac{k}{D}\right) \qquad (6.2\text{-}14)$$

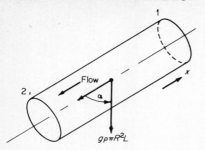

Fig. 6.2-2 System for steady flow in inclined pipe.

The group k/D is the dimensionless pipe roughness. In theory, it is supposed to represent the ratio of some sort of *average* roughness height to the pipe diameter. In fact, data were obtained by gluing particles of known uniform size to a pipe wall, plotting this data (for which k/D is simply the ratio of particle diameter to pipe diameter), then taking data for various types of pipe and assigning values of k/D to the pipes depending upon the data for artificially roughened pipe to which they correspond. Figure 6.2-3 is a plot obtained in this manner.

This plot is called a friction factor chart or Stanton diagram. It is based on data of Moody. The data can be plotted in such a way as to give either a Fanning friction factor or a Darcy (or Blasius) friction factor. This is confusing but is, unfortunately, a fact. The values of relative roughness for most commercial types of pipe are given in Fig. 6.2-4.

We can obtain an *analytic* expression for the friction factor in *laminar* flow. Making a force balance on an element of fluid gave us, just above,

$$\Delta p = -\frac{2fL\rho\langle v \rangle^2}{D} \tag{6.2-15}$$

But the Hagen-Poiseuille equation, Eq. (5.3-9), gives us a relation between Δp and other variables in laminar flow

$$\Delta p = -\pi R^2 \langle v \rangle \frac{8\mu L}{\pi R^4} = -\frac{32 \langle v \rangle \mu L}{D^2} \tag{6.2-16}$$

Notice that, in laminar flow, fluid density does not affect pressure drop. Substituting

$$\frac{32 \langle v \rangle \mu L}{D^2} = \frac{2fL\rho\langle v \rangle^2}{D} \tag{6.2-17}$$

or

$$f = \frac{16}{\text{Re}} \tag{6.2-18}$$

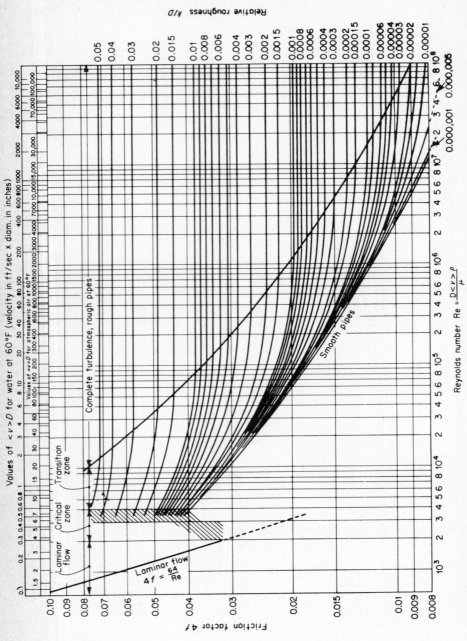

Fig. 6.2-3 Friction factor chart. (*Reprinted from the Pipe Friction Manual, 3d ed., copyright (c) 1961 by the Hydraulic Institute, 122 East 42d St., New York, N. Y. 10017.*)

Values of $\langle v \rangle D$ for water at 60°F (velocity in ft/sec × diam. in inches)

Values of $\langle v \rangle D$ for atmospheric air at 60°F

Relative roughness k/D

Friction factor $4f$

Reynolds number $Re = \dfrac{D\langle v \rangle \rho}{\mu}$

Laminar flow

Critical zone

Transition zone

Complete turbulence, rough pipes

Smooth pipes

Laminar flow $4f = \dfrac{64}{Re}$

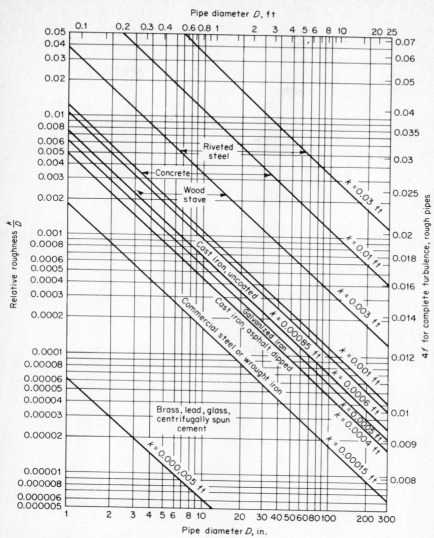

Fig. 6.2-4 Relative roughness for new clean pipes. (*Reprinted from the Pipe Friction Manual, 3d ed., copyright (c) 1961 by the Hydraulic Institute, 122 East 42d St., New York, N. Y. 10017.*)

Note that f (for laminar flow) is not a function of pipe roughness. This line is labeled on Fig. 6.2-3.

For the turbulent region, f gradually approaches a fairly constant value—in other words, f *no longer* depends on Re but is primarily a function of pipe roughness.

A qualitative explanation of the behavior of f is that at low Reynolds numbers the flow is streamlined and roughness contributes only skin friction, not form drag. At higher Reynolds numbers the roughness protrudes beyond the viscous sublayer and the form drag becomes important, overriding the skin friction contribution. By controlling flow very carefully and seeing that no external disturbances are imposed upon the flow stream, it is sometimes possible to maintain laminar flow up to a Reynolds number of about 10,000. In ordinary pipes, however, there are sufficient external disturbances to trigger turbulence far below this point. It is not possible in going to laminar from turbulent flow to maintain turbulent flow much below a Reynolds number of about 2,100.

The universal velocity distribution, Eq. (5.5-23), can be used to fit the turbulent region of the friction factor chart. By substituting the universal velocity profile into the definition of the friction factor, and going through a great deal of algebraic manipulation, one can show that, in essence, the data of the friction factor chart for hydraulically smooth pipes are curve-fit by

$$\frac{1}{\sqrt{f}} = 4.06 \log (\text{Re } \sqrt{f}) - 0.6 \tag{6.2-19}$$

A portion of the turbulent region can also be represented by an empirical curve fit using the Blasius relation:

$$f = \frac{0.0791}{\text{Re}^{1/4}} \qquad 2.1 \times 10^3 < \text{Re} < 10^5 \tag{6.2-20}$$

For rough pipes the universal velocity distribution yields the equation

$$\frac{1}{\sqrt{f}} = 4.06 \log \frac{R}{k} + 3.36 \tag{6.2-21}$$

In many practical situations it is more convenient to use the friction factor chart than the above algebraic forms. However, with computer calculations of complicated flow problems the various equations may be more convenient to use.

In a general flow system, the pipe may not be of a uniform size and/or there may be various kinds of fittings, valves, and changes in pipe diameter in the system which will cause losses. The lost work term in the mechanical energy balance must include these losses. The latter losses can be estimated from the kinetic energy term in the energy balance. For most calculations, losses caused by fittings and valves are treated by a model that replaces the fitting or valve with an equivalent length of straight pipe—a pipe length that would give the same pressure loss as the fitting or valve at the same flow rate. Equivalent lengths for standard-size

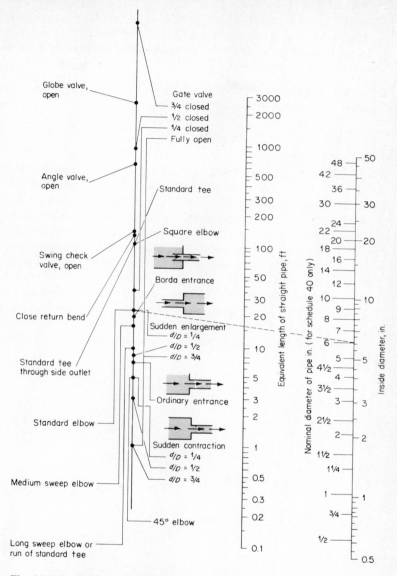

Fig. 6.2-5 Equivalent lengths for friction losses. (*Crane Co.*)

fittings and valves are given in the nomograph of Fig. 6.2-5. This nomograph will also give expansion and contraction losses for most cases except for sudden expansions into large regions, e.g., tanks. This type of situation is treated as detailed in Example 6.2-1.

Example 6.2-1 Calculation of expansion losses The nomograph in Fig. 6.2-5 does not give losses for sudden enlargements with diameter ratio

greater than 4/1. Develop an approximate formula for calculation of enlargement losses and use it to calculate the lost work involved in pumping
water at 25 gal/min from a 1-in. schedule 40 steel pipe into the side of a
large tank.

Solution We obtain the lost work by writing a mechanical energy balance for the system shown in Fig. 6.2-6.

$$\frac{\Delta p}{\rho} + \frac{\Delta \langle v \rangle^2}{2} = -\widehat{lw}$$

We do not know Δp in this relation and so we use the momentum balance
to solve for Δp in terms of velocities. In doing this, we observe that the
actual flow expands fairly slowly to fill the larger pipe and a sizable regime
of eddying fluid exists, as shown in Fig. 6.2-6. Since the radial velocity is
not large compared to the axial velocity, we conclude that radial pressure
gradients are small with respect to axial gradients. This permits us to assume uniform pressure across the entire (here, larger) pipe at points 1 and
2. We further know from experiment that the length of pipe required
to permit the flow to "fill" the larger pipe is short enough that the wall
stress will be a small contribution compared to the momentum change and
pressure terms in the momentum balance, and so we will neglect this
force. With these two assumptions, a momentum balance on the system
yields

$$\Sigma F_x = \underbrace{p_1 A_2 - p_2 A_2}_{\text{force}} = \underbrace{\underbrace{\rho A_2 \langle v_2 \rangle}_{\substack{\text{mass} \\ \text{flowing}}} \underbrace{[\langle v_2 \rangle - \langle v_1 \rangle]}_{\substack{\text{velocity} \\ \text{change}}}}_{\text{momentum change}}$$

Solving for $p_1 - p_2 = -\Delta p$:

$$-\Delta p = \rho \frac{A_2}{A_2} \langle v_2 \rangle (\langle v_2 \rangle - \langle v_1 \rangle)$$

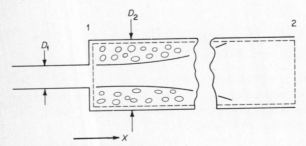

Fig. 6.2-6 Sudden enlargement.

(Note: We have assumed plug flow at velocity $\langle v \rangle$, so that we have pretty well limited ourselves to turbulent flow, as discussed in Chap. 2.)

Substituting in the mechanical energy balance:

$$-\widehat{lw} = -\langle v_2 \rangle \left(\langle v_2 \rangle - \langle v_1 \rangle \right) + \frac{\langle v_2 \rangle^2 - \langle v_1 \rangle^2}{2}$$

But a mass balance shows

$$\langle v_1 \rangle A_1 \rho = \langle v_2 \rangle A_2 \rho$$

or

$$\langle v_2 \rangle = \langle v_1 \rangle \frac{A_1}{A_2}$$

and so

$$-\widehat{lw} = -\langle v_1 \rangle \frac{A_1}{A_2} \left(\langle v_1 \rangle \frac{A_1}{A_2} - \langle v_1 \rangle \right) + \left(\frac{\langle v_1 \rangle^2}{2} \frac{A_1^2}{A_2^2} - \frac{\langle v_1 \rangle^2}{2} \right)$$

$$-\widehat{lw} = -\langle v_1 \rangle^2 \left[\left(\frac{A_1}{A_2} \right)^2 - \frac{A_1}{A_2} - \frac{1}{2} \frac{A_1^2}{A_2^2} + \frac{1}{2} \right]$$

$$\widehat{lw} = \frac{\langle v_1 \rangle^2}{2} \left(1 - \frac{A_1}{A_2} \right)^2$$

or

$$\widehat{lw} = \frac{\langle v_1 \rangle^2}{2} \left[1 - \left(\frac{D_1}{D_2} \right)^2 \right]^2$$

the required relationship.

Applying this result to the numerical problem stated,

$$\frac{D_1}{D_2} \cong 0$$

Therefore

$$\widehat{lw} = \frac{1}{2} \langle v_1 \rangle^2$$

$$= \frac{1}{2} \left[\left(\frac{25 \text{ gal}}{\text{min}} \right) \left(\frac{\text{ft}^3}{7.48 \text{ gal}} \right) \left(\frac{\text{min}}{60 \text{ sec}} \right) \left(\frac{1}{0.006 \text{ ft}^2} \right) \right]^2 \left(\frac{\text{lbf sec}^2}{32.2 \text{ lbm ft}} \right)$$

$$= 1.34 \text{ ft lbf/lbm}$$

□ □ □

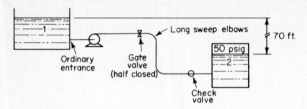

Fig. 6.2-7 Sketch for Example 6.2-2.

Example 6.2-2 Friction loss in a piping system   A distilled oil is transferred from a tank at 1 atm absolute pressure to a pressure vessel at 50 psig by means of the piping arrangement shown above. The oil flows at a rate of 23,100 lb$m$/hr through a 3-in. schedule 40 steel pipe. The total length of straight pipe in the system is 450 ft. Calculate the minimum horsepower input to a pump having an efficiency of 60 percent. The oil has a density of 52 lb$m$/ft³ and viscosity of 3.4 cp. (See Fig. 6.2-7.)

Solution   We first evaluate the terms in the mechanical energy balance using a system which is from liquid surface 1 to liquid surface 2:

$$\frac{\Delta p}{\rho} = \left[(50 - 0)\,\frac{\mathrm{lb}f}{\mathrm{in.}^2}\right]\left(144\,\frac{\mathrm{in.}^2}{\mathrm{ft}^2}\right)\left(\frac{\mathrm{ft}^3}{52\ \mathrm{lb}m}\right) = 138\ \mathrm{ft\ lb}f/\mathrm{lb}m$$

$$g\,\Delta z = \left(\frac{32.2\ \mathrm{ft}}{\mathrm{sec}^2}\right)\left[\frac{(0-70)\ \mathrm{ft\ lb}f\ \mathrm{sec}^2}{32.2\ \mathrm{lb}m\ \mathrm{ft}}\right] = -70\ \mathrm{ft\ lb}f/\mathrm{lb}m$$

$$\frac{\Delta \langle v \rangle^2}{2} = 0$$

The equivalent length of pipe in the system is

$$L_{eq} = \underset{\text{(straight)}}{450} + \underset{\text{(entry)}}{4.5} + \underset{\text{(gate)}}{50} + \underset{\text{(elbows)}}{(2)(5)} + \underset{\text{(check)}}{20} = 535\ \mathrm{ft}$$

$$\langle v \rangle = \frac{Q}{A} = \left(\frac{23{,}100\ \mathrm{lb}m}{\mathrm{hr}}\right)\left(\frac{\mathrm{ft}^3}{52\ \mathrm{lb}m}\right)\left(\frac{\mathrm{hr}}{3{,}600\ \mathrm{sec}}\right)\left(\frac{1}{0.0513\ \mathrm{ft}^2}\right)$$
$$= 2.41\ \mathrm{ft/sec}$$

$$\mathrm{Re} =$$
$$\left(\frac{3.068\ \mathrm{in.\ ft}}{12\ \mathrm{in.}}\right)\left(\frac{2.41\ \mathrm{ft}}{\mathrm{sec}}\right)\left(\frac{52\ \mathrm{lb}m}{\mathrm{ft}^3}\right)\left[\frac{\mathrm{cp\ ft\ sec}}{(3.4\ \mathrm{cp})(6.72 \times 10^{-4}\ \mathrm{lb}m)}\right]$$
$$= 14{,}000$$

Using the friction factor plot and pipe roughness chart:

$$k/D = 0.0006$$
$$f = 0.007$$

The total lost work is the sum of the loss calculated using the friction factor chart plus the loss from the sudden expansion into tank 2 (which we cannot get from the friction factor approach, and therefore calculate as in Example 6.2-1).

$$\widehat{lw} = \frac{2fL\langle v \rangle^2}{D} + \frac{\langle v \rangle^2}{2}\left[1 - \left(\frac{D_1}{D_2}\right)^2\right]^2$$

$$= \left[\frac{(2)(0.007)(535)\ \text{ft}\ (2.41)^2}{3.068\ \text{in.}}\right]\left(\frac{\text{ft}^2}{\text{sec}^2}\right)\left(\frac{12\ \text{in.}}{\text{ft}}\right)\left(\frac{\text{lbf sec}^2}{32.2\ \text{lbm ft}}\right)$$

$$+ \left[\frac{(2.41)^2}{2}\right]\left(\frac{\text{ft}^2}{\text{sec}^2}\right)\left(\frac{\text{lbf sec}^2}{32.2\ \text{lbm ft}}\right)$$

$$= 5.30 + 0.0903 = 5.31\ \text{ft lbf/lbm}$$

Notice that the sudden expansion loss is negligible for this problem; it was included simply to illustrate the method of its inclusion. Substituting in the mechanical energy balance,

$$138 - 70 + 0 = -5.3 + \widehat{W}$$

$$\widehat{W} = -73.3\ \text{ft lbf/lbm}$$

The sign is negative, indicating that the work is done *on* the fluid and not *by* the fluid. Since $\widehat{W}$ is the actual work done on the fluid, the pump must have an input of

$$0.6\widehat{W}_{\text{to pump}} = -73.3$$

$$\widehat{W}_{\text{to pump}} = -122\ \text{ft lbf/lbm}$$

or

$$\text{Power to pump} =$$

$$-\left(\frac{122\ \text{ft lbf}}{\text{lbm}}\right)\left(\frac{\text{hp min}}{33,000\ \text{ft lbf}}\right)\left(23,100\ \frac{\text{lbm}}{\text{hr}}\right)\left(\frac{\text{hr}}{60\ \text{min}}\right)$$

$$\text{Power to pump} = -1.42\ \text{hp}$$

□ □ □

Example 6.2-3 Pressure loss calculations Benzene (density = 55 lbm/ft³, viscosity = 0.65 cp) is stored in a vented tank and is to be pumped to the top of a vented absorption column at a controlled rate of 120 lbm/min. You have a pump with a ¹/₂-hp motor, which operates with an 80 percent total efficiency, and which has an outlet for 1-in. steel pipe.

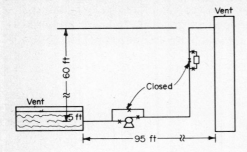

Fig. 6.2-8 Sketch for Example 6.2-3.

1. Can 1-in. schedule 40 steel pipe be used throughout?
2. Could a smaller standard pipe size (schedule 40 steel) be used throughout?

The system is shown in Fig. 6.2-8. The rotameter has flow resistance equivalent to 20 ft of pipe; standard elbows, standard tees, and gate valves are to be used in the system. (Neglect kinetic energy changes.)

Solution   From the mechanical energy balance and Figs. 6.2-3 to 6.2-5:

		Equivalent length, ft
1	Ordinary entrance	1.5
2	Runs of tee	3.2
4	Standard elbows	11.2
2	Tees (side outlet)	10
4	Open gate valves	2.4
1	Rotameter	20
	Straight pipe	160 (we neglect the expansion loss)
		208.3 Total

$$\langle v \rangle = \left(\frac{w}{\rho A}\right) = \left(120 \ \frac{\text{lb}m}{\text{min}}\right)\left(\frac{\text{min}}{60 \ \text{sec}}\right)\left(\frac{\text{ft}^3}{55 \ \text{lb}m}\right)\left(\frac{1}{0.006 \ \text{ft}^2}\right)$$

$$= 6.06 \ \text{ft/sec}$$

$$\text{Re} =$$
$$\left(\frac{1.049 \ \text{in. ft}}{12 \ \text{in.}}\right)\left(6.06 \ \frac{\text{ft}}{\text{sec}}\right)\left(\frac{55 \ \text{lb}m}{\text{ft}^3}\right) \ \left(\frac{1}{0.65 \ \text{cp}}\right)\left(\frac{\text{cp ft sec}}{6.72 \times 10^{-4} \ \text{lb}m}\right)$$

$$= 6.68 \times 10^4$$

for 1-in. pipe, $k/D = 0.0018$. Then

$f = 0.0063$ (Note: This is *not* more accurate than perhaps ± 0.0001.)

Applying the mechanical energy balance:

$$-\widehat{W} = \widehat{lw} + g\,\Delta z$$

$$-\widehat{W} = \frac{2fL\langle v \rangle^2}{D} + g\,\Delta z = [(2)(0.0063)(208)\text{ ft}]\left[(6.06)^2\,\frac{\text{ft}^2}{\text{sec}^2}\right]$$

$$(1.049 \text{ in.})\left(\frac{12 \text{ in.}}{\text{ft}}\right)\left(\frac{\text{lbf sec}^2}{32.2 \text{ lbm ft}}\right) + \left(\frac{32.2 \text{ ft lbf sec}^2}{32.2 \text{ sec}^2 \text{ lbm ft}}\right)(60 \text{ ft})$$

Therefore the energy required is

$$\widehat{W} = -\left(33.4\,\frac{\text{lbf}}{\text{lbm}}\right) - \left(60\,\frac{\text{ft lbf}}{\text{lbm}}\right) = -93.4 \text{ ft lbf/lbm}$$

The energy available *to the fluid* is

$$\widehat{W}_{\text{avail}} = \left(\frac{-0.8}{2}\text{ hp}\right)\left(\frac{33,000 \text{ ft lbf}}{\text{hp min}}\right)\left(\frac{\text{min}}{120 \text{ lbm}}\right) = -110 \text{ ft lbf/lbm}$$

Thus we can use 1-in. pipe throughout.

If we go to the next smaller size of standard pipe, 3/4 in., the equivalent length in *feet* will change but the equivalent length in *diameters* will be about the same, and so

$$L_{\text{eq}} = 160 + \left(\frac{3/4}{1}\right)(48) = 186 \text{ ft}$$

(We use nominal diameter with sufficient accuracy.) Also

$$\langle v \rangle_2 = 6.06\,\frac{A_1}{A_2} = 6.06\,\frac{0.006 \text{ ft}^2}{0.00371 \text{ ft}^2} = 9.8 \text{ ft/sec}$$

The Reynolds number is

$$\text{Re} = (6.68 \times 10^4)\left(\frac{0.92 \text{ in.}}{1.049 \text{ in.}}\right)\left(\frac{9.8 \text{ ft/sec}}{6.06 \text{ ft/sec}}\right) = 9.46 \times 10^4$$

$$\frac{k}{D} = \frac{0.00015}{3/4}\,(12) = 0.0024$$

$$f = 0.0061$$

and so

$$-\widehat{W} = \frac{2fL\langle v\rangle^2}{D} + g\,\Delta z = \frac{(2)(0.0061)(186)(9.8)^2(12)}{(0.92)(32.2)}$$
$$+ \frac{(32.2)(60)}{(32.2)}$$

$$-\widehat{W} = \left(89\,\frac{\text{ft lbf}}{\text{lb}\,m}\right) + \left(60\,\frac{\text{ft lbf}}{\text{lb}\,m}\right) = \left(149\,\frac{\text{ft lbf}}{\text{lb}\,m}\right)$$

Since only 110 ft lbf/lbm is available, 3/4-in. pipe is too small.

□ □ □

In problems where the velocity is unknown, but the pressure losses are known, it is necessary to do a trial and error calculation, since the velocity is necessary to find f. Problems of this kind are simplified by plotting the friction factor f as a function of the Karman number.

$$\text{Re}\,\sqrt{f} = \frac{D\rho}{\mu}\sqrt{\frac{\widehat{lw}D}{2L}} = \text{Karman number} = \lambda \qquad (6.2\text{-}22)$$

The motivation to construct such a plot comes from the fact that, in evaluating lost work when velocity is unknown, neither the ordinate nor the abscissa of the ordinary friction factor plot is known. When flow rate (i.e., velocity) is given, we normally first calculate from the chart the friction factor, which is basically

$$f = \phi_1(\langle v\rangle) \qquad (6.2\text{-}23)$$

and then substitute f and $\langle v\rangle$ in the relation for lost work

$$\widehat{lw} = \frac{2fL\langle v\rangle^2}{D} = \phi_2(f,\langle v\rangle) \qquad (6.2\text{-}13)$$

When pressure drop (i.e., lost work) is given, however, we cannot solve the above set of equations by successive substitution because *two* unknowns ($\langle v\rangle$ and f) appear in the relationship given in graphical form. We therefore rearrange the second equation to read

$$\frac{D\widehat{lw}}{2L\langle v\rangle^2} = f \qquad (6.2\text{-}24)$$

and multiply both sides by $(D\langle v\rangle\rho/\mu)^2$ to obtain

$$\frac{D^2\rho^2}{\mu^2}\frac{D\widehat{lw}}{2L} = \text{Re}^2 f \qquad (6.2\text{-}25)$$

or

$$\frac{D\rho}{\mu} \sqrt{\frac{\widehat{lw}D}{2L}} = \mathrm{Re}\ \sqrt{f}$$ (6.2-26)

This equation involves only one unknown (f), since we know $\widehat{lw}$ from the mechanical energy balance if we are given Δp (assuming no large kinetic energy changes). We can easily plot the relation between f and $\mathrm{Re}\ \sqrt{f}$ from the friction factor plot by picking values of Re, reading f, calculating $\mathrm{Re}\ \sqrt{f}$ and plotting. Such a plot is given in Fig. 6.2-9. One uses such a plot by evaluating $\widehat{lw}$ using the mechanical energy balance. If the pipe is of the same diameter at the entrance and exit of the system, the velocity is not needed to evaluate the Karman number and f may be found directly. (If the velocity change is large enough to be a significant part of $\widehat{lw}$, even Fig. 6.2-9 will not avoid the trial and error solution.) Knowing f and $\widehat{lw}$, we can easily get $\langle v \rangle$ from

$$\widehat{lw} = \frac{2fL\langle v \rangle^2}{D}$$ (6.2-13)

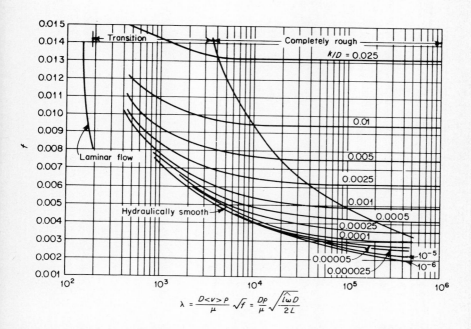

Fig. 6.2-9 Friction factor versus Karman number. (*C. O. Bennett, and J. E. Myers, "Momentum, Heat, and Mass Transfer," McGraw-Hill Book Company, New York, 1962.*)

Example 6.2-4 Calculation of flow rate when pressure drop is known
Water at 110°F flows through a horizontal 2-in. schedule 40 pipe with
equivalent length of 100 ft. A pressure gage at the upstream end reads 30
psi, while one at the downstream end reads 15 psi. What is the flow rate in
gallons per minute?

Solution Applying the mechanical energy balance:

$$\frac{\Delta p}{\rho} + \frac{\Delta \langle v \rangle^2}{2} + g\Delta z = -\widehat{lw}$$

$$-\widehat{lw} = \left[(15-30) \frac{lbf}{in.^2} \right] \left(\frac{ft^3}{61.8 \ lbm} \right) \left(\frac{ft^3}{61.8 \ lbm} \right)$$

$$\widehat{lw} = 35 \ ft \ lbf/lbm$$

Calculating the Karman number:

$$\lambda = Re \ \sqrt{f} = \frac{D\rho}{\mu} \sqrt{\frac{\widehat{lw}D}{2L}}$$

$$\lambda = \frac{[(2.067 \ in. \ ft)/(12 \ in.)](61.8 \ lbm/ft^3)}{(0.62 \ cp)(6.72 \times 10^{-4} \ lbm/ft \ sec \ cp)}$$

$$\sqrt{\frac{(35) \ ft \ lbf/lbm \ [(2.067 \ ft/12)] \ (32.2 \ lbm \ ft/lbf \ sec^2)}{(2)(100) \ ft}}$$

$$\lambda = 2.52 \times 10^5$$

From Fig. 6.2-9, with $k/D = 0.0009$,

$$f = 0.0048$$

Then

$$\widehat{lw} = \frac{2fL\langle v \rangle^2}{D}$$

$$35 \ \frac{ft \ lbf}{lbm} = \left\{ \frac{[(2)(0.0048)(100) \ ft] \langle v \rangle^2}{2.067/12 \ ft} \right\} \left(\frac{lbf \ sec^2}{32.2 \ lbm \ ft} \right)$$

$$\langle v \rangle = 14.2 \ ft/sec$$

□ □ □

Pipe networks
The calculation of flow in piping systems containing branches (sometimes
called pipe networks) is complicated and often requires trial and error

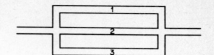

Fig. 6.2-10 Parallel pipe system.

solutions best done on a computer. When several pipes of different sizes or properties (such as roughness) are connected such that fluid flows first through one pipe, then the next, etc., they are said to be connected in series. For such a system a mass balance shows that the mass flow rate is constant, and one simply applies the mechanical energy balance, finding a lost work term for each section of pipe. The approach in the examples we have solved so far is adequate for solving this problem. When two or more pipes are connected as in Fig. 6.2-10, the flow divides among the pipes (they are connected in parallel). Adding pipes in parallel is a common way of adding flow capacity to a system. In a parallel pipe system such as is illustrated in Fig. 6.2-10

$$\left(\frac{\Delta p}{\rho}\right)_1 = \left(\frac{\Delta p}{\rho}\right)_2 = \left(\frac{\Delta p}{\rho}\right)_3 \qquad (6.2\text{-}27)$$

and

$$Q = Q_1 + Q_2 + Q_3 \qquad (6.2\text{-}28)$$

Example 6.2-5 Flow in a parallel piping system A pipeline 20 miles long delivers 5,000 barrels of petroleum per day. The pressure drop over the line is 500 psig. Find the new capacity of the total system if a parallel and identical line is laid along the last 12 miles. The pressure drop remains at 500 psig and flow is laminar. (See Fig. 6.2-11.)

Solution A mass balance yields

$$Q_T = Q_A + Q_B$$

Neglecting the effect of the short connecting length

$$Q_A = Q_B$$

$$\therefore Q_B = \frac{Q_T}{2}$$

Fig. 6.2-11 Sketch for Example 6.2-5.

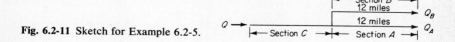

The overall pressure drop we denote as

$$\Delta p_T = 500 \text{ psig}$$

$$\left(\frac{\Delta p}{\rho}\right)_A = \left(\frac{\Delta p}{\rho}\right)_B$$

(a) $\therefore \Delta p_T = \Delta p_C + \Delta p_B = \Delta p_C + \Delta p_A$

But if we assume laminar flow

(b) $-\Delta p = \dfrac{32\mu L \langle v \rangle}{D^2} = \dfrac{32\mu L Q}{D^2 A} = \alpha L Q$

We do not have physical properties, and so we evaluate α using the information given for the original line.

$$\left(144 \frac{\text{in.}^2}{\text{ft}^2}\right) \left(500 \frac{\text{lbf}}{\text{in.}^2}\right) = \alpha(20\text{miles})\left(5{,}280 \frac{\text{ft}}{\text{mile}}\right)$$

$$\left(5{,}000 \frac{\text{barrels}}{\text{day}}\right)\left(1 \frac{\text{day}}{24 \text{ hr}}\right)\left(\frac{1 \text{ hr}}{3{,}600 \text{ sec}}\right)\left(\frac{42 \text{ gal}}{\text{barrel}}\right)\left(\frac{\text{ft}^3}{7.48 \text{ gal}}\right)$$

$$\alpha = 2.1 \ \text{lbf sec/ft}^6$$

Substituting in equation (a) using equation (b):

$$(500)\,(144) = (2.1)\,(8)\,(5{,}280)\,Q + (2.1)\,(12)\,(5{,}280)\frac{1}{2}\,Q$$

$$Q = \frac{(500)\,(144)}{(2.1)\,(5{,}280)\,(14)}\left(\frac{\text{ft}^3}{\text{sec}}\right)\left(7.48 \frac{\text{gal}}{\text{ft}^3}\right)\left(\frac{\text{barrel}}{42 \text{ gal}}\right)\left(\frac{3{,}600 \text{ sec}}{\text{hr}}\right)\left(\frac{24 \text{ hr}}{\text{day}}\right)$$

$$Q = 7{,}200 \text{ barrels/day}$$

□ □ □

Such pipelines do not normally operate in laminar flow, of course. The problem could be solved for turbulent flow by expressing Δp as a function of friction factor and fitting the friction factor chart with an analytic expression. This would complicate matters, but the attack would remain the same.

Example 6.2-6 Input of additional fluid to an existing pipe network It is desired to add an additional plant to the 24-in. pipeline network designed in Example 6.2-5. The plant will produce 1,880 barrels/day and will be located at point 3. (See Fig. 6.2-12.) The pipeline network is to supply equal quantities of product to the plants at points 4 and 5. What pressure must a throttling valve at 5 maintain in order that $Q_A = Q_B$(p_4 remains at 0 psig and the pressure at 1 is changed in such a way that the flow in Section C is maintained at 7,200 barrels/day).

The Hagen-Poiseuille equation yields as in Example 6.2-5

$$\Delta p = \frac{-32\mu L \langle v \rangle}{D^2} = \frac{-32\mu L Q}{D^2 A}$$

$$\Delta p \ \frac{lbf}{ft^2} = -\alpha L Q = \left(-21 \ \frac{lbf \ sec}{ft^6}\right) (LQ \ ft) \left(\frac{ft^3}{sec}\right)$$

or, converting the units of α so that we can use Δp in psi, L in miles, and Q in barrels/day,

$$\Delta p \ \frac{lbf}{in.^2} = (-5 \times 10^{-3}) \left(\frac{psi \ day}{miles \ barrel}\right) LQ \left(\frac{miles \ barrel}{day}\right)$$

We now write down the equations describing our system. First, we apply the Hagen-Poiseuille law to each section of pipe:

Section C:
(a) $\quad p_2 - p_1 = (-5 \times 10^{-3})(8)(7,200)$
Section D:
(b) $\quad p_3 - p_2 = (-5 \times 10^{-3})(4) Q_D$
Section B:
(c) $\quad 0 - p_2 = (-5 \times 10^{-3})(12) Q_B$
Section A:
(d) $\quad p_5 - p_3 = (-5 \times 10^{-3})(8) Q_A$

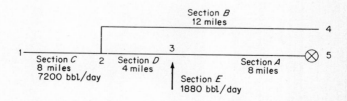

Fig. 6.2-12 Sketch for Example 6.2-6.

Mass balances around points 2 and 3 yield

> Point 2:
> (e) $7{,}200 - Q_D - Q_B = 0$
> Point 3:
> (f) $1{,}880 + Q_D - Q_A = 0$

(Note: We must assume a direction of flow for stream D in equations (d) and (f)—if the direction assumed is wrong, the value calculated will simply be negative—but we *must assume* a *consistent* direction in our equations.)

We also have the imposed restriction

> (g) $Q_A = Q_B$

Rewriting equations (a) to (g):

(a') $-p_1 + p_2$ $\qquad\qquad\qquad\qquad\qquad\qquad\qquad\qquad\; + \quad 288 = 0$

(b') $\quad\;\; -p_2 + p_3 \qquad\qquad\qquad\qquad\qquad + 2 \times 10^{-2} Q_D \qquad\qquad = 0$

(c') $\quad\;\; -p_2 \qquad\qquad\qquad + 6 \times 10^{-2} Q_B \qquad\qquad\qquad\qquad\quad = 0$

(d') $\qquad\qquad -p_3 + p_5 + 4 \times 10^{-2} Q_A \qquad\qquad\qquad\qquad\qquad = 0$

(e') $\qquad\qquad\qquad\qquad\qquad\qquad\;\; -Q_B \qquad\quad -Q_D \qquad +7{,}200 = 0$

(f') $\qquad\qquad\qquad -Q_A \qquad\qquad\qquad\qquad\qquad\quad +Q_D \qquad +1{,}800 = 0$

(g') $\qquad\qquad\qquad\quad Q_A \qquad\qquad -Q_B \qquad\qquad\qquad\qquad\qquad\quad = 0$

Equations (a') to (g') are a set of seven linearly independent equations in the seven unknowns p_1, p_2, p_3, p_5, Q_A, Q_B, and Q_D. They may be solved by elementary matrix methods, but this set is also simply solved as follows:

Use equation (g') to substitute for Q_A in equation (f'), then add the result to equation (e') to yield

$$-2 Q_B + 9{,}080 = 0$$

$$Q_B = 4{,}540$$

Therefore

$$Q_A = 4{,}540$$

and equation (e') will give

$$Q_D = 2{,}660$$

Equation (c') will then give

$$p_2 = 272$$

Equation (b') will give

$$p_3 = 219$$

following which (a') gives

$$p_1 = 560$$

and equation (d') gives

$$p_5 = 37 \text{ psig}$$

Notice that the pump at point 1 must now supply 560 psig, and that the fluid added at point 3 must come from a pump which will supply 219 psig. Both of these requirements would probably be feasible in a real world situation.

This problem illustrates the general attack on pipe network problems: Write down the independent equations describing the system; if these are fewer than the number of unknowns the system is underdetermined and some of the unknowns must be chosen; if there are fewer unknowns than independent equations, the system is overdetermined and some variables that have been fixed will have to be considered unknowns.

□ □ □

Thus far, we have been discussing only single-phase flow. Many important industrial problems involve the flow of more than a single phase: typically, the flow of a gas phase and a liquid phase, the flow of two liquid phases, or the flow of a solid phase in a gas or liquid. In particular, gas-liquid flows are very important. The addition of multiple phases to our flow problem introduces a problem in flow regime over and above the simple classification of laminar and turbulent flow. The flow regime must be known to describe precisely the mechanism of momentum transfer. Shown in Fig. 6.2-13a is a highly empirical but useful chart which permits rough approximation of flow regime for horizontal two-phase gas-liquid flow in pipes. This chart is called the Baker plot. (The corresponding flow regimes are shown in Fig. 6.2-13b.) We illustrate its use by the following example.

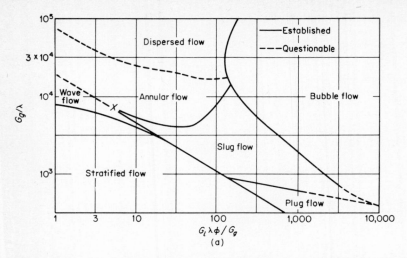

Fig. 6.2-13a Area plot for two-phase horizontal flow. $\lambda = [(\rho_g/0.075)(\rho_l/62.4)]^{1/2}$ and $\phi = (73/\sigma)[\mu_l(62.3/\rho_l)^2]^{-(1/3)}$, where $\sigma =$ surface tension, dynes/cm; $\rho_l =$ density of the liquid, lb/ft^3; $\mu_l =$ viscosity of the liquid, centipoise; and $\rho_g =$ density of the gas, lb/ft^3. [*O. Baker, Gas J.*, **53**: *185 (1954).*]

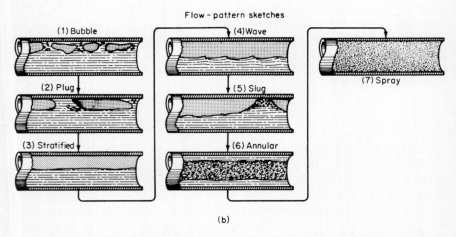

Fig. 6.2-13b Two-phase gas-liquid flow regimes. [*G. E. Alves, Chem. Eng. Progr.*, **50**: *449 (1954).*]

Example 6.2-7 Calculation of flow regime for two-phase flow from a Baker plot Use the Baker plot to determine flow regime for the flow of a mixture of natural gas and water through a 4-in. schedule 40 pipe. The gas flows at 250 ft^3/min and the water flows at 0.2 ft^3/min. The following physical properties may be used:

$$\rho_L = 60 \text{ lb} m/\text{ft}^3 \qquad \rho_G = 0.04 \frac{\text{lb}m}{\text{ft}^3}$$

$$\mu_L = 0.3 \text{ cp} \qquad \mu_G = 0.01 \text{ cp}$$

$$\sigma = 60 \text{ dynes/cm}$$

Solution We calculate the ordinate and abscissa for the Baker plot as follows:

$$\lambda = \left(\frac{\rho_G}{0.075}\frac{\rho_L}{62.4}\right)^{1/2} = \left(\frac{0.04}{0.075}\frac{60}{62.4}\right)^{1/2} = 0.716$$

$$\phi = \frac{73}{\sigma}\left[\mu_L\left(\frac{62.3}{\rho_L}\right)^2\right]^{-1/3}$$

$$= \frac{73}{60}\left[0.3\left(\frac{62.3}{60}\right)^2\right]^{-1/3} = 1.76$$

$$G_G = \left(250\ \frac{\text{ft}^3}{\text{min}}\right)\left(0.04\ \frac{\text{lb}m}{\text{ft}^3}\right)\left(60\ \frac{\text{min}}{\text{hr}}\right)\left(\frac{1}{0.0884\ \text{ft}^2}\right)$$

$$= 6.8 \times 10^3\ \text{lb}m/\text{hr ft}^2$$

$$G_L = \left(0.2\ \frac{\text{ft}^3}{\text{min}}\right)\left(60\ \frac{\text{lb}m}{\text{ft}^3}\right)\left(60\ \frac{\text{min}}{\text{hr}}\right)\left(\frac{1}{0.0884\ \text{ft}^2}\right)$$

$$= 8.16 \times 10^3\ \frac{\text{lb}m}{\text{hr ft}^2}$$

$$\frac{G_G}{\lambda} = \frac{6.8 \times 10^3}{0.716} = 9.5 \times 10^3$$

$$\frac{G_L}{G_G}\lambda\phi = \frac{8.16 \times 10^3}{6.8 \times 10^3}(0.716)(1.76) = 1.52$$

From Fig. 6.2-13a we see we have the stratified flow regime or the wave flow regime. The lines on this chart really should not be drawn so sharply since the boundaries are quite fuzzy. This means that in many cases, as here, we will not be able to decide among two or more possible flow regimes.

□ □ □

A similar approach to the Baker plot can be taken for other two-phase flow situations.

Despite the fact that a detailed knowledge of regime is necessary to make precise flow calculations in multiphase flow, it is possible to use certain overall correlations which provide useful answers without knowledge of regime. The classical example of such a correlation is the Lockhart-Martinelli approach for gas-liquid pipe flow.

The Lockhart-Martinelli correlation rests on two basic assumptions: (1) The static pressure drops for the gas and for the liquid are the same, and (2) the volume of the gas plus the volume of the liquid equals the pipe volume. The method of calculation is as follows:

1. Calculate the pressure drop per unit length for liquid as if the pipe carried only the liquid flow. This we call $(\Delta p/\Delta L)_l$, and the Reynolds number used to obtain it we call Re_l.
2. Repeat using the gas phase to obtain $(\Delta p/\Delta L)_g$ and Re_g.
3. Calculate

$$\chi^2 = \frac{(\Delta p/\Delta L)_l}{(\Delta p/\Delta L)_g} \tag{6.2-29}$$

4. Use Table 6.2-2 to decide which curve on Fig. 6.2-14 to use. The curves are labeled with three subscripts. The first, l or g, refers to liquid or gas. The second, v or t, refers to whether the *liquid* is in "viscous" or "turbulent" flow. The third, v or t, refers to whether the *gas* is in "viscous" or "turbulent" flow. The table gives the last two of these three subscripts as a function of Re_l and Re_g.
5. Using the appropriate curve, calculate the two-phase pressure drop per unit length, $(\Delta p/\Delta L)_{TP}$, by either

$$\left(\frac{\Delta p}{\Delta L}\right)_{TP} = \phi_{l..}^2 \left(\frac{\Delta p}{\Delta L}\right)_l \tag{6.2-30}$$

or

$$\left(\frac{\Delta p}{\Delta L}\right)_{TP} = \phi_{g..}^2 \left(\frac{\Delta p}{\Delta L}\right)_g \tag{6.2-31}$$

Table 6.2-2   Flow regime for Lockhart-Martinelli correlation[a]

	t-t	v-t	t-v	v-v
$N_{Re,l}$	> 2,000	< 1,000	> 2,000	< 1,000
$N_{Re,g}$	> 2,000	> 2,000	< 1,000	< 1,000

[a] R. W. Lockhart and R. C. Martinelli, *Chem. Eng. Progr.*, **45**:39 (1949).

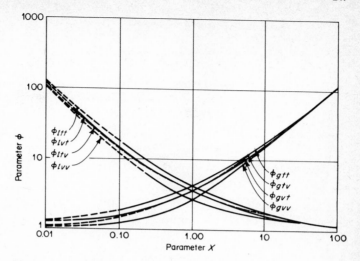

Fig. 6.2-14 Lockhart-Martinelli correlation. [*R. W. Lockhart, and R. C. Martinelli, Chem. Eng. Progr.*, **45**: *39 (1949)*.]

where the dots simply indicate the appropriate missing subscripts, t or v (this notation is also used in tensor manipulations to indicate summation—such is *not* the case here). Equivalent answers are obtained regardless of which of the last two equations is used.

Example 6.2-8   Calculation of two-phase pressure drop using Lockhart-Martinelli approach   Calculate $(\Delta p/\Delta L)_{TP}$ for the conditions given in Example 6.2-7.

Solution   Following the procedure outlined above:

1. $\mathrm{Re}_l = \dfrac{Dv\rho}{\mu} = D \dfrac{Q}{A} \dfrac{\rho}{\mu}$

$$= \left(\frac{4.026}{12}\ \text{ft}\right) \left(0.2\ \frac{\text{ft}^3}{\text{min}}\right) \left(\frac{1}{(0.0884)\ \text{ft}^2}\right) \left(60\ \frac{\text{lbm}}{\text{ft}^3}\right)$$
$$\left(\frac{1.0}{0.3\ \text{cp}}\right) \left(\frac{\text{cp}}{(6.72 \times 10^{-4})}\right) \left(\frac{\text{ft sec}}{\text{lbm}}\right) \left(\frac{\text{min}}{60\ \text{sec}}\right)$$

$= 3{,}780$

From the friction factor chart with $k/D = 0.00042$,
$f = 0.01$ (Note: This is really in the transition region—the value of 0.01 is conservative.)

$$\Delta p = \frac{2fL\rho \langle v \rangle^2}{D}$$

Since the symbol L means the same as ΔL:

$$\left(\frac{\Delta p}{\Delta L}\right) = \frac{2f\rho\langle v\rangle^2}{D}\left[(2)(0.01)(60)\,\frac{\text{lbm}}{\text{ft}^3}\right]\left(\frac{(12)}{(4.026)\,\text{ft}}\right)$$

$$\left(\frac{\text{lbf sec}^2}{32.2\,\text{lbm ft}}\right)\left[\left(\frac{0.2\,\text{ft}^3}{\text{min}}\right)\left(\frac{1.0}{0.0884\,\text{ft}^2}\right)\left(\frac{\text{min}}{60\,\text{sec}}\right)\right]^2$$

$$\left(\frac{\Delta p}{\Delta L}\right)_l = 1.59\times10^{-4}\,(\text{lbf/ft}^2)/\text{ft}$$

2. $\text{Re}_g = D\dfrac{Q}{A}\dfrac{\rho}{\mu}$

$$= \left(\frac{4.026\,\text{ft}}{12}\right)\left(250\,\frac{\text{ft}^3}{\text{min}}\right)\left(\frac{1}{0.0884\,\text{ft}^2}\right)\left(0.04\,\frac{\text{lbm}}{\text{ft}^3}\right)$$

$$\left(\frac{\text{cp ft sec}}{0.01\,\text{cp}\,(6.72\times10^{-4})\,\text{lbm}}\right)\left(\frac{\text{min}}{60\,\text{sec}}\right)$$

$\text{Re}_g = 9.45\times10^4$

and

$f = 0.046$

yielding

$$\left(\frac{\Delta p}{\Delta L}\right)_g = \frac{(2)(0.046)(0.04)(12)}{(32.2)(4.026)}\left[\frac{(250)}{(0.0884)(60)}\right]^2$$

$$= 0.757\,(\text{lbf/ft}^2)/\text{ft}$$

3. $\chi^2 = \dfrac{1.59\times10^{-4}}{0.757} = 2.1\times10^{-4}$

$\chi = 1.45\times10^{-2}$

4. From the table we see that we are in the t-t region. Therefore, from Fig. 6.2-14,

$\phi_g = 1.5$

5. $\left(\dfrac{\Delta p}{\Delta L}\right)_{TP} = \phi_{gtt}^2\left(\dfrac{\Delta p}{\Delta L}\right)_g = (1.5)^2(0.757)$

$$= 1.7\,\frac{\text{lbf/ft}^2}{\text{ft}} = 0.0118\,\text{psi/ft}$$

We could also have used $\phi_{ltt} = 100$ and the equation based on the liquid phase:

$$\left(\frac{\Delta p}{\Delta L}\right)_{TP} = \phi_{ltt}^2 \left(\frac{\Delta p}{\Delta L}\right)_l$$

$$= (100)^2 (1.59 \times 10^{-4})$$

$$= \frac{1.59 \ \text{lbf/ft}^2}{\text{ft}} = 0.011 \ \text{psi/ft}$$

Note: The agreement is reasonable. This correlation gives answers that may differ from the true answer by more than 100 percent occasionally. It is useful, however, for order-of-magnitude calculations.

□ □ □

The friction factors we have dealt with thus far have been applied to the flow of a Newtonian liquid. The determination of this friction factor will change when we use non-Newtonian rather than Newtonian fluids in our pipes. One way that we can account for non-Newtonian behavior is to introduce a correction to the Newtonian fluid Reynolds number. An example of the general approach used is detailed in the example below.

The approach below is that of Metzner and Reed.[1] They defined a modified Reynolds number as

$$\text{Re}_m = \frac{D^n \langle v \rangle^{2-n}}{\gamma} \rho \tag{6.2-32}$$

where n and γ are properties of the particular non-Newtonian fluid. This definition of Re permits the usual chart for Fanning friction factor to be used, e.g., in the laminar regime $f = 16/\text{Re}_m$. Obviously for a Newtonian fluid $n = 1$ and $\gamma = \mu$.

The parameters n and γ are characteristic of the particular non-Newtonian fluid and can be determined by doing flow measurements in a capillary tube.[2] We can incorporate the diameter and velocity into an "apparent" viscosity by defining

$$\mu_{ap} = \gamma \left(\frac{\rho}{\langle v \rangle}\right)^{1-n} \tag{6.2-33}$$

[1] A. B. Metzner and J. C. Reed, Flow of non-Newtonian Fluids—Correlation of the Laminar, Transition, and Turbulent-flow Regions, *A.I.Ch.E. J.*, **1** (4): 434 (1955)
[2] R. J. Brodkey, "The Phenomena of Fluid Motions," Addison-Wesley Publishing Company, Inc., Reading, Mass., 1967.

and, substituting, this permits us to write the Reynolds number in the familiar form:

$$\text{Re} = \frac{D\langle v \rangle \rho}{\mu_{\text{ap}}} \qquad (6.2\text{-}34)$$

Note that the apparent viscosity is not a simple property of the fluid but depends on flow rate and geometry.

The use of this type of approach is typical, but far from unique. The example below simply illustrates this method; many methods exist for non-Newtonian flow problems and this single example should not be generalized without reference to the literature.

Example 6.2-9 Friction factor for non-Newtonian flow Molten polyethylene is to be pumped through a schedule 40 steel pipe at a rate of 0.25 gal/min. Calculate the pressure drop per unit length.

Properties:

$$n = 0.49 \qquad \rho = 58 \ \text{lb}m/\text{ft}^3$$

$$\mu_{\text{ap}} = 22{,}400 \ \text{poise}$$

Solution The Reynolds number is

$$\text{Re}_m = \frac{D\langle v \rangle \rho}{\mu_{\text{ap}}} = D \, \frac{Q}{A} \, \frac{\rho}{\mu_{\text{ap}}}$$

$$= \left(\frac{1.049}{12} \ \text{ft}\right) \left[\left(\frac{0.25 \ \text{gal}}{\text{min}}\right) \left(\frac{\text{ft}^3}{7.48 \ \text{gal}}\right) \left(\frac{1}{0.00211 \ \text{ft}^2}\right) \right]$$

$$\left[\frac{\text{ft sec poise}}{(2.2 \times 10^4 \ \text{poise})(6.72 \times 10^{-2} \ \text{lb}m)} \right] \left(58 \ \frac{\text{lb}m}{\text{ft}^3}\right) \left(\frac{\text{min}}{60 \ \text{sec}}\right)$$

$$= 9.05 \times 10^{-4}$$

$$f = \frac{16}{\text{Re}} = \frac{16}{9.05 \times 10^{-4}} = 1.77 \times 10^4$$

$$\Delta p = \frac{2 f L \rho \langle v \rangle^2}{D}$$

$$\frac{\Delta p}{L} = \frac{2 f \rho \langle v \rangle^2}{D} = \left[(2)(1.77 \times 10^4)(54) \ \frac{\text{lb}m}{\text{ft}^3}\right] \left(\frac{12}{1.049 \ \text{ft}}\right)$$

$$\left(\frac{\text{lb}f \ \text{sec}^2}{32.2 \ \text{lb}m \ \text{ft}}\right) \left[\left(\frac{0.25 \ \text{gal}}{\text{min}}\right) \left(\frac{\text{ft}^3}{7.48 \ \text{gal}}\right) (0.00211 \ \text{ft}^2) \left(\frac{\text{min}}{60 \ \text{sec}}\right) \right]^2$$

$$= 1.93 \times 10^5 \; \frac{\mathrm{lb}f/\mathrm{ft}^2}{\mathrm{ft}}$$

$$= 1{,}340 \; \mathrm{psi/ft}$$

The apparent viscosity and therefore the pressure drop decreases with increasing flow rate.

□ □ □

6.3 NONCIRCULAR CONDUITS

The hydraulic radius is a useful empiricism for calculating *turbulent* flow in noncircular conduits. One assumes that the resistance for a circular pipe of some equivalent diameter is the same as the conduit in question. The hydraulic radius is defined as the flow cross-sectional area of the conduit in question divided by the wetted perimeter.

$$R_H = \frac{S}{Z} \tag{6.3-1}$$

The equivalent diameter of the pipe for use in determining friction factors and Reynolds numbers is

$$D_{eq} = 4R_H \tag{6.3-2}$$

Note that the equivalent diameter is *four* times the hydraulic radius, not twice the hydraulic radius. The force due to friction balances the pressure force

$$-F = \Delta p \, A_{xs} = -\frac{1}{2}\,\rho\langle v \rangle^2 A_w f \tag{6.3-3}$$

where A_{xs} = cross-sectional area, A_w = wall area.

$$\Delta p \pi R^2 = -\frac{1}{2}\,\rho\langle v \rangle^2 2\pi RLf \tag{6.3-4}$$

$$\frac{\Delta p}{L} = -\frac{1}{2}\,\rho\langle v \rangle^2 \frac{2}{R}f \tag{6.3-5}$$

$$\frac{\Delta p}{L} = -\frac{\rho\langle v \rangle^2}{R}f \tag{6.3-6}$$

Since $R_H = \dfrac{1}{2} R$

$$\frac{\Delta p}{L} = - \frac{\rho \langle v \rangle^2}{2 R_H} f \tag{6.3-7}$$

$$\frac{\Delta p}{L} = - \frac{1}{2} \rho \langle v \rangle^2 \frac{f}{R_H} \tag{6.3-8}$$

and the appropriate Reynolds number is

$$\mathrm{Re_{eq}} = \frac{4 R_H \langle v \rangle \rho}{\mu} \tag{6.3-9}$$

Pressure drop is determined as in the previous sections using the pipe friction factor chart, except that one enters the chart with Reynolds number as defined by Eq. (6.3-9). The pressure loss is determined from the resultant friction factor and Eq. (6.3-8).

Example 6.3-1 Pressure drop in an annulus Water at 70°F is flowing at 10 ft/sec in the annulus between a 1-in.-OD tube and 2-in.-ID tube of a concentric tube heat exchanger. If the exchanger is 100 ft long, what is the pressure drop in the annulus? Neglect end effects.

(Note that if we are going to manufacture an actual exchanger for a plant application, we certainly would not make one which was 100 ft long, but rather would make it in several sections and place the sections side by side.)

$$R_H = \frac{\pi D_2^2/4 - \pi D_1^2/4}{\pi D_2 + \pi D_1} = \frac{D_2^2 - D_1^2}{4(D_2 + D_1)} = \frac{(D_2 - D_1)(D_2 + D_1)}{4(D_2 + D_1)}$$

$$= \frac{D_2 - D_1}{4}$$

$$\mathrm{Re} = \frac{D_{eq}\langle v \rangle \rho}{\mu} = \frac{4 R_H \langle v \rangle \rho}{\mu}$$

$$= \frac{[(4/_4)(2/_{12} - 1/_{12})\text{ ft}](10\text{ ft/sec})(62.4\text{ lb}m/\text{ft}^3)}{(1\text{ cp})(6.72 \times 10^{-4}\text{ lb}m)/\text{cp ft sec}}$$

$$= 7.75 \times 10^4$$

$$\frac{k}{D} = 0.0018 \qquad \text{(based on } D_2 - D_1\text{)}$$

$$f = 0.0062$$

$$\Delta p = -\frac{2fL\rho\langle v\rangle^2}{D_{eq}}$$

$$= [(2)(0.0062)(100)\text{ft}]\left(\frac{62.4\ \text{lb}m}{\text{ft}^3}\right)$$

$$\left\{(10^2)\frac{\text{ft}^2}{\text{sec}^2[(4/4)(2/12 - 1/12)\ \text{ft}]}\right\}\left(\frac{\text{lb}f\ \text{sec}^2}{32.2\ \text{lb}m\ \text{ft}}\right)\}$$

$$\Delta p = -2,880\ \text{lb}f/\text{ft}^2$$

$$= -20\ \text{psi}$$

Note that what we have done above is consistent with our equations for pipe, because the limit as D_1 approaches zero gives $D_2/4$ as the hydraulic radius. Four times the hydraulic radius is the pipe diameter D_2.

□ □ □

Example 6.3-2 Error using hydraulic radius for laminar flow If one has laminar flow of Newtonian fluid in an annulus it is easy to show[1] that the expression for pressure drop (neglecting body forces) is

(a) $\qquad \Delta p = \frac{-8\mu L\langle v\rangle}{R_1^2}\frac{1}{C}$

where

$$C = \frac{1}{1 - K^2}\left[(1 - K^4) - \frac{(1 - K^2)^2}{\ln 1/K}\right]$$

and

$R_1 =$ inside radius of annulus

$K =$ ratio of outside radius of annulus to inside radius

Notice that the first factor on the right-hand side of equation (a) is just the Hagen-Poiseuille expression.

If, however, we attempt to obtain equation (a) by using the concept of hydraulic radius we have

$$\Delta p = -\frac{2fL\rho\langle v\rangle^2}{D_{eq}}$$

[1]R. B. Bird, W. E. Stewart, and E. N. Lightfoot, "Transport Phenomena," John Wiley & Sons, Inc., New York, 1960.

and

$$f = \frac{16}{\text{Re}} = \frac{16\mu}{D_{eq}\langle v \rangle \rho}$$

so that

$$(b) \qquad \Delta p = \frac{-(2)(16)\mu}{D_{eq}\langle v \rangle \rho} \frac{L\rho \langle v \rangle^2}{D_{eq}} = \frac{-32\mu L \langle v \rangle}{D_{eq}^{\,2}}$$

But

$$D_{eq} = 4R_H$$

and for an annulus

$$R_H = \frac{D_2 - D_1}{4} = \frac{R_2 - R_1}{2}$$

so that

$$D_{eq} = 2(R_2 - R_1) = 2R_1(K - 1)$$

Substituting in equation (b)

$$\Delta p = \frac{-32\mu L \langle v \rangle}{4R_1^{\,2}(K-1)^2} = \frac{-8\mu L \langle v \rangle}{R_1^{\,2}} \frac{1}{(K-1)^2}$$

which, when compared with equation (a), gives $C = (K-1)^2$ rather than the correct result.

□ □ □

6.4 FLOW THROUGH POROUS MEDIA Many fluid flow operations involve porous media. Fixed bed reactors, packed absorption columns, and filters are common examples of such operations. A very important operation, that of removing petroleum and natural gas from underground reservoirs, is mainly a problem of flow in porous media.

The packed bed problem is usually attacked via the friction factor route. One assumes that the porous medium can be modeled as a bundle of capillary tubes that has equivalent flow resistance. With the problem of filters and underground reservoirs, we usually use an empirical equation, Darcy's law, to estimate pressure drops and flow rates.

The laminar flow of fluids in porous media can be represented by Darcy's law

$$v_\infty = \frac{-k}{\mu} \frac{dp}{dx} \tag{6.4-1}$$

where k is the permeability and represents the conductivity to flow of the porous medium.

A friction factor for flow in a packed bed is defined in a parallel fashion to that for a pipe, except that a different characteristic length and velocity are used [compare with Eq. (6.2-5)]:

$$\frac{\Delta p}{L} = -\frac{1}{2} \rho v^2 \frac{4f}{D_{eq}} \tag{6.4-2}$$

where v is the interstitial velocity and D_{eq} is the equivalent diameter of the packing. The velocity of approach is sometimes called the superficial velocity and is the velocity obtained by taking the total volumetric flow and dividing it by the total cross section normal to flow. Rather than using the cross-sectional area divided by the wetted perimeter for hydraulic radius, we use void *volume* divided by the wetted *surface*, thus extending our two-dimensional definition to the third dimension. For packed beds, the hydraulic radius is defined

$$R_H = \frac{\text{void volume of bed}}{\text{surface area of packing}} \tag{6.4-3}$$

Packed beds are porous media where the interstitial spaces are quite large and the particles that make up the porous medium are quite large.

If the particle volume is V_p and the surface area of a particle is S_p the specific surface is defined as

$$S = \frac{S_p}{V_p} \tag{6.4-4}$$

For a spherical particle it readily follows that

$$S = \frac{6}{D} \tag{6.4-5}$$

If ϵ is the void *fraction* of the bed, then

$$\frac{\epsilon}{1 - \epsilon} = \frac{\text{(void volume)/(total volume)}}{\text{(non-void volume)/(total volume)}}$$

and

$$\frac{\epsilon}{1-\epsilon} \times \text{(volume of particles)}$$

is the void volume, and N is the number of particles, so

$$R_H = \frac{\epsilon V_P N/(1-\epsilon)}{S_P N} = \frac{V_p}{S_p}\frac{\epsilon}{1-\epsilon} \qquad (6.4\text{-}6)$$

For nonspherical particles an effective particle diameter is defined

$$D_p = \frac{6}{S} \qquad (6.4\text{-}7)$$

and

$$R_H = \frac{D_p}{6}\frac{\epsilon}{1-\epsilon} \qquad (6.4\text{-}8)$$

The velocity of approach is related to the interstitial velocity by a mass balance

$$v_\infty A_{tot}\rho = vA_{void}\rho \qquad (6.4\text{-}9)$$

and as long as the porosity in terms of area ratio is the same as the volumetric porosity

$$v_\infty = \epsilon v \qquad (6.4\text{-}10)$$

where v is the interstitial velocity.

The Reynolds number associated with the velocity of approach and the effective particle diameter from Eqs. (6.4-8) and (6.4-10) is

$$\text{Re} = \frac{4R_H v_\infty \rho}{\epsilon\mu} = \frac{2}{3(1-\epsilon)}\frac{D_p v_\infty \rho}{\mu} \qquad (6.4\text{-}11)$$

The equation for the friction factor in terms of the hydraulic radius and the effective particle diameter is

$$\frac{\Delta p}{L} = -\frac{1}{2}\rho v^2 \frac{4f}{D_{eq}} = -\frac{3\rho v_\infty^2(1-\epsilon)}{D_p \epsilon^3}f \qquad (6.4\text{-}12)$$

Since the friction factor and Reynolds number for flow in a packed bed are based on empirical arguments (experimentally verified), it is simpler to define the friction factor and Reynolds number *without the numerical constants*.

$$\text{Re}_p = \frac{D_p v_\infty \rho}{\mu(1 - \epsilon)} \tag{6.4-13}$$

$$\frac{\Delta p}{L} = \frac{-\rho v_\infty^2 (1 - \epsilon) f_p}{D_p \epsilon^3} \tag{6.4-14}$$

We obtain f from experimental data for flow in packed beds, which for $\text{Re}_p < 1$ show

$$f_p = \frac{150}{\text{Re}_p} \tag{6.4-15}$$

This is the Blake-Kozeny equation. For completely turbulent flow one would expect the friction factor to be constant, and further experiments show that

$$f_p = 1.75 \tag{6.4-16}$$

This is the Burke-Plummer equation and is applied above Re_p about 10^4. For intermediate Reynolds numbers:

$$f_p = \frac{150}{\text{Re}_p} + 1.75 \tag{6.4-17}$$

This is the Ergun equation. It is extremely important that the reader recognize that all the above discussion has presumed a uniform packing with no channels of low flow resistance. Such channels do, in fact, exist in most packed beds. Although they yield lower pressure drop they also permit some fluid to pass through the bed more quickly than is desired; this is important, for example, if the bed is a catalytic reactor.

It is instructive to consider the problem of calculating flow rate through a packed bed when pressure drop is given, since this is a problem which in straight pipe led to a trial and error procedure which we circumvented by introducing the Karman number. For calculation of flow in a packed bed given the pressure drop, one combines Eqs. (6.4-17), (6.4-14), and (6.4-13) to yield

$$\frac{\Delta p}{L} = -\frac{\rho v_\infty^2 (1 - \epsilon)}{D_p \epsilon^3} \left[\frac{150 \, \mu(1 - \epsilon)}{D_p \rho v_\infty} + 1.75 \right] \tag{6.4-18}$$

Rearranging:

$$1.75v_\infty^2 + \frac{150v_\infty\mu(1-\epsilon)}{D_p\rho} + \frac{\Delta p}{L}\frac{D_p\epsilon^3}{\rho(1-\epsilon)} = 0 \qquad (6.4\text{-}19)$$

or

$$v_\infty = \frac{\dfrac{-150\mu(1-\epsilon)}{D_p\rho} \pm \sqrt{\left[\dfrac{150\mu(1-\epsilon)}{D_p\rho}\right]^2 - 4(1.75)\left(\dfrac{\Delta p}{L}\right)\dfrac{D_p\epsilon^3}{\rho(1-\epsilon)}}}{(2)(1.75)}$$

$$(6.4\text{-}20)$$

Notice that for a packed bed, we can solve explicitly for v_∞ without trial and error. This is because the relationship between friction factor and Reynolds number is available in simple analytical rather than graphical form, as was the case for straight pipe. (If the friction factor were available in complicated analytical form, the resulting equation would possibly have to be solved by trial and error anyhow.)

We can use the explicit solution shown to calculate flow velocity. We can *also* solve the problem by use of a Karman plot as we did for straight pipe. We now show this approach to illustrate the parallel between the pipe problem and the packed bed problem. Rearranging Eq. (6.4-14):

$$f_p = -\frac{\Delta p}{L}\frac{D_p\,\epsilon^3}{\rho v_\infty^2(1-\epsilon)} \qquad (6.4\text{-}21)$$

and so

$$\lambda_p = \mathrm{Re}_p\sqrt{f_p}$$

becomes

$$\lambda_p = \frac{D_p v_\infty \rho}{\mu}\left[\frac{-\Delta p}{L}\frac{D_p\epsilon^3}{\rho v_\infty^2(1-\epsilon)}\right]^{1/2} \qquad (6.4\text{-}22)$$

or

$$\lambda_p = \left[\frac{-\Delta p}{L}\left(\frac{\rho}{\mu}\right)^2\left(\frac{D_p\epsilon}{1-\epsilon}\right)^3\right]^{1/2} \qquad (6.4\text{-}23)$$

which permits us to calculate λ without a knowledge of v_∞.

We can prepare a plot of Re_p versus λ_p using

$$\lambda_p = \mathrm{Re}_p \sqrt{f_p} = \mathrm{Re}_p \sqrt{\frac{150}{\mathrm{Re}_p} + 1.75} \qquad (6.4\text{-}24)$$

and by choosing values of Re_p plot a figure like Fig. 6.4-1. This figure is the analogous plot to the one for straight pipe (Fig. 6.2-9). To use it we calculate λ_p using Eq. (6.4-23), read Re_p, and calculate v_∞ from the definition of Re_p, Eq. (6.4-13). This procedure is completely equivalent to using the explicit solution, Eq. (6.4-20).

Example 6.4-1 Pressure drop for air flowing in porous media   Calculate the pressure drop for air at 100°F and 1 atm, flowing at 2.1 lb*m*/sec through a bed of 1/2-in. spheres. The bed is 4 ft in diameter and 8 ft high.

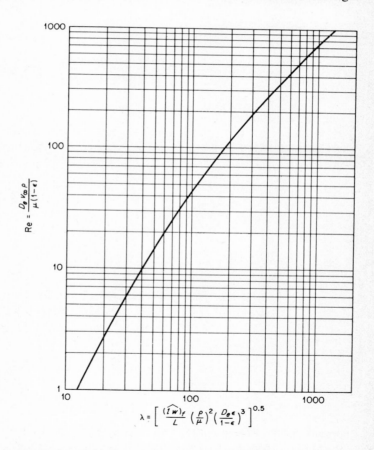

$$\lambda = \left[\frac{(\widehat{l\,w})_f}{L} \left(\frac{\rho}{\mu} \right)^2 \left(\frac{D_\theta \epsilon}{1-\epsilon} \right)^3 \right]^{0.5}$$

Fig. 6.4-1 Karman plot for Ergun correlation. (*K. C. Chao & D. Adams, Hydrocarbon Process, Petrol. Refiner,* **45** (8): 165 (1966). *Copyright (c) 1966 by Gulf Publishing Co., Houston, Tex.*)

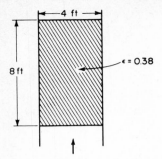

Fig. 6.4-2 Sketch for Example 6.4-1.

The porosity (void fraction) of the bed is 0.38. (See Fig. 6.4-2.) The Reynolds number is

$$\text{Re}_p = \frac{D_p v_\infty \rho}{\mu(1 - \epsilon)} = \frac{D(w/A)}{\mu(1 - \epsilon)}$$

$$= \frac{(^1/_{24}\,\text{ft})\,(2.1\;\text{lb}m/\text{sec})}{(0.182\;\text{cp})\,(6.72 \times 10^{-4})\;\text{lb}m/\text{cp ft sec }\pi\,(^{16}/_4\;\text{ft}^2)\,1.38}$$

$$= \frac{5.7 \times 10^2}{0.62} = 920$$

The Ergun equation yields

$$f_p = \frac{150}{920} + 1.75 = 1.91$$

We then have

$$\Delta p = \frac{-L\rho v_\infty^2 (1 - \epsilon)}{D_p \epsilon^3} f_p = \frac{(-L\rho w^2/A^2)\,(1 - \epsilon)}{D_p \epsilon^3} f_p$$

$$= \frac{-8\;\text{ft}\,[\,(2.1)^2\;\text{lb}m^2/\text{sec}^2\,]\,[\,\text{lb}f\;\text{sec}^2/(32.2\;\text{lb}m\;\text{ft})\,]\,(1 - 0.38)}{(^1/_{24}\;\text{ft})\,\{\pi(4)^2/4^2\;\text{ft}^4\}\,[\,(0.0808)(492/560)\;\text{lb}m/\text{ft}^3\,]\,(0.38)^3}\;1.91$$

$$= 50\;\frac{\text{lb}f}{\text{ft}^2}$$

$$= 0.35\;\text{psi}$$

□ □ □

Example 6.4-2 Permeability of a sandstone A core (from an underground oil reservoir) 10 cm long and 2.5 cm in diameter is placed in a Hasler core holder. Water flows through this core at the rate of 1 cm³/sec under a pressure drop of 1 atm. What is the average permeability of the unconsolidated sand?

Solution   Integration of the Darcy equation for constant k, μ, and v_∞ and rearranging yields

$$k = \frac{-v_\infty \mu L}{\Delta p} = \frac{-Q}{A} \frac{\mu L}{\Delta p}$$

$$= \left(10 \, \frac{cm^3}{sec} \right) \frac{(1 \, cp) \, (10 \, cm)}{\{ [\pi \, (2.5)^2 / 4] \, cm^2 \} \, (1 \, atm)}$$

$$= 2.04 \, cm^2 \, cp/sec \, atm$$

The unit used to measure permeability is the *darcy*. A porous medium with a permeability of 1 darcy will permit a fluid of 1 cp viscosity to flow at a rate of 1 cm³/sec per cm² of cross section under a pressure drop of 1 atm. The darcy therefore has units as above—cm² cp/sec atm—and our core has a permeability of 2.04 darcies. Since the darcy is a rather large unit for most practical problems, the millidarcy (1/1,000 darcy) is commonly used.

□ □ □

Filters

An important application of the principles of flow in porous media is the operation and design of filters. Both the filter medium (cloth and/or precoat) and the resultant filter cake are porous media. The laminar steady-state velocity at any time through the filter medium and filter cake often can be described by Darcy's law. The resistance of the filter cake will be a function of time or volume throughput since the thickness of the cake is increasing during operation of the filter. The pressure drop across a filter with an incompressible cake can be calculated using Darcy's law as

$$v_\infty = \frac{-k}{\mu} \frac{dp}{dx} \tag{6.4-25}$$

Multiplying and dividing by A:

$$Q = \frac{-k}{\mu} A \frac{dp}{dx} \tag{6.4-26}$$

Integrating in the x direction for constant permeability, area, etc.,

$$Q = \frac{-k}{\mu} A \frac{\Delta p}{\Delta x} \tag{6.4-27}$$

$$\Delta p = -\left(\frac{\mu \, \Delta x}{kA}\right) Q \tag{6.4-28}$$

The term in parentheses represents the resistance to flow.

Since the cake thickness Δx is proportional to V_f (volume of filtrate collected) and the medium (that is, the support, precoat, etc.) has a *constant* resistance, this may be written as

$$\Delta p = (K_1 V_f + K_2) Q \tag{6.4-29}$$

but by definition

$$Q = \frac{dV}{dt} \tag{6.4-30}$$

Thus at constant *pressure drop*:

$$\frac{dV}{dt} = \frac{\Delta p}{K_1 V + K_2} \tag{6.4-31}$$

Integrating from an initial volume $V = 0$ to a final volume V_f, and from $t = 0$ to $t = t_f$,

$$t_f = \frac{1}{2\Delta p} V_f^2 + \frac{K_2}{\Delta p} V_f \tag{6.4-32}$$

Data can be taken for t_f and V_f and t_f / V_f plotted versus V_f to obtain K_1 and K_2, since

$$\frac{t_f}{V_f} = \frac{K_1}{2\,\Delta p} V_f + \frac{K_2}{\Delta p} \tag{6.4-33}$$

is the equation of a straight line with slope $K_1/2\,\Delta p$ and intercept $K_2/\Delta p$.

Considering a constant-*rate* operation:

$$\frac{dV}{dt} = Q = \text{constant} \tag{6.4-34}$$

Therefore integration from $V = 0$ to V_f and from $t = 0$ to t_f yields

$$V_f = t_f Q \tag{6.4-35}$$

and

$$\Delta p = K_1 Q^2 t_f + K_2 Q \tag{6.4-36}$$

A plot of Δp versus t for a given rate Q permits calculation of K_1 and K_2.

Most filter operations have both a constant-rate and a constant-pressure period. At the beginning of a normal cycle the rate is held constant until the pressure builds up to a maximum determined by the equipment. Then the pressure drop is held constant until the rate drops below an economical limit. The cake is then washed at constant pressure, and so the entire cycle consists of a sum of times for the constant-rate operation, constant-pressure operation, washing operation, and dumping of the filter cake.[1]

Example 6.4-3 Operation of a filter at constant rate and at constant pressure A filter is run at constant rate for 20 min. The amount of filtrate collected is 30 gal. The initial pressure drop is 5 psig and the pressure drop at the end of the 20 min is 50 psig. If the filter is now operated for an additional 20 min at a constant pressure of 50 psig, how much filtrate will be collected?

Solution

    CONSTANT RATE

$$Q_0 = \frac{30 \text{ gal}}{20 \text{ min}} = 1.5 \text{ gal/min}$$

At constant rate

$$\Delta p = K_1 Q_0 t_f + K_2 Q_0$$

At $t = 0$, $\Delta p = 5$ psig

$$5 = K_1 (1.5)^2(0) + K_2(1.5)$$

$$K_2 = \frac{5}{1.5} = 3.33$$

At $t = 20$, $\Delta p = 50$ psig

$$50 = K_1 (1.5)^2 (20) + (3.33)(1.5)$$

$$K_1 = \frac{50 - 5}{45} = 1$$

---

[1] For a more extensive treatment of filtration including, for example, continuous filters and compressible cakes, the reader is referred to unit operations texts. An excellent discussion of filtration may be found in W. L. McCabe and J. C. Smith, "Unit Operations of Chemical Engineering," (2d ed.). McGraw-Hill, New York, 1967.

Now we wish to apply the equation for constant pressure drop:

$$t_f = \frac{K_1' V_f^2}{2 \Delta p} + \frac{K_2' V_f}{\Delta p}$$

We must remember that we obtained this equation by integrating the rate relationship assuming that at $t = 0$, $V = 0$. This means that K_2' in this equation represents not just the resistance of the medium, but the resistance of the medium *and the cake deposited during the constant-rate part of the cycle*. It will therefore be different in numerical value from the K_2 for the constant-pressure part of the cycle. The constant K_1' will not differ from the K_1 obtained because the additional resistance is still proportional to V_f in the same manner as before. We append the primes to distinguish the two cases.

We find K_2' using the fact that at the end of the constant-rate period, and therefore at the beginning of the constant-pressure cycle,

$$\frac{dV}{dt} = Q_0 = 1.5 \frac{\text{gal}}{\text{min}} = \frac{\Delta p}{K_1' V_f + K_2'}$$

but $V_f = 0$ and $\Delta p = 50$ psig at the beginning of the constant-pressure portion, and so

$$K_2' = \left(50 \frac{\text{lbf}}{\text{in.}^2}\right)\left(\frac{\text{min}}{1.5 \text{ gal}}\right) = 33.3 \text{ lbf min./in.}^2 \text{ gal}$$

Substituting in the constant-pressure expression gives

$$20 = \frac{1 V_f^2}{(2)(50)} + \frac{33.3 V_f}{50}$$

$$0.01 V_f^2 + 0.667 V_f - 20 = 0$$

$$V_f = \frac{-0.667 \pm \sqrt{0.445 + 0.8}}{0.02}$$

$$V_f = 20.4 \text{ gal} \qquad \text{(The other root is negative)}$$

Thus the total filtrate is

$$30 \text{ gal} + 20.4 \text{ gal} = 50.4 \text{ gal}$$

□ □ □

PROBLEMS

6.1 Calculate the total drag force on a bridge cable which is 8 in. in diameter and extends a distance of 1,000 ft between support towers when a wind is blowing across the line with a velocity of 40 mph. Use properties of air at 60°F.

6.2 Calculate the force acting on a circular bridge pier 5 ft in diameter which is immersed 60 ft in a river which is traveling at 3 mph. River temperature is at 70°F.

6.3 In calculating the drag force on a wing, we might consider it as two flat plates plus half of a cylinder on the leading edge. Estimate the drag force for a wing traveling at 1,000 ft/sec in air at 32°F. Data for air at 32°F, 1 atm: $\rho = 0.081$ lbm/ft³; $\mu = 0.023$ cp.

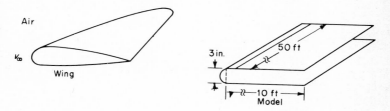

6.4 For the situation sketched, find the required power input to the fluid from the pump.

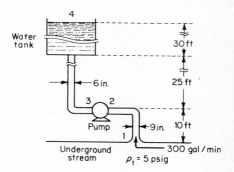

6.5 Water is flowing in a pipe network shown below. The water flows through a process in branch C which for the purpose of calculating pressure drop can be considered as 300 ft of straight pipe. A bypass section is provided with a gate valve so that the water flow to the process can be adjusted. For the system shown, find p_2, p_3, $\langle v_A \rangle$, $\langle v_B \rangle$, and $\langle v_C \rangle$. $\rho = 62.4$; $\mu = 1$ cp.

Branch	Length, ft
A	400
B	400
C	300
D	200

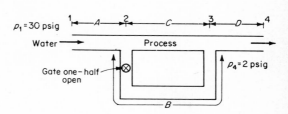

All pipe is horizontal 3-in. schedule 40 steel.

6.6   A 25-hp pump is employed to deliver an organic liquid ($\rho = 50$ lbm/ft^3) through a piping system, shown in the figure below. Under the existing conditions, the liquid flow rate is 4,488 gal/min.

(*a*) Calculate the frictional losses ("lost work") in the piping system between points 1 and 2 in ft lbf/sec.

(*b*) Assuming that the frictional losses vary as the square of the flow rate, calculate the required pump horsepower if the delivery rate is to be doubled.

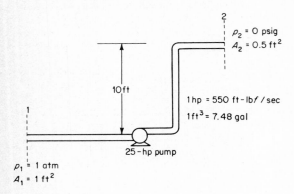

6.7   A fluid with viscosity of 100 cp and density 77.5 lbm/ft^3 is flowing through a 100-ft length of vertical pipe (schedule 40 steel). If the pressure at the bottom of the pipe is 20 psi greater than at the top, what is the mass flow rate?

6.8   Water at 60°F (viscosity 1.05 cp) is being pumped at a rate of 185 gal/min for a total distance of 200 ft from a lake to the bottom of a tank in which the water surface is 50 ft above the level of the lake. The flow from the lake first passes through 100 ft of 3-1/2-in. schedule 40 pipe (ID = 3.548 in.). This section contains the pump, one open globe valve, and three standard elbows. The remaining distance to the tank is through two parallel 100-ft lengths of 2-1/2-in. schedule 40 pipe (ID = 2.469), each including four standard elbows. A reducing tee is used for the junction between the 3-1/2-in. line and the two 2-1/2-in. lines. Calculate the lost work in the piping system.

6.9   Water at 60°F is flowing from a very large constant head reservoir to an irrigation ditch through a 1-1/2-in. schedule 40 cast iron pipe as shown below. All elbows are 90° standard elbows. Find the water flow rate in gallons per minute.

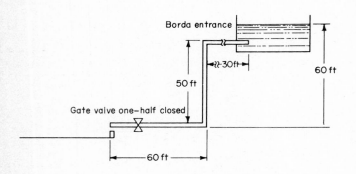

6.10 A large constant head water reservoir feeds two irrigation lines as shown below. If we neglect any kinetic energy change due to an area change of the piping and any losses which are associated with the tee joint at point A, find the bulk velocities in each line.

D_I = 1-1/2-in. schedule 40 commercial pipe

D_{II} = 2-in. schedule 40 commercial pipe

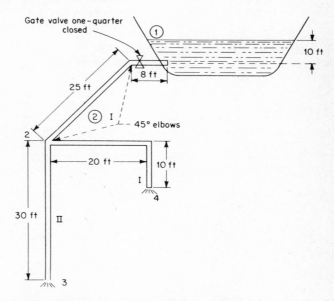

6.11 In many nuclear reactors, the nuclear fuel is assembled in solid form in the form of a lattice through which coolant is pumped. This assembly is usually built up of fuel plates which are thin rectangular prisms. In one such assembly, the openings for coolant are 2.620 in. wide by 0.118 in. The plates are 25 in. long. Manifolding at the ends is such that an infinite diameter may be assumed. Calculate the pressure drops in psi between the inlet and outlet manifolds of the fuel assembly for both water at an average temperature of 100°F and sodium at 400°F, assuming that the surface roughness of the plates is equivalent to drawn tubing and that fluid velocities of 20 ft/sec are used.

μ_{Na} = 0.428 cp

ρ_{Na} = 0.901 g/cm³

6.12 A hypodermic syringe is used to inject acetone at 70°F into a laboratory equilibrium still operating at 5.0 psig. The syringe has a 1.5-in. barrel ID and is connected to the still by 5.0 ft of special hypodermic tubing having an ID of 0.084 in. What injection rate in gallons per minute would you expect for a force of 15 lbs applied to the syringe plunger? State your assumptions.

Barometer = 750 mm Hg

Viscosity = 0.35 cp

Sp.gr.= 0.7

6.13 An industrial furnace uses natural gas (assume 100% CH_4) as a fuel. The natural gas is burned with 20% excess air, and is converted completely to CO_2 and water. The combustion products leave the furnace at 500°F and pass into the stack, which is 150 ft high and 3.0 ft ID. The temperature of the air outside of the stack is 65°F and barometric pressure is 745 mm Hg.

(*a*) If the gases in the stack were standing still, what would be the stack draft (difference in pressure between inside and outside the bottom of the stack)?

(*b*) If the gases were moving upward in the stack at a rate of 10 ft/sec, what would be the stack draft, neglecting friction?

(*c*) Including the effect of friction, estimate the expected flow rate. (The 10 ft/sec given previously no longer applies.) Make whatever simplifying assumptions you may feel necessary.

6.14 Water at 68°F flowing at an average velocity of 9.7 ft/sec through an old rusty pipe having an ID of 4.026 in. gave a net loss of head of 12.4 ft for a 51.5-ft test length. How does the experimental friction factor compare with that calculated for new commercial pipe?

6.15 Shown is an open waste-surge tank with diameter of 30 ft. This tank is fed by a pump supplying a steady 100 gal/min. When the pump is first turned on, the level in the tank is 10.0 ft. The tank drains to the ocean through 3-in. cast iron pipe as shown. Will the tank empty, overflow, or reach a constant level? Support your answer with calculations.

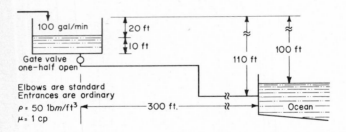

6.16 Calculate the pump horsepower required for the system below if the pump operates at an efficiency of 70 percent.

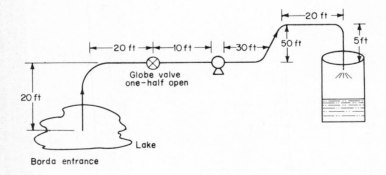

Data: The fluid is water; $\rho = 62.4$ lb$m$/ft^3, $\mu = 1$ cp. The pipe is galvanized iron and is 3-in. ID. $Q = 300$ gal/min. All elbows are standard.

6.17 Water, at 60°F, is to be pumped at 200 gal/min for a total distance of 200 ft from a lake to the bottom of a tank in which the water surface is 50 ft above the level of the lake. The flow from the lake first passes through 100 ft of 3 1/2-in. schedule 40 commercial steel pipe (ID = 3.548-in.). This section contains the pump, one open globe valve, and three standard elbows. The remaining distance to the tank is through *two* parallel 100-ft lengths of 2 1/2-in. schedule 40 cast iron pipe (ID = 2.469 in.) *each* including four standard elbows. A reducing tee is used for the junction between the 3 1/2- and the two 2 1/2-in. lines. Calculate the lost work in the system.

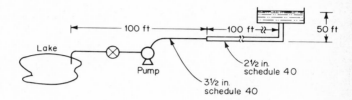

6.18 An experiment is performed to check the equivalent length of a standard elbow in a galvanized iron pipe system. The pressure drop $p_1 - p_2$ between two taps 10 ft on either side of the elbow is found to be 11 psig when the mass flow rate of water is 26 lb*m*/sec. Calculate the equivalent length of the elbow.

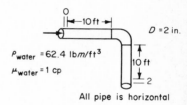

All pipe is horizontal

6.19 Calculate the equivalent diameter of the annular space between a 1 1/2- and a 2-in. schedule 40 pipe.

6.20 A rectangular enclosed channel of asphalted cast iron is used to transport water at 30 ft/sec. Find the pressure drop for each foot of channel length.

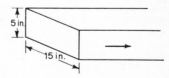

6.21 A conduit whose cross reaction is an equilateral triangle (with side 5 in.) has a pressure drop of 1 psi/ft. How much air at standard conditions will the conduit deliver?

6.22 Calculate the pressure drop for water at 100°F and 1 atm flowing at 500 lb/hr through a bed of 1/2-in. spheres. The bed is 4 in. in diameter and 8 in. high. The porosity (void fraction) of the bed is 0.38.

6.23 A filter is to be operated as follows:

    (*a*) At a constant rate of 2 gal/min until the pressure drop reaches 5 psi, and then

    (*b*) At a constant pressure drop of 5 psi to the end of the run. How long will it take to collect 500 gal of filtrate operating as above? You may make appropriate assumptions. The following data have been taken at a constant rate of 1 gal/min.

t	Δp
100 sec	1 psi
200 sec	3 psi

7
Momentum Transfer Applications

This chapter, like the parallel chapters in heat and mass transfer, is divided into two sections. The first section is a relatively detailed presentation of the determination of pressure-drop data in pipes and development of the friction factor chart. The data are compared with the calculations and equations given in Chap. 6. In addition, a brief discussion of the allied topic, drag reduction, is initiated.

In the second section, we discuss momentum transfer applications (fluid flow) with reference to the natural gasoline plant which is introduced in Chap. 1. Several applications which might be associated with the plant are presented in the form of examples.

As with the parallel chapters, a series of plates is included at the end of the chapter to provide a look at equipment associated with fluid flow.

7.1 EXPERIMENTAL DETERMINATION OF FRICTION FACTORS

We do a great deal with momentum transfer coefficients (that is, friction factors), heat transfer coefficients, and mass transfer coefficients. In

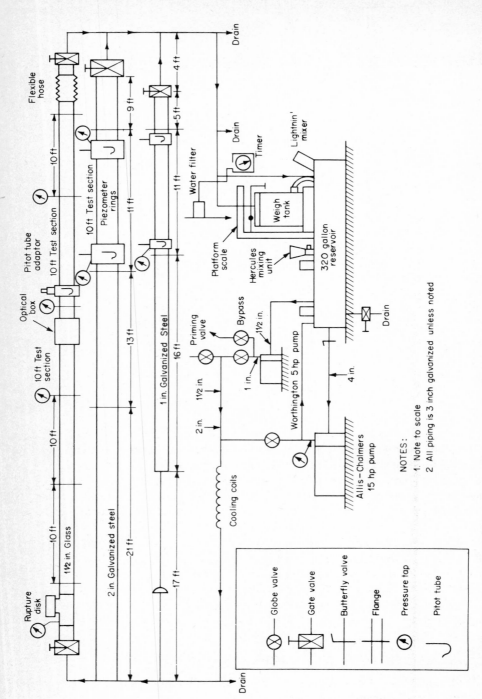

Fig. 7.1-1 Flow loop schematic.

order to give some insight into how these coefficients might be determined in the laboratory, we will include in each appropriate chapter an example of how one experimentally could determine one of these coefficients. By extending the procedures we have shown in these examples, one could develop data for a complete design equation such as we discuss in Chaps. 6, 9, and 12. Below we consider the first of these three examples, how to determine friction factors in a pipe.

In order to develop data for the friction factor chart, we know we must vary Reynolds numbers and relative roughness. Relative roughness could be varied either by producing an artificially rough pipe with known scale[1] or by using different pipes of various but unknown roughness and comparing this data with pipes of known roughness. In other words, we vary the relative roughness by varying the pipe itself.

If we believe dimensional analysis, we can vary the Reynolds number by varying any or all of the following: density or viscosity of the fluid, pipe diameter, or flow rate. For example, we would have precisely the same effect by doubling the flow velocity that we would by maintaining flow velocity and changing the operating fluid to a fluid with half the original viscosity, because the Reynolds number would be the same for each case. Obviously, we vary the thing which is easiest to vary: in this case, the velocity.

To determine friction factor for plotting on a friction factor chart or to determine the constants in one of the design equations one finds the necessary pressure drop to pump a given flow rate of a fluid of known properties through known lengths of various-size pipes. A flow loop for this purpose is shown schematically in Fig. 7.1-1.[2]

Water flows into the storage tank through a filter. The mixing unit is for mixing various additives to the water to study the drag coefficients of more complex solutions. Two centrifugal pumps are used in the system to provide a very wide flow rate range from nearly 0 to 300 gal/min (for water Re = 150 to 250,000). Control of the flow was accomplished by bypassing fluid into the mixing tank.

Table 7.1-1 gives the pipe dimensions. The length of the test section in each case was 10 to 11 ft preceded by a calming section to ensure developed flow. The 1- and 2-in. pipes are steel—schedule 40. The 1 1/2-in. pipe is glass and is smooth. Flow rates were determined by direct timed measurement of the effluent in the weigh tank. The pressure drop over

[1] J. Nikuradse, *VDI-Forschungsh.* **361**:1 (1933).
[2] Friction factors were determined for water (and other fluids) in the system of Fig. 7.1-1 by T. R. Sifferman under the direction of the authors; the discussion of this section is based on these data. (See T. R. Sifferman, "Drag Reduction for Rheologically Complex Fluid in Smooth and Rough Pipes," doctoral dissertation, Purdue University, Lafayette, Ind., 1970.)

Table 7.1-1 Pipe dimensions

Nominal size, in.	Actual ID, in.	Actual OD, in.	Entrance length approx., ft	$L/D_{ent.}$ approx.	Exit length approx., ft	L/D_{exit} approx.
1	1.049	1.315	16	180	5	50
$1\frac{1}{2}$	1.5	1.844	20	160	10	80
			(30)	(240)		
2	2.067	2.375	34	200	9	50

the test section was metered using manometers or pressure transducers depending on the flow rate range. (The piezometer rings in Fig. 7.1-1 are used to provide an average pressure tap.)

Velocity profiles were obtained for the turbulent regime using a pitot tube assembly constructed from a 16-gage (0.065-in.) hypodermic needle. The pitot tube was positioned in the pipe with a micrometer.

Two fluids were used to obtain the data discussed here, water and a white mineral oil. The oil had a specific gravity of 0.89 and a viscosity of 200 cp at 77°F. (All experiments were run at 77°F.)

Tables 7.1-2 to 7.1-4 give the computer results for one experiment in each pipe summarizing the pressure-drop data collected on the flow loop for water in all three pipe sizes. Tables 7.1-5 and 7.1-6 give the computer output summarizing the pressure-drop data for the oil. Figures 7.1-2 to 7.1-4 give the pressure gradient versus flow rate for water. Figure 7.1-5 presents the data for water plotted on a friction factor chart. The line is the smooth pipe reference line from the chart of Fig. 6.2-3. Figures 7.1-6 to 7.1-8 are the pressure gradient versus flow rate data for the oil.

Figures 7.1-9 and 7.1-10 show the power-law velocity profile in the 1-in. pipe at three different Reynolds numbers. The solid line uses Eq. (6.2-19) as the model. The data are replotted in Figs. 7.1-11 and 7.1-12 at all velocities with the solid line using Eq. (6.2-19) as the model. Finally, the data are plotted on Figs. 7.1-13 and 7.1-14 including both rough and smooth pipe where the solid line uses Eq. (6.2-19) as the model.

Drag reduction

An interesting topic is the phenomenon of drag reduction. This phenomenon is of great interest in such diverse applications as the pumping of sewage, the pumping of solid waste, the pumping of concrete, the fracturing of oil wells, the performance of hydrofoils, and the pipeline transport of various materials such as oil products and slurries of solid such as coal. By drag reduction we mean modification of the system in such a way that flow at the same volumetric throughput is obtained but with less pressure drop. A variety of methods is used to accomplish drag reduction. Two examples are the use of Newtonian fluids in two-phase flow, and the incorporation of polymers in single-phase flow, which produces non-Newtonian behavior.

An example of drag reduction unrecognized for many years is that dolphins swim unreasonably fast. Despite the fact that they have extremely streamlined shapes, they move faster through the water than one would normally predict. One explanation that has been given is that apparently dolphins secrete a substance through their skin which modifies the boundary layer surrounding them in such a way that drag reduction is achieved. Such information is obviously of great interest to the Navy, which can envision increasing the top speed of ships by many knots. It is also of interest for solution of such problems as the increasing load of sewage or for increasing amounts of products run through existing pipelines, because, by addition of polymer, one could perhaps increase the capacity of lines without having to construct additional lines.

Table 7.1-2 Experimental data
Exp. No. 51. Diam. = 1.049; the fluid system is water.

Flow rate, gal/min	Press. grad., psi/10 ft	Reynolds no.	Friction factor	Press. instr.
0.05	0.001	155	0.7860	a
0.78	0.004	2,502	0.0148	b
0.78	0.004	2,522	0.0142	a
1.61	0.011	5,258	0.0099	a
2.16	0.018	7,000	0.0090	a
2.84	0.066	8,977	0.0912	b
2.97	0.031	9,564	0.0083	a
4.31	0.107	13,622	0.0136	b
7.16	0.191	22,613	0.0088	b
9.32	0.289	29,429	0.0078	b
10.86	0.392	33,171	0.0078	c
10.87	0.369	34,312	0.0074	b
24.63	1.515	75,243	0.0059	c
34.97	2.887	106,838	0.0056	c
45.30	4.918	138,419	0.0057	c
51.82	6.254	158,336	0.0055	c
67.00	10.976	204,732	0.0058	c
74.26	12.918	226,893	0.0055	c
79.15	14.112	231,853	0.0053	c
82.32	16.250	241,527	0.0057	c

[a] Early calibration of 1 in. of water transducer.
[b] 1.3 sp. gr. oil under water.
[c] Mercury manometer.

Table 7.1-3 Experimental data

Exp. No. 52. Diam. = 2.067; the fluid system is water.

Flow rate, gal/min	Press. grad., psi/10 ft	Reynolds no.	Friction factor	Press. instr.
1.76	0.001	2,945	0.0134	a
2.89	0.001	4,833	0.0103	a
6.85	0.005	11,467	0.0076	a
9.93	0.009	16,426	0.0067	a
11.93	0.050	19,491	0.0244	b
14.50	0.017	23,994	0.0057	b
14.65	0.019	24,238	0.0063	a
16.68	0.025	27,605	0.0062	a
20.71	0.033	34,268	0.0054	a
21.64	0.033	35,353	0.0050	b
42.14	0.099	68,845	0.0039	b
48.86	0.096	79,814	0.0028	b
49.69	0.196	80,246	0.0056	c
62.12	0.146	101,487	0.0027	b
72.42	0.426	116,942	0.0057	b
75.85	0.270	122,481	0.0033	b
90.07	0.426	145,446	0.0037	b
94.19	0.520	152,101	0.0041	b
94.19	0.552	152,101	0.0044	c
137.18	1.123	210,198	0.0042	c
160.05	1.604	245,231	0.0044	c
194.65	2.619	298,254	0.0048	c
244.14	3.421	374,081	0.0040	c
271.78	4.383	416,430	0.0042	c
277.00	4.437	424,438	0.0041	c

a Early calibration of 1 in. of water transducer.
b 1.3 sp. gr. oil under water.
c Mercury manometer.

Table 7.1-4 Experimental data

Exp. No. 53. Diam. = 1.500; the fluid system is water.

Flow rate, gal/min	Press. grad., psi/10 ft	Reynolds no.	Friction factor	Press. instr.
1.84	0.077	3,732	0.3235	b
1.84	0.039	3,732	0.1635	b
2.12	0.004	4,325	0.0130	a
2.12	0.008	4,462	0.0247	a
7.91	0.030	16,166	0.0069	a
8.52	0.032	17,301	0.0062	b
8.52	0.032	17,421	0.0063	a
16.32	0.107	31,206	0.0057	c
19.28	0.140	39,192	0.0053	a
23.60	0.303	45,121	0.0077	c
23.60	0.270	45,121	0.0068	b
30.49	0.321	58,313	0.0049	b
30.49	0.374	58,313	0.0057	c
36.26	0.445	69,329	0.0048	c
49.29	0.766	94,259	0.0044	c
57.34	0.998	109,655	0.0043	c
71.61	1.461	136,933	0.0040	c
80.86	1.728	154,627	0.0037	c
81.80	1.853	158,181	0.0036	c
82.72	1.764	159,096	0.0041	c
83.20	2.013	160,880	0.0039	c
85.67	2.067	163,831	0.0040	c
93.46	2.299	178,724	0.0037	c
100.65	2.619	192,472	0.0036	c
104.30	3.029	199,446	0.0039	c

a Early calibration of 1 in. of water transducer.
b 1.3 sp. gr. oil under water.
c Mercury manometer.

Table 7.1-5   Experimental data
Exp. No. 131. Diam. = 1.049; the system is oil.

Total flow, gal/min	Oil flow, gal/min	Press. grad., psi/10 ft	Percent oil (by vol.)	Press. instr.
10.67	10.67	3.564	100.00	a
18.77	18.77	6.806	100.00	a
23.15	23.15	9.123	100.00	a

[a] Mercury manometer.

Table 7.1-6   Experimental data
Exp. No. 132. Diam. = 2.067; the system is oil.

Total flow, gal/min	Oil flow, gal/min	Press. grad., psi/10 ft	Percent oil (by vol.)	Press. instr.
73.40	73.40	1.550	100.00	a
102.66	102.66	2.049	100.00	a
120.95	120.95	2.940	100.00	a
137.47	137.47	2.655	100.00	a
168.83	168.83	3.688	100.00	a
190.54	190.54	4.312	100.00	a

[a] Mercury manometer.

This is an extremely interesting area of research. At this time, we still do not know the mechanism by which drag reduction is achieved. Further, we do not even know how to correlate the data in any sort of general form.

Another example of the same problem is in the oil business, where it is possible to wrap an annular ring of water around a high-viscosity oil and achieve pumping rates which were completely unattainable in pumping the oil alone. This drag reduction does not depend upon the non-Newtonian character of the fluids, but upon the mechanism of two-phase flow.[1]

[1] A. R. Sinclair, "Rheology of Viscous Fracturing Fluids," presented at the Society of Petroleum Engineers of AIME 44th Annual Fall Meeting, Denver, Colorado, October, 1969, SPE 2623.

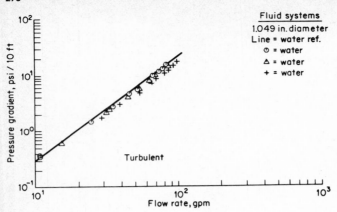

Fig. 7.1-2 Pressure drop for water in 1-in. pipe.

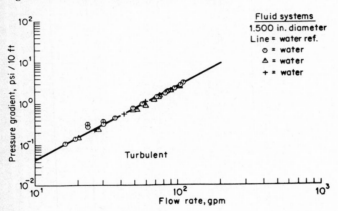

Fig. 7.1-3 Pressure drop for water in 1 1/2-in. pipe.

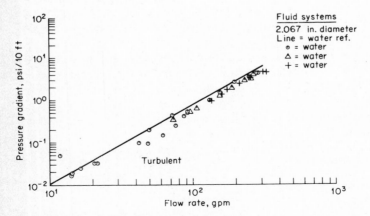

Fig. 7.1-4 Pressure drop for water in 2-in. pipe.

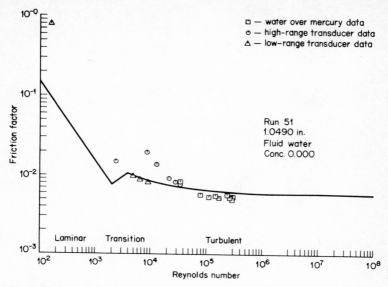

Fig. 7.1-5a Water data—1-ft pipe.

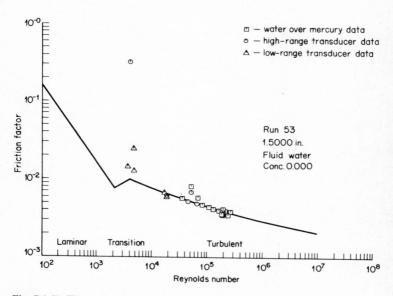

Fig. 7.1-5b Water data —1 1/2-in. pipe.

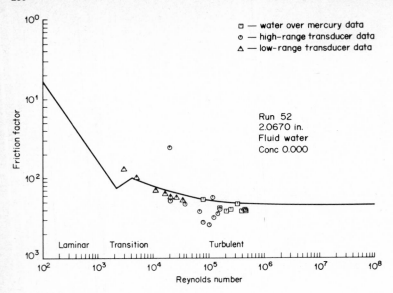

Fig. 7.1-5c Water data—2-in. pipe.

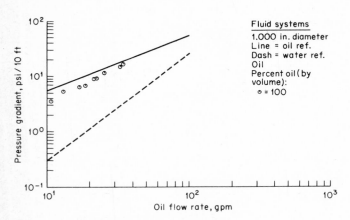

Fig. 7.1-6 Pressure drop for oil in 1-in. pipe.

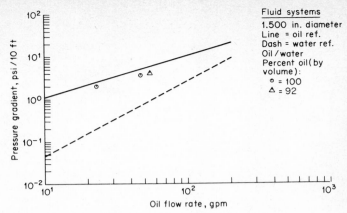

Fig. 7.1-7 Pressure drop for oil in 1 1/2-in. pipe.

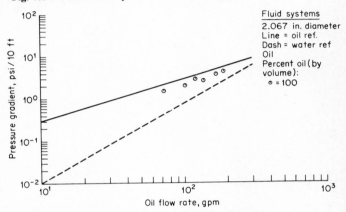

Fig. 7.1-8 Pressure drop for oil in 2-in. pipe.

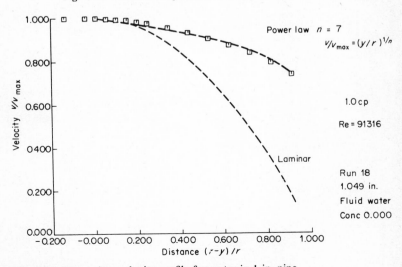

Fig. 7.1-9 Power-law velocity profile for water in 1-in. pipe.

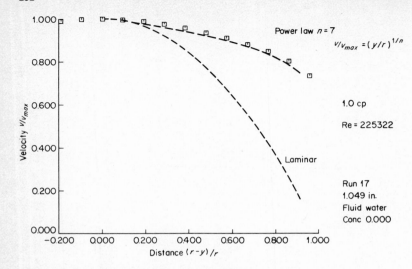

Fig. 7.1-10 Power-law velocity profile for water in 1-in. pipe.

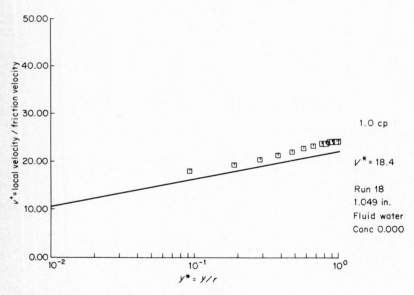

Fig. 7.1-11 Composite velocity profile for water in 1-in. pipe.

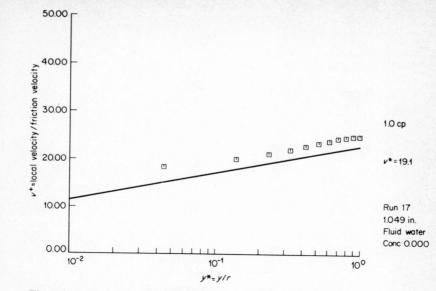

Fig. 7.1-12 Composite velocity profile for water in 1-in. pipe.

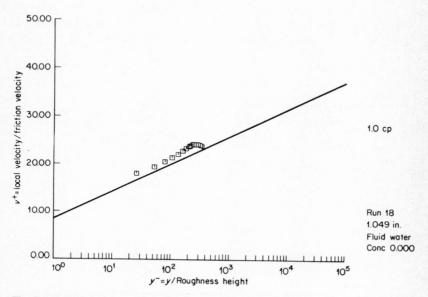

Fig. 7.1-13 Roughness velocity profile for water in 1-in. pipe.

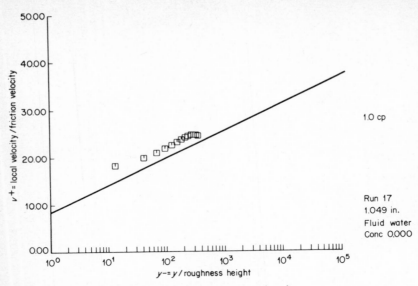

Fig. 7.1-14 Roughness velocity profile for water in 1-in. pipe.

We have taken some pressure-drop data in the flow loop of Fig. 7.1-1 using dilute polymer solutions and oil with an annular ring. Tables 7.1-7 to 7.1-9 summarize some of these data.[1] The data are plotted in

Table 7.1-7 Experimental data

Exp. No. 57. Diam. = 1.049; the fluid system is CMC; conc. 0.3 percent.

Flow rate, gal/min	Press. grad., psi/10 ft	Reynolds no.	Friction factor	Press. instr.
0.71	0.085	2,055	0.4012	a
1.57	0.121	4,389	0.1157	a
4.74	0.194	13,270	0.0203	a
13.46	0.489	37,632	0.0064	a
16.14	0.731	45,136	0.0066	b
35.90	2.174	100,402	0.0040	b
58.52	4.223	163,663	0.0029	b
77.39	6.539	212,201	0.0026	b
77.39	8.802	212,201	0.0035	b
115.15	10.994	315,755	0.0020	b

[a] 1.3 sp. gr. oil under water.

[b] Mercury manometer.

[1] T. R. Sifferman, "Drag Reduction for Rheologically Complex Fluid in Smooth and Rough Pipes," doctoral dissertation, Purdue University, Lafayette, Ind., 1970.

Table 7.1-8   Experimental data

Exp. No. 70.   Diam. = 2.067; the fluid system is Jaguar; conc. 0.030 percent.

Flow rate, gal/min	Press. grad., psi/10 ft	Reynolds no.	Friction factor	Press. instr.
3.14	0.001	4,755	0.0048	a
4.22	0.002	6,392	0.0064	a
8.53	0.007	12,935	0.0068	a
19.57	0.032	29,665	0.0058	a
22.09	0.033	33,487	0.0048	b
62.08	0.215	94,111	0.0039	b
68.22	0.232	96,291	0.0035	c
95.38	0.445	144,594	0.0034	c
96.66	0.446	146,535	0.0033	b
171.37	1.016	241,873	0.0024	c
266.57	2.227	376,247	0.0022	c
276.83	2.013	390,718	0.0018	c
306.28	2.708	432,284	0.0020	c

[a] 1 in. of water transducer.
[b] 1-psi transducer.
[c] Mercury manometer.

Table 7.1-9   Experimental data

Exp. No. 69.   Diam = 1.049; the fluid system is Jaguar; conc. 0.030 percent.

Flow rate, gal/min	Press. grad., psi/10 ft	Reynolds no.	Friction factor	Press. instr.
0.31	0.001	929	0.0137	a
1.03	0.004	3,035	0.0086	a
1.82	0.014	5,326	0.0102	a
2.52	0.024	7,365	0.0090	a
7.18	0.153	21,006	0.0070	b
9.52	0.267	26,179	0.0070	c
12.70	0.389	37,171	0.0057	b
18.80	0.709	53,345	0.0044	c
19.40	0.695	55,028	0.0047	b
23.00	0.958	67,335	0.0043	b
35.27	1.764	93,584	0.0033	c
54.72	3.777	145,180	0.0030	c
71.29	6.254	192,840	0.0028	c
72.68	6.308	208,671	0.0029	c
88.29	8.873	234,547	0.0027	c
95.94	11.208	254,548	0.0029	c
101.34	12.651	268,889	0.0029	c

[a] 1 in. of water transducer.
[b] 1-psi transducer.
[c] Mercury manometer.

Figs. 7.1-15 to 7.1-18. If we compare these figures with the corresponding data of Sec. 7.1 on water and the water reference line we see the evidence of reduction of the drag coefficient.

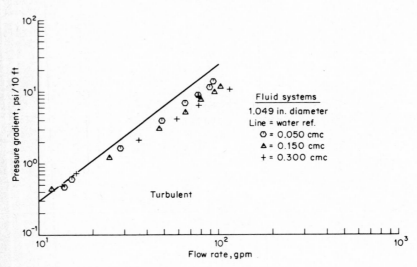

Fig. 7.1-15 Pressure drop for CMC in 1-in. pipe.

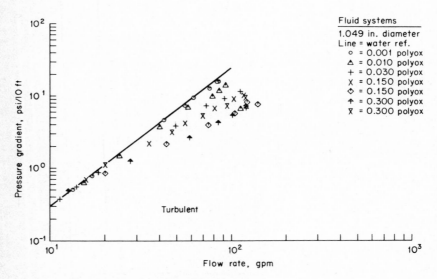

Fig. 7.1-16 Pressure drop for POLYOX in 1-in. pipe.

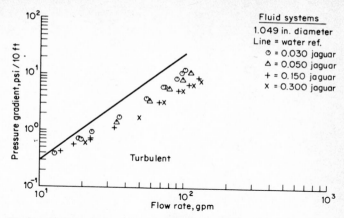

Fig. 7.1-17 Pressure drop for Jaguar in 1-in. pipe.

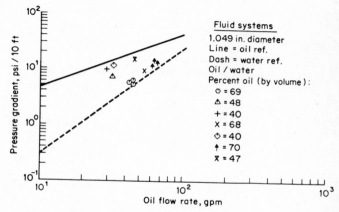

Fig. 7.1-18 Pressure drop for oil/water in 1-in. pipe.

7.2 SELECTED APPLICATIONS

Example 7.2-1 Pipe network There are two general classes of problems that often face the engineer: first, the design of new equipment, and second, the modification of existing equipment to perform a new function. A frequent problem which faces the person who operates a process plant is how to modify the existing plant (for which one can no longer change the basic design) in light of changing market demands and to improve plant operations. In Fig. 1.3-1 it would not be unlikely that equal-size lines might have been installed at one time to supply cooling water to the lean oil cooler and the product cooler. In examining the flowsheet, we see that the heat load on the product cooler is only one-tenth that of the lean oil

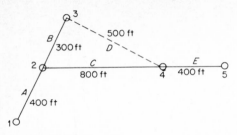

Fig. 7.2-1 Pipe network.

cooler. Since these units run more or less around the clock, it would be desirable that they be supplied with lines which have more nearly appropriate capacity. One could, of course, simply throttle one of the existing lines; however, this means a continual friction loss across the valve. There are obviously a number of possible ways that the capacity of one line could be increased at the expense of another. One of these is to add a parallel line to the network. (See Example 6.2-6.) Another is to add a new line to the network as below. If one were faced with this situation in practice, one would evaluate many alternatives and not the single one which we evaluate below. The decision would eventually be an economic one involving the balance between the capital cost of the installation and the operating cost savings.

Problem statement (Refer to Fig. 7.2-1)   A horizontal pipeline system, composed of 6-in.-ID pipe, carries cooling water at about 60°F for the lean oil cooler and the product cooler. The product cooler needs much less cooling water than does the lean oil cooler. (Lengths given are equivalent lengths; that is, they include bends, fittings, etc.) Outlets exist at points 3 (lean oil cooler) and 5 (product cooler) and the outlet pressures may both be taken to be 0 psig. The inlet is supplied by a pump capable of maintaining a 30-psig pressure. It has become necessary to increase the flow supplied at point 3, with less flow required at point 5. If a 6-in. line is installed from point 4 to point 3 as indicated, what will be the flow rate in this branch? Assume $f = 0.015$ over the entire system.

Solution   We can apply the mechanical energy balance to each section of pipe:

$$\frac{p_2 - p_1}{\rho} = (-\widehat{lw})_A = \frac{-2fL_A \langle v_A \rangle^2}{D}$$

$$\frac{p_3 - p_2}{\rho} = (-\widehat{lw})_B = \frac{-2fL_B \langle v_B \rangle^2}{D}$$

$$\frac{p_4 - p_2}{\rho} = (-\widehat{lw})_C = \frac{-2fL_C \langle v_C \rangle^2}{D}$$

$$\frac{p_3 - p_4}{\rho} = (-\widehat{lw})_D = \frac{-2fL_D \langle v_D \rangle^2}{D}$$

$$\frac{p_5 - p_4}{\rho} = (-\widehat{lw})_E = \frac{-2fL_E \langle v_E \rangle^2}{D}$$

This is a system of five independent equations with seven unknowns $\langle v_A \rangle$, $\langle v_B \rangle$, $\langle v_C \rangle$, $\langle v_D \rangle$, $\langle v_E \rangle$, p_2, p_1. We obtain the other two equations required by applying the total mass balance first around points 2 and 4,

(a) $\langle v_A \rangle A\rho - \langle v_B \rangle A\rho - \langle v_D \rangle A\rho - \langle v_E \rangle A\rho = 0$

and then around point 4, giving

(b) $\langle v_C \rangle A\rho - \langle v_D \rangle A\rho - \langle v_E \rangle A\rho = 0$

(We could replace either of the above equations by a balance around point 2 but only two of the three equations are independent—(a) minus (b) yields the balance around point 2.)

We now have a system of seven equations in seven unknowns. The equations are nonlinear algebraic equations and may be solved by some reasonably sophisticated numerical procedures. However, we can proceed for this case simply by trial and error. The equations reduce to

(c) $(p_2 - 30) = -0.162 \langle v_A \rangle^2$

(d) $(0 - p_2) = -0.121 \langle v_B \rangle^2$

(e) $(p_4 - p_2) = -0.324 \langle v_C \rangle^2$

(f) $(0 - p_4) = -0.202 \langle v_D \rangle^2$

(g) $(0 - p_4) = -0.162 \langle v_E \rangle^2$

(a) $\langle v_A \rangle - \langle v_B \rangle - \langle v_D \rangle - \langle v_E \rangle = 0$

(b) $\langle v_C \rangle - \langle v_D \rangle - \langle v_E \rangle = 0$

Assume $p_2 = 10$, $p_4 = 10$.

(c) $\rightarrow \quad -20 = -0.162 \langle v_A \rangle^2$

$\quad\quad \langle v_A \rangle = \sqrt{123} = 11.1$

(d) $\rightarrow \quad -10 = -0.121 \langle v_B \rangle^2$

$\quad\quad \langle v_B \rangle = \sqrt{82.7} = 9.1$

$(e) \rightarrow \quad 0 = \langle v_C \rangle$

$(f) \rightarrow \quad -10 = -0.202 \, \langle v_D \rangle^2$
$\qquad \langle v_D \rangle = 7.04$

$(g) \rightarrow \quad -10 = -0.162 \langle v_E \rangle^2$
$\qquad \langle v_E \rangle = 7.85$

$(a) \rightarrow \quad 11.1 - 9.1 - 7.04 - 7.85 \overset{?}{=} 0$
$\qquad\qquad\qquad\qquad\qquad -12.89 \overset{?}{=} 0$

$(b) \rightarrow \quad 0 - 7.04 - 7.85 \overset{?}{=} 0$
$\qquad\qquad\qquad\qquad -14.89 \overset{?}{=} 0$

Assume $p_2 = 10$, $p_4 = 2$.

$(c) \quad 10 - 30 = -0.162 \langle v_A \rangle^2$
$\qquad \langle v_A \rangle = 11.1$

$(d) \quad \langle v_B \rangle = 9.1$

$(e) \quad 2.10 = -0.324 \langle v_C \rangle^2$
$\qquad \langle v_C \rangle = 4.97$

$(f) \quad -2 = -0.202 \langle v_D \rangle^2$
$\qquad \langle v_D \rangle = 3.15$

$(g) \quad -2 = -0.162 \langle v_E \rangle^2$
$\qquad \langle v_E \rangle = 3.5$

$(a) \quad 11.1 - 9.1 - 3.15 - 3.5 \overset{?}{=} 0$
$\qquad\qquad\qquad\qquad -4.55 \overset{?}{=} 0$
$(b) \quad 4.97 - 3.15 - 3.5 \overset{?}{=} 0$
$\qquad\qquad\qquad\qquad -1.68 \overset{?}{=} 0$

Assume $p_2 = 6$, $p_4 = 2$.

$(c) \quad -24 = -0.162 \langle v_A \rangle^2$
$\qquad \langle v_A \rangle = 12.2$

$(d) \quad -6 = -0.121 \langle v_B \rangle^2$
$\qquad \langle v_B \rangle = 7.05$

$(e) \quad -4 = -0.325 \langle v_C \rangle^2$
$\qquad \langle v_C \rangle = 3.5$

$(f) \quad -2 = -0.202 \langle v_D \rangle^2$
$\qquad \langle v_D \rangle = 3.15$

$(g) \quad -2 = -0.162 \langle v_E \rangle^2$
$\qquad \langle v_E \rangle = 3.52$

(a) $12.2 - 7.05 - 3.15 - 3.52 \overset{?}{=} 0$
$$- 1.52 \overset{?}{=} 0$$

(b) $3.5 - 3.15 - 3.52 \overset{?}{=} 0$
$$- 3.17 \overset{?}{=} 0$$

Assume $p_2 = 6$, $p_4 = 1$.

(c) $\langle v_A \rangle = 12.2$

(d) $\langle v_B \rangle = 7.05$

(e) $-5 = -0.324 \langle v_C \rangle^2$
$\langle v_C \rangle = 3.93$

(f) $-1 = -0.202 \langle v_D \rangle^2$
$\langle v_D \rangle = 2.23$

(g) $-1 = 0.162 \langle v_E \rangle^2$
$\langle v_E \rangle = 2.52$

(a) $12.2 - 7.05 - 2.23 - 2.52 \overset{?}{=} 0$
$$+ 0.4 \overset{?}{=} 0$$

(b) $3.93 - 2.23 - 2.52 \overset{?}{=} 0$
$$- 0.82 \overset{?}{=} 0$$

This is sufficiently accurate for our purposes.

$Q_D = \langle v_D \rangle A$

$$= \left(2.23 \, \frac{\text{ft}}{\text{sec}}\right) \left[\frac{\pi \left(\frac{1}{2}\right)^2}{4} \, \text{ft}^2\right] \left(\frac{60 \, \text{sec}}{\text{min}}\right) \left(\frac{7.48 \, \text{gal}}{\text{ft}^3}\right) = 197 \, \text{gal/min}$$

$$\left(Q_A = \frac{12.2}{2.23} \times 197 = 1{,}080 \, \text{gal/min}\right)$$

□ □ □

In order to evaluate all the alternatives that would increase capacity in one branch of a pipe network with respect to another, one would have to consider a large number of cases and consider the capital cost required to construct pipe paths for each of these alternatives. It cannot be considered here, but in practice one of the things which would determine the pipeline path is the number of units which were in the way and the cost of supporting or the cost of burying the pipeline. Therefore, construction costs per foot of pipe for pipe and supports will not be constant for different paths.

Also in real systems the friction loss in cooling water piping systems increases over the life of the system due to scaling, fouling, and tuberculation of the pipe and this would have to be taken into account.

Example 7.2-2 Pump characteristics of a centrifugal pump One method of supplying a relatively constant pressure to water lines used for fire fighting, drinking water, sanitary use, etc.—that is, potable water—is to maintain a rather large reservoir of such water at sizable elevation in a standpipe, more commonly known as a water tower. The idea in using a standpipe is to use a smaller pump which runs more or less continuously to fill the standpipe although the use of water will vary widely at different times during the day—plus the fact that the standpipe will also furnish a constant pressurized supply which is independent of any pump in the system so that water for fire fighting is independent of electrical power.

The kind of pump one would probably utilize for such an application is a centrifugal pump, which is typical of the class of pumps most widely used in plants.   Applications include booster pumps on the heat exchangers, product transfer pumps, etc. A centrifugal pump consists of an impeller with vanes mounted on a shaft. As the impeller rotates, the centrifugal forces developed in the water contained in the case or volute hurl the water from its entrance at the center of the impeller to the periphery where it exits into the pipe it supplies. One of the advantages of the centrifugal pump is that one can throttle the outlet. As a consequence of throttling the fluid tends to recirculate within the casing of the pump and heat up to some degree. However, fairly extensive degrees of throttling are possible without this becoming a serious problem.   The result of throttling a centrifugal pump is expressed in the form of a "char-

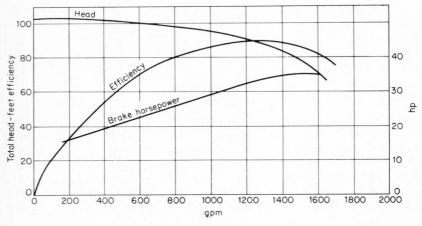

Fig. 7.2-2 Characteristic curves for typical centrifugal pump. (*Ingersoll-Rand Co.*)

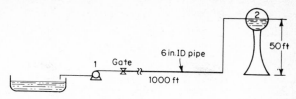

Fig. 7.2-3 Pumping system. Length given is equivalent length.

acteristic curve" of the pump, which gives the output in terms of head (or pressure at outlet divided by the density of the fluid being pumped) versus the volumetric throughput of the liquid. Figure 7.2-2 is a characteristic curve of a centrifugal pump. A centrifugal pump does not supply a constant volumetric flow of water but rather the volumetric flow supplied depends upon the hydraulic resistance of the system to which the water is supplied. An example follows of how one uses the characteristic curve of a centrifugal pump in determining the flow rate in a pipe network which the pump supplies.

Problem statement Water is supplied to a standpipe by a pump with characteristic curve as indicated in Fig. 7.2-2. If the pipeline is as shown in Fig. 7.2-3 what will the rate of water be if the gate valve is full open?

Solution Assume $\mu = 1$ cp, $\rho = 62.4$ lb$m$/ft^3. A mechanical energy balance from 1 to 2 yields

$$\Delta \frac{\langle v \rangle^2}{2} + g\,\Delta z + \frac{\Delta p}{\rho} = -\widehat{lw} - \widehat{W}$$

$$(50 \text{ ft})\left(32.2\,\frac{\text{ft}}{\text{sec}^2}\right)\left(\frac{\text{lb}f\,\text{sec}^2}{32.2\,\text{lb}m\,\text{ft}}\right)$$

$$+ \left(\frac{0-p_1}{62.4}\right)\left(\frac{\text{lb}f\,\text{ft}^3}{\text{in.}^2\,\text{lb}m}\right)\left(144\,\frac{\text{in.}^2}{\text{ft}^2}\right)$$

$$= -(2f)(1{,}000 \text{ ft})\,v^2\left(\frac{\text{ft}^2}{\text{sec}^2}\right)\left(\frac{\text{lb}f\,\text{sec}^2}{32.2\,\text{lb}m\,\text{ft}}\right)\left(\frac{1}{\frac{1}{2}\,\text{ft}}\right)$$

(a) $50 - 2.31\,p_1 = -124.4\,fv^2$
(b)   Determine p_1 from Fig. 7.2-2 as a function of $Q(\langle v \rangle A)$
(c)   Determine f from Fig. 6.2-3 as a function of v

(Assume pipe is smooth.)

Therefore we have three equations in three unknowns.

Proceed as follows: (1) Assume $\langle v_1 \rangle$
              (2) Determine Re and f from Fig. 6.2-3

(3) Determine p_1 from Fig. 7.2-2

(4) Put in equation (a) to see if it checks.

Thus: (1) Assume $\langle v \rangle = 10$ ft/sec

(2) $\text{Re} = \dfrac{(1/2)\ (10)\ (62.4)}{(1)\ (6.72 \times 10^{-4})} = 4.64 \times 10^5 \quad f = 0.0033$

from Fig. 6.2-3

(3) $Q = \langle v \rangle A = \left(10\ \dfrac{\text{ft}}{\text{sec}}\right) \left(\dfrac{\pi}{4}\right)\left[\left(\dfrac{1}{2}\right)^2\ \text{ft}^2\right] \left(7.48\ \dfrac{\text{gal}}{\text{ft}^3}\right) \left(60\ \dfrac{\text{sec}}{\text{min}}\right)$

$= 800$ gal/min

$\dfrac{\Delta p}{\rho} = 95$ ft lbf/lbm from Fig. 7.2-2

$\Delta p = \left(95\ \text{ft}\ \dfrac{\text{lbf}}{\text{lbm}}\right) \left(62.4\ \dfrac{\text{lbm}}{\text{ft}^3}\right) \left(\dfrac{\text{ft}^2}{144\ \text{in.}^2}\right) = 41.2\ \text{lbf/in.}^2$

(4) $50 - (2.31)\ (41.2) = -(124.4)\ (0.0033)\ (100)$

$- 45.2 \neq -41.2$

Making a new assumption:

(1) Assume $\langle v_1 \rangle = 5$ ft/sec

(2) $\text{Re} = 2.3 \times 10^5 \quad f = 0.0037$ from Fig. 6.2-3

(3) $Q = 800\ \dfrac{5}{10} = 400$ gal/min

$\dfrac{\Delta p}{\rho} = 102$ ft lbf /lbm from Fig. 7.2-2

$\Delta p = 41.2\ \dfrac{102}{75} = 44.2\ \text{lbf/in.}^2$

(4) $50 - (2.31)\ (44.2) = -(124.4)\ (0.0037)\ (100)$

$- 52 \neq -46.0$

Making a new assumption:

(1) Assume $\langle v_1 \rangle = 3$ ft/sec

(2) $\text{Re} = 1.39 \times 10^5 \quad f = 0.0042$ from Fig. 6.2-3

(3) $Q = 800\ \dfrac{4}{10} = 320$ gal/min

$\Delta p = 41.2\ \dfrac{103}{95} = 44.6$

(4) $50 - (2.31)(44.6) = -(124.4)(0.0042)(100)$

$-53 \cong -52.7$

$\therefore Q = 320$ gal/min

□ □ □

There are many types of pumps which might be selected for the same application; however, a centrifugal would be by and large the most common choice. One could consider positive displacement pumps as exemplified by the piston pumps, gear pumps, lobe pumps, etc. However, these pumps cannot be throttled, and they would be less likely to be used because in general they would cost more and require more maintenance.

The natural gasoline plant of Fig. 1.3-1 uses a recirculating cooling water system where water is pumped into a storage basin, cooled in a cooling tower, pumped through the heat exchangers, back over the cooling tower, and then disposed. The water may be treated or if it is brackish (salt), it may be disposed in underground wells.

Example 7.2-3 Pressure drop for gate versus globe valve One of the problems that face engineers when they pump things through pipes is that somehow they must always control the flow rate and pressure. An initial reaction might be to control the flow rate by controlling the speed of the pump; however, this is almost never done because most motors are designed for constant speed. Rather, some sort of valve is usually introduced in the line which, by adding frictional loss to the fluid, will control the flow rate and pressure obtained from the pump (this must never, of course, be done for positive displacement pumps). Many valves are used mostly for on-off service; that is, the valve will be either completely open or completely closed—as, for example, for isolation of a heat exchanger or some other unit in order to maintain it by changing gaskets, cleaning, etc. Some valves such as globe valves, needle valves, etc., are designed for throttling use. (See the plates at the end of the chapter.) However, valves designed for throttling almost always have a larger pressure drop when full open than do valves designed for on-off service. In the example below, we compare the consequences of using a globe valve and a gate valve for open service.

Problem statement A side stream of product from the natural gasoline plant, Fig. 1.3-1, at 90°F is flowing through a horizontal 1 1/2-in. schedule 40 steel pipe at the rate of 40 gal/min. The line contains two standard-radius 90° elbows, one tee (used as an elbow with the liquid entering the run), and two valves. The length of straight pipe in the system is 50 ft.

The density of the product at 90°F is 53 lb$m$/ft^3 and the viscosity is 0.5 cp. Compute the pressure loss in the pipe when gate valves are installed. Compare with the pressure loss when globe valves are installed. (The valves are fully open.)

Solution

For 1 1/2-in. schedule 40 pipe $\quad D = 1.61$ in. $= 0.134$ ft

From Table 6.2-1 $\qquad\qquad\qquad A = 0.01414$ ft^2

$$\langle v \rangle = \left(40 \; \frac{\text{gal}}{\text{min}}\right)\left(\frac{\text{ft}^3}{7.48 \; \text{gal}}\right)\left(\frac{\text{min}}{60 \; \text{sec}}\right)\left(\frac{1}{0.01414 \; \text{ft}^2}\right) = 6.3 \; \text{ft/sec}$$

$$\text{Re} = \frac{D\langle v \rangle \rho}{\mu} = \frac{(0.134)(6.3)(53)}{(0.5)(6.72 \times 10^{-4})} = 1.34 \times 10^5$$

$$\frac{k}{D} = 0.0011 \; \text{from Fig. 6.2-4}$$

$f = 0.0055$ from Fig. 6.2-3

Equivalent lengths of fittings, from Fig. 6.2-5

WITH GATE VALVES

2 elbows (2)(4)	= 8
1 tee (cross)	= 8
2 gate valves (open)(2)(1)	= 2
Equivalent length	= 18 ft

$L = 50 + 18 = 68$ ft

$$\widehat{lw} = \frac{2fL\langle v \rangle^2}{D} = \frac{(2)\,(0.0055)\,(68)\;\text{ft}\;(6.31)^2\;\text{ft}^2}{0.134\;\text{ft}} \frac{\text{lb}f\;\text{sec}^2}{32.2\;\text{lb}m\;\text{ft}}$$

$$= 6.9 \; \text{ft lb}f/\text{lb}m$$

Mechanical energy balance: $\Delta \dfrac{\langle v \rangle^2}{2} + g\,\Delta z + \dfrac{\Delta p}{\rho} = -\widehat{lw} - \widehat{W}$

$$\Delta p = -\widehat{lw}\rho = -(6.9)\,(53) = -365 \; \text{lb}f/\text{ft}^2 = -2.53 \; \text{lb}f/\text{in.}^2$$

WITH GLOBE VALVES

2 globe valves (open) (2)(40.5) = 81

Equivalent length = $81 + 18 - 2 = 97$ ft

$L = 50 + 97 = 147$ ft

$$\widehat{lw} = 6.9 \; \frac{147}{68} = 14.9 \; \text{ft lb}f/\text{lb}m$$

$$\Delta p = -(14.9)(53) = -795 \text{ lb}f/\text{ft}^2 = -5.52 \text{ lb}f/\text{in.}^2$$

$$\Delta p \text{ increase} = \frac{795 - 365}{365} \ 100 \text{ percent} = 118 \text{ percent}$$

□ □ □

Notice that we get a significant increase in pressure drop by using the "wrong" type of valve in the line. This increased pressure drop might or might not be economically significant depending upon a number of things: the cost of power, the possibility that one might wish in the future to throttle the valve, whether there are suspended solids in the flowing stream (in which case the tortuous path through the globe valve might very well plug up where the gate valve would provide satisfactory service), etc. It is not good practice to throttle with a gate valve for several reasons: the valve wears, the pressure drop is high, and the valve opening versus stem position is nonlinear, making control awkward.

Example 7.2-4 Optimization of pipe diameter Since the cost of process piping is frequently a large percentage of the cost of a process plant, in design it is frequently worthwhile to worry about optimizing the pipe size used. For example, the larger the pipe diameter, the less the pressure drop involved in pumping fluid through the pipe; however, the greater the capital and installation cost for installation of the pipe. Obviously, there must be some sort of compromise between these two factors. Specific optimization for the case of very large piping systems, such as cross-country pipelines or pipe systems made of high-cost materials of construction, is usually calculated on a digital computer using a specific optimization routine for the particular system. However, in many cases generalized charts can be used to select the economic pipe diameter. In Example 7.2-2 where we have a 6-in. pipe which runs a thousand feet, there will be a significant investment involved in the purchase of the pipe. There will also be a significant day-to-day operating cost for pumping. In that example, we made no effort to optimize the pipe diameter. However, based on one of the simpler rule-of-thumb optimizations which is presented in the nomograph of Fig. 7.2-4,[1] let us consider what an economic pipe diameter for that case might be. The nomograph is developed from a fairly complicated equation.

[1] R. H. Perry, C. H. Chilton, and S. D. Kirkpatrick (eds.), "Chemical Engineers Handbook," 4th ed., p. 5–29, McGraw-Hill Book Company, New York, 1963.

Problem statement   Determine the economic pipe diameter for the pipe in Example 7.2-2, assuming the same flow rate. (Use the nomograph, Fig. 7.2-4.)

Solution   From the figure the optimum diameter is 7.6 in.; therefore we would probably use an 8-in. pipe.   Notice that calculations from the nomograph are relatively crude.   A next step would be to take into account current material and labor costs as well as current power costs.   It may also be necessary to consider the effect of fouling on friction factors. The combination of optimization when changing a pipe network such as that of Example 7.2-2 becomes very complex and would be solved by computer.

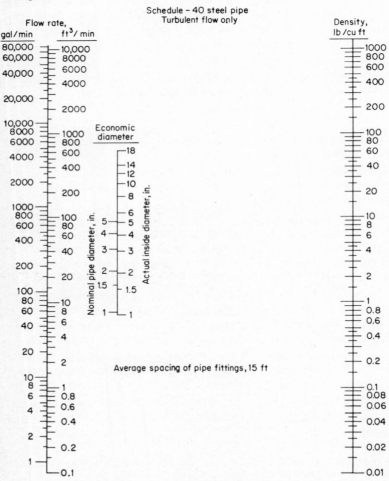

Fig. 7.2-4 Economic pipe diameters. (*E. I. duPont de Nemours and Co.*)

□ □ □

APPENDIX 7A

Fig. 7A-1 Centrifugal pump. (*Allis-Chalmers.*)

Fig. 7A-2 Centrifugal pump. (*Allis-Chalmers.*)

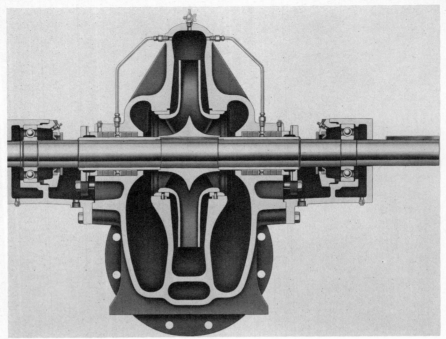

Fig. 7A-3 Centrifugal pump (section view). (*Allis-Chalmers.*)

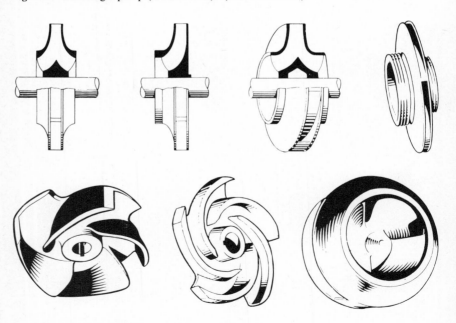

Fig. 7A-4 Centrifugal pump (impellers). (*Allis-Chalmers.*)

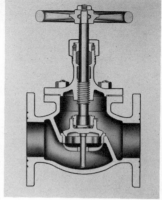

Fig. 7A-5 Globe valve. (*Crane Co.*)

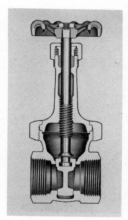

Fig. 7A-6 Gate valve. (*Crane Co.*)

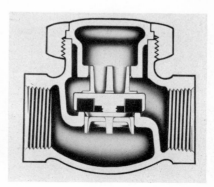

Fig. 7A-7 Check valve. (*Crane Co.*)

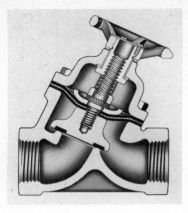

Fig. 7A-8 Diaphragm valve. (*Crane Co.*)

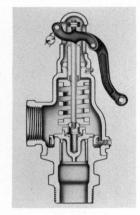

Fig. 7A-9 Relief valve. (*Crane Co.*)

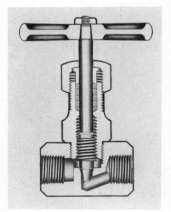

Fig. 7A-10 High-pressure valve. (*Crane Co.*)

8
Heat Transfer

In construction of the overall energy balance, we defined a term called "heat" which was energy transferred as a result of a temperature difference. In the treatment we did not make any statement about how the *rate* of transfer of the various types of energy (work, heat, internal, potential, and kinetic) related to the physical parameters of the situation—we only specified that the total energy must be constant, that is, that energy be a *conserved* quantity. By the physical parameters of the system we mean such things as velocity, temperature, and concentration gradient. Since we add and remove energy with respect to the system at rates which depend on these parameters, the overall mass, energy, and momentum balances are seldom *in themselves* sufficient conditions to give a unique answer to our problem.

To define our solution uniquely we must develop further equations which relate the rate of transfer of each particular type of energy to the physical parameters of our system. In Chap. 5 we related the conversion of mechanical energy into either heat or internal energy to the concept of "lost work," which was related to *momentum* transfer using Newton's law of viscosity:

$$\tau_{zy} = -\mu \frac{dv_z}{dy} \tag{8-1}$$

We now proceed to develop an analogous equation for *heat* transfer which is (in its one-dimensional form)

$$\left(\frac{Q}{A}\right)_y = q_y = -k \frac{dT}{dy} \tag{8-2}$$

where Q is heat flow and q is the heat flux. The above equation is known as Fourier's law, and is a phenomenological law; that is, it is true not because of our ability to derive it from theory, but because we define the thermal conductivity as the number which makes the equation true, and, conveniently, by doing this the thermal conductivity is relatively constant. In fact, some efforts[1] have been made to find higher-order dependence of heat transfer upon the temperature gradient, but such efforts to date have been unsuccessful. The isotropic thermal conductivity k, as will be mentioned later, is a function of pressure and (primarily) of temperature.[2]

Equation (8-2) describes heat transfer by *conduction*. Elementary physics teaches us that heat transfers by three mechanisms: conduction, convection, and radiation.

Conduction is the transfer of energy on a molecular scale independent of net bulk flow of the material. For example, in a solid the molecules are held in a more or less rigid structure, so that there is no large amount of molecular translation from point to point. As long as the molecules have significant amounts of thermal energy, however, they do vibrate, with the vibrations increasing in amplitude with the thermal energy level. These vibrations can be transmitted from one layer of molecules to an adjacent layer of molecules at a lower thermal energy level, which is one of the ways in which heat conduction takes place. Another mechanism of heat conduction on a molecular level involves the so-called free electrons which are more abundant in metals than other substances. The "free" electrons are electrons which are very loosely bound to atoms and are relatively free to move about in the solid. Sometimes these electrons are referred to as "electron gas." These electrons are excellent carriers of both thermal energy and electrical energy and, therefore, substances

---

[1] R. W. Flumerfeld and J. C. Slattery, *A. I. Ch. E. J.*, **15**:291 (1969).

[2] For cases where the thermal conductivity varies with direction in the material (anisotropy) the thermal conductivity becomes a tensor; we will treat in this text only materials which may be considered isotropic. An example of an anisotropic material is a fibrous material such as wood, where the resistance to heat conduction varies depending on whether one is going along the grain or across the grain.

which have large numbers of free electrons are good conductors of heat as well as good conductors of electricity, for example, silver and aluminum. In liquids, molecules also vibrate as a result of thermal energy. In liquids, however, they are not held in a rigid lattice structure and the molecules can undergo a process of molecular diffusion, which is the interchange of molecules on a molecular scale. In this process, energy is not transferred directly from one molecule to the next, but rather by the interchange of a high-energy molecule with a low-energy molecule. If one looks at the problem microscopically, that is to say, on a molecular level, some of these mechanisms of conduction look much like what we define below as convection. The question is really one of *scale*. In convection we are concerned with *groups* of molecules which contain literally many millions of molecules, and so we simply do not examine the problem at the molecular level. A classical example of conduction of energy in a solid is that of transfer of energy down a poker thrust into a fire; a more literary example is a cat on a hot tin roof.

Convection is the transfer of energy from place to place by a bulk flow of matter from a hotter region to a colder, or vice versa. Convection is usually subdivided into two categories: *free convection* and *forced convection*. By forced convection, we mean convection which is induced by the introduction of some prime mover such as a pump, fan, blower, etc., into the system. By free convection, we mean convection which results from density differences within the system. This convection, for example, can result from thermal gradients or concentration gradients. Each of these mechanisms of convection can give rise to either laminar or turbulent flow, although turbulent flow is perhaps more common in forced convection problems than in free convection problems. Again, the classification of heat transfer as by convection is largely a question of scale. In fact, if we wish to be rigorous, what we refer to as heat transfer by convection is a sequential process. In a three-step process, a mass of material receives energy by conduction, then moves by bulk flow from a hotter region to a colder region, at which point it gives up the acquired energy by conduction to the colder surrounding fluid. Typical examples illustrating convective heat transfer are forced air furnaces in homes and transfer of energy on the earth by large-scale movements of the atmosphere.

Heat transfer by *radiation* is accomplished by the emission and subsequent absorption of electromagnetic radiation. A heat lamp transfers heat primarily by radiation, as does a campfire. The energy from the sun arrives via radiation. Heat transfer by radiation does not fall into the convenient parallelism with mass and momentum transfer that conductive and convective heat transfer do. There is no mass or momentum transfer process analogous to that of radiation for cases of interest here. We know, of course, from physics that radiation is associated with an equivalent amount of mass, and therefore when radiation is emitted mass is transferred. This amount of mass, however, is insignificant in any present-day engineering process.

Convective and conductive heat transfer are intimately related, because heat transferred by convection ultimately involves conduction. Perhaps a crude analogy will make this clearer. Suppose we have a group of people assembled, and someone in the back of the group wishes to communicate with someone in the front. One approach to the problem is for the person to write a note and hand it to his neighbor and have the neighbor pass it to his neighbor to the front, etc. This is analogous to *conduction*. Another alternative is for the person in the back row to walk to the front of the group (convection) and hand the note to the individual in front (conduction). Here, although the mechanism is convection, the ultimate transfer is still by conduction. The *radiation* analog is for the person standing at the rear of the group to shout the message, which utilizes yet another medium (air) to carry the message—one in which the message moves as waves, although of a very different nature from electromagnetic radiation.

8.1 HEAT TRANSFER BY CONDUCTION

In the last section we wrote the Fourier equation in a special form (one-dimensional Cartesian coordinates) to show that it is parallel to our previous treatment of momentum transfer. The *general* form of the Fourier equation for isotropic media is

$$\mathbf{q} = - k\nabla T \qquad\qquad\qquad (8.1\text{-}1)$$

where $\mathbf{q}$ = heat transfer in energy per unit time and unit area (typically, Btu/hr ft²), a *vector* quantity

k = thermal conductivity in units of energy per unit time, unit area, and unit temperature gradient (typically, Btu/hr ft² °F/ft)

∇T = gradient of the temperature field: a vector whose dimensions are temperature per unit length (typically, °F/ft)

The negative sign in Eq. (8.1-1) is there because the heat flow is in the *opposite* direction to the temperature gradient: the heat runs downhill, as it were. Remember also that the gradient operator has the forms listed in Table 8.1-1 in various coordinate systems.

Typical values for thermal conductives of common materials (in Btu/hr ft °F) are

1. Metals   25–250

Table 8.1-1   Fourier equations in various coordinates

Coordinate system	Components of q/A	∇ operator
Rectangular Cartesian	$q_x = k\dfrac{\partial T}{\partial x}$	$\left(\dfrac{\partial}{\partial x}\right)$
	$q_y = k\dfrac{\partial T}{\partial y}$	$\left(\dfrac{\partial}{\partial y}\right)$
	$q_z = k\dfrac{\partial T}{\partial z}$	$\left(\dfrac{\partial}{\partial z}\right)$
Cylindrical	$q_r = -k\dfrac{\partial T}{\partial r}$	$\left(\dfrac{\partial}{\partial r}\right)$
	$q_\theta = -k\dfrac{1}{r}\dfrac{\partial T}{\partial \theta}$	$\left(\dfrac{1}{r}\dfrac{\partial}{\partial \theta}\right)$
	$q_z = -k\dfrac{\partial T}{\partial z}$	$\left(\dfrac{\partial}{\partial z}\right)$
Spherical	$q_r = -k\dfrac{\partial T}{\partial r}$	$\left(\dfrac{\partial}{\partial r}\right)$
	$q_\theta = -k\dfrac{1}{r}\dfrac{\partial}{\partial \theta}$	$\left(\dfrac{1}{r}\dfrac{\partial}{\partial \theta}\right)$
	$q_\phi = -k\dfrac{1}{r\sin\theta}\dfrac{\partial T}{\partial \phi}$	$\dfrac{1}{r\sin\theta}\dfrac{\partial}{\partial \phi}$

2. Nonmetallic solids   0.025–2.0
3. Liquids   0.08–50
4. Gases   0.008–0.13

Table 8.1-2 shows relative values of thermal conductivity for many substances.

The thermal conductivity is a fairly weak function of pressure and a much stronger one of temperature. In our work we will not be greatly concerned with the effect of pressure; however, we will treat cases in which an average thermal conductivity cannot be assumed satisfactorily. Data are available or may be estimated by a variety of equations based on molecular principles.[1-3]

[1] R. B. Bird, W. E. Stewart, and E. N. Lightfoot, "Transport Phenomena," pp. 247ff, John Wiley & Sons, Inc., New York, 1960.
[2] S. T. Hsu, "Engineering Heat Transfer," pp. 9ff, 569ff, D. Van Nostrand Company, Inc., Princeton, N.J. 1963.
[3] R. H. Perry, C. H. Chilton, and S. D. Kirkpatrick (eds.), "Chemical Engineers Handbook," 4th ed., p. 3-224, McGraw-Hill Book Company, New York, 1963.

Table 8.1-2   Relative valves of thermal conductivity[a]

Btu/ft hr °F	Solids, 32° F	Btu/ft hr °F	Liquids, 68° F	Btu/ft hr °F	Gases, 32° F
250	Silver				
200	Copper				
150	Aluminum				
100	Magnesium				
	Zinc				
50	Nickel				
	Iron				
	Platinum				
30	Lead				
	Uranium				
	Monel metal				
	Brick (SiC)				
10					
	Quartz		Mercury		
	Ice				
	Sandstone				
1	Brick (masonry)	1	Water		
	Plaster		Ammonia		
	Fiber		Glycerine		
0.2	Oak	0.2	Castor oil		
	Maple		Ethyl alcohol		
0.1	Leather	0.1		0.1	Hydrogen
	Pine		Benzol		
	Rubber				Helium
	Asbestos				
	Sawdust				
	Felt				
0.03	Cork				Methane
	Cotton				Oxygen
0.01	Wool				Nitrogen
					Air
					Carbon Monoxide
0.001					Carbon dioxide
					Sulfur dioxide

[a] From S. T. Hsu, "Engineering Heat Transfer," copyright (c) 1963 by Litton Educational Publishing, Inc., by permission of D. Van Nostrand Company, Inc.

The higher thermal conductivity of metals is largely due to the great number of "free" electrons—the so-called electron gas. These electrons are responsible for electrical conductivity as well as thermal conductivity, and so these two properties are highly correlated. Nonconducting (electrically) solids must rely mostly on lattice vibrations to conduct energy, except for some cases where radiation is exchanged between molecules.

One-dimensional steady-state conduction in rectangular coordinates

If we assume our x axis to be in the direction of heat flow, and further assume that no heat is being conducted in the y or z directions (that is, no temperature gradients exist in those directions), the Fourier equation reduces to

$$q_x = -k \frac{dT}{dx} \qquad (8.1\text{-}2)$$

If we consider the case of a single homogeneous solid as shown in Fig. 8.1-1 where the area normal to heat flow is constant and the faces are maintained at T_1 and T_2, we may integrate the equation as follows:

$$q_x \int_{x_1}^{x_2} dx = - \int_{T_1}^{T_2} k \, dT \qquad (8.1\text{-}3)$$

$$q_x (x_2 - x_1) = - \int_{T_1}^{T_2} k \, dT \qquad (8.1\text{-}4)$$

An energy balance shows us that $Q = $ constant, and so since $A = $ constant, q_x is constant and may be removed from the integrand above. If we know k as a function of T, we can integrate the right-hand side analytically, numerically, or graphically.

The *simplest* case is that of constant k, where Eq. (8.1-4) integrates to

$$q_x (x_2 - x_1) = -k (T_2 - T_1) \qquad (8.1\text{-}5)$$

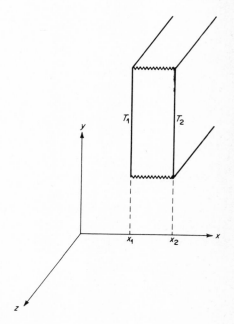

Fig. 8.1-1 Homogeneous solid.

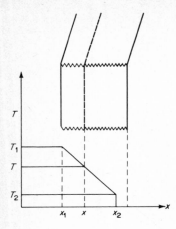

Fig. 8.1-2 Temperature profile in heat transfer by conduction.

If the limits on the integration are taken rather as between x_1 (where we have T_1) and some *arbitrary* point x in the solid (where we have T) we obtain the temperature profile:

$$q_x(x - x_1) = - k(T - T_1) \tag{8.1-6}$$

This is the equation of a straight line. Therefore, for conduction problems where (1) we have steady-state, (2) the cross-sectional area normal to the flow of heat is constant, and (3) k is constant, the temperature profile is linear, as shown in Fig. 8.1-2.

The above equation is satisfactory for a single solid; however, we frequently have a series of these solids, as in the case where we have a composite wall (for example, a layer of insulation plus the wall itself).

Before we discuss this case we must briefly treat a very important point: the interface condition between solids. In this text we will make the *assumption* that there is *no thermal resistance* between solids in *direct contact*. Another way to state this assumption is to say that the faces of solids I and II in contact have the same surface temperature—T_3 in Fig. 8.1-3.

For clean solids in intimate contact this is a good assumption. Dirt at the interface or a poor contact leaving a void space may make it a poor assumption, however. When interfacial resistance is present, one way it can be handled is to obtain an effective thickness and thermal conductivity for the resistance. In effect, one simply introduces another solid in the series of solids in the conduction path. (See Sec. 9.2, i.e., fouling coefficient.)

We could calculate the heat flow through the wall in Fig. 8.1-3 by applying Eq. (8.1-5) to *either* the wall *or* the insulation—either approach would give the same answer because

$$q_x = \frac{-k_{\mathrm{I}}(T_3 - T_1)}{x_3 - x_1} \tag{8.1-7}$$

and

$$q_x = \frac{-k_{\mathrm{II}}(T_2 - T_3)}{x_2 - x_3} \tag{8.1-8}$$

However, to do the problem in such a fashion requires a knowledge of T_3, and this temperature is hard to obtain. For example, suppose instead of insulation that solid II is merely a coat of paint—how does one measure T_3?

The usual information we have at our disposal is the *overall* temperature drop, that is, $T_1 - T_2$. (In fact, we usually do not even have the *surface* temperatures T_1 and T_2 but rather the temperatures somewhere in the fluid phase adjoining the wall. More about this when we study convection; for the moment assume that we know T_1 and T_2.) Let's see if we can develop an equation for heat flow in terms of the *overall* temperature drop $T_1 - T_2$, the thickness, and the thermal conductivities.

In other words, we want an equation which gives

$$q_x = (\text{something})(T_1 - T_2) \tag{8.1-9}$$

Thus far we only know how to relate q_x to the individual temperature drops. Observe that

$$(\Delta T)_{\text{overall}} = T_1 - T_2 = -[(T_2 - T_3) + (T_3 - T_1)] \tag{8.1-10}$$

and that by rewriting Eq. (8.1-5),

$$\Delta T = -\frac{q_x \Delta x}{k} \tag{8.1-11}$$

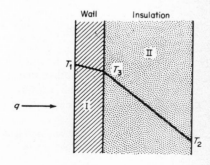

Fig. 8.1-3 Conduction through two solids in contact.

Substituting Eq. (8.1-11) in Eq. (8.1-10) and the result in Eq. (8.1-9), we have

$$q_x = - \text{(something)} \left[\frac{-q_x(x_3 - x_1)}{k_I} - \frac{q_x(x_2 - x_3)}{k_{II}} \right] \qquad (8.1\text{-}12)$$

This yields

$$\text{(something)} = \frac{+1}{(x_3 - x_1)/k_I + (x_2 - x_3)/k_{II}} \qquad (8.1\text{-}13)$$

Substituting in Eq. (8.1-9)

$$q_x = \left[\frac{1}{(x_3 - x_1)/k_I + (x_2 - x_3)/k_{II}} \right] (T_1 - T_2) \qquad (8.1\text{-}14)$$

If we regard this equation as

$$\text{Flux} = \text{conductance} \times \text{driving force} \qquad (8.1\text{-}15)$$

the bracketed term is simply the *equivalent conductance* of the system. Since the conductance is equal to the reciprocal of the resistance, the *equivalent resistance* is the denominator of the bracketed term in Eq. (8.1-14). The resistance is seen to be the sum of two resistances, one for solid I and one for solid II, since the first term involves only properties of solid I and the second only of solid II. The problem is equivalent to the electrical circuit shown, where temperature corresponds to the voltage imposed.[1] (See Fig. 8.1-4.)

Observe that although this example was done for two solids, the same procedure is valid for *any* number of layers. We simply add up the individual temperature drops to get the overall ΔT, substitute using Eq. (8.1-11), and obtain (minus because x_1 corresponds to T_1, etc.),

$$q_x = \frac{-1}{\displaystyle\sum_{i=I,II,...} (\Delta x/k)_i} (\Delta T)_{\text{overall}} = \left(\frac{Q}{A} \right)_x \qquad (8.1\text{-}16)$$

Example 8.1-1 Conduction through layers A thermocouple attached to the inside wall of a furnace indicates 1500°F. The inside wall is firebrick, 3 in. thick, and is to be in intimate contact with a 1/4-in. steel wall. For

---

[1] Series resistance is the common problem; however, it is possible to have heat transfer in parallel, for example, heat flow through two slabs side by side. (See Fig. 8.1-4.)

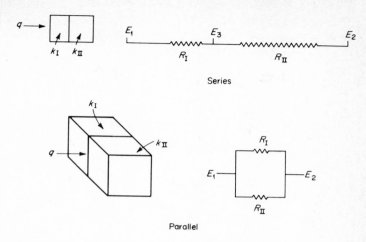

Fig. 8.1-4 Equivalent electrical circuit for heat transfer.

economic reasons the loss must not exceed 5 Btu/hr ft². The outside wall temperature is 150°F. How thick a layer of magnesia insulation is necessary to meet this requirement?

Solution To solve we use the fact that Q/A is the same for each of the three solids:

(a) $\quad q_x = \left[\dfrac{-k}{\Delta x} (T_2 - T_1) \right]_{\text{fire}}$

(b) $\quad q_x = \left[\dfrac{-k}{\Delta x} (T_3 - T_2) \right]_{\text{steel}}$

(c) $\quad q_x = \left[\dfrac{-k}{\Delta x} (T_4 - T_3) \right]_{\text{mag}}$

The above is a set of three equations in three unknowns $(\Delta x)_{\text{mag}}$, T_2, and T_3:

$$5 \text{ Btu/hr ft}^2 = [\, (-0.6 \text{ Btu/hr ft °F}) \frac{4}{1 \text{ ft}} (T_2 - 1500)°\text{F}]$$

$$5 = -\frac{26}{1} (4)(12)(T_3 - T_2)$$

$$5 = -\frac{0.03}{x} (150 - T_3)$$

From equation (*a*)

$$T_2 = -2.08 + 1500 = 1498$$

Substituting in equation (*b*)

$$\frac{5}{(26)(4)(12)} = 1498 - T_3$$

$$T_3 = 1498 + 0.004 \cong 1498$$

Substituting in equation (*c*)

$$x = \frac{-0.03}{5}(1498 - 150) = 8.08 \text{ ft}$$

This means that the insulation is unrealistically thick, and is hardly practical.

□ □ □

One-dimensional steady-state conduction in cylindrical coordinates

We consider a derivation similar to that of the preceding section now for the case of cylindrical coordinates. This time we consider conduction only in the *r* direction as shown in Fig. 8.1-5.

The Fourier equation, Eq. (8.1-1), reduces, using Table 8.1-1, to

$$q_r = -k \frac{dT}{dr} \tag{8.1-17}$$

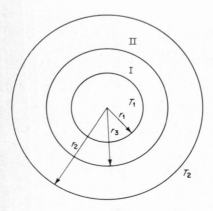

Fig. 8.1-5 Conduction in a composite cylinder wall.

Our integration now must account for the fact that A (and therefore q) is not constant as in the previous (rectangular) case but is now a function of r. An energy balance, however, tells us that Q_r is constant. Considering the case of constant k, we have from the form in Table 8.1-1

$$Q_r \int_{r_1}^{r_2} \frac{dr}{A} = -k \int_{T_1}^{T_2} dT \qquad (8.1\text{-}18)$$

but $A = 2\pi r L$, and so

$$\frac{Q_r}{2\pi L} \ln \frac{r_2}{r_1} = -k(T_2 - T_1) \qquad (8.1\text{-}19)$$

This corresponds to Eq. (8.1-6) but is not the equation used most by engineers. Many engineers prefer to say, "I intend to apply the rectangular coordinate *form* [Eq. (8.1-6)] to the case where the area is changing." Very well, let us put Eq. (8.1-19) in this form; that is,

$$\frac{\text{Heat transfer} \times \text{distance}}{\text{Area}}$$
$$= -\text{thermal conductivity} \times \text{temperature drop} \qquad (8.1\text{-}20)$$

Looking at Eq. (8.1-19), we see we lack (1) the appropriate area and (2) the distance. To introduce the distance, we multiply and divide Eq. (8.1-19) by $(r_2 - r_1)$, and rearrange slightly:

$$\left[\frac{Q_r(r_2 - r_1)}{2\pi(r_2 - r_1)L/\ln(r_2/r_1)} \right] = -k(T_2 - T_1) \qquad (8.1\text{-}21)$$

In fact, we can rewrite as

$$\frac{2\pi(r_2 - r_1)L}{\ln(r_2/r_1)} = \frac{2\pi r_2 L - 2\pi r_1 L}{\ln(2\pi r_2 L/2\pi r_1 L)} \equiv A_{\text{lm}} = \frac{A_2 - A_1}{\ln(A_2/A_1)} \qquad (8.1\text{-}22)$$

and *define* this quantity as the logarithmic mean area. We then may write Eq. (8.1-20) as

$$\frac{Q_r(r_2 - r_1)}{A_{\text{lm}}} = -k(T_2 - T_1) \qquad (8.1\text{-}23)$$

which corresponds directly to Eq. (8.1-6). Referring to Fig. 8.1-5 and using the same argument for a series of resistances (now *cylinders*) in series as we used above, we write the sum of the individual temperature differences as

$$(\Delta T)_{\text{overall}} = -Q_r \sum_{i=\text{I},\text{II},\ldots} \left(\frac{\Delta r}{kA_{\text{lm}}}\right)_i \tag{8.1-24}$$

and finally

$$Q_r = \frac{-1}{\sum_i (\Delta r/kA_{\text{lm}})_i}(\Delta T)_{\text{overall}} \tag{8.1-25}$$

Again, the denominator of the left-hand side is the thermal resistance term.

Example 8.1-2 Conduction through an insulated pipe A 3-in. glass-lined steel pipe carries steam which keeps the inside surface of the glass at 350°F. The glass coating is 1/32-in. thick and the pipe is coated with magnesia insulation 1-in. thick on the outside. Calculate the heat loss per foot of pipe if the outside surface temperature of the insulation may be assumed to be 120°F. (See Fig. 8.1-6.)

ID $= 3$ in.

OD $= 3.5$ in.

$k_{\text{glass}} = 0.5$ Btu/hr ft °F

$k_{\text{steel}} = 26$ Btu/hr ft °F

$k_{\text{mag}} = 0.03$ Btu/hr ft °F

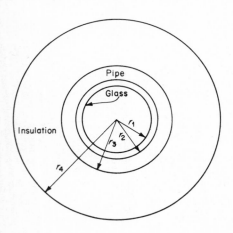

Fig. 8.1-6 Sketch for Example 8.1-2.

Solution The formula for conduction through a series of cylindrical resistances is

(a) $\quad -Q_r = \dfrac{(\Delta T)_{\text{overall}}}{\displaystyle\sum_i \dfrac{\Delta r}{kA_{\text{lm}}}}$

For the log mean areas, assuming $r = 1$ ft

$$(A_{\text{lm}})_{\text{mag}} = \frac{2\pi r_4(1) - 2\pi r_3(1)}{\ln(2\pi r_4/2\pi r_3)}$$

$$= \frac{[2\pi(4.5)(1) - 2\pi(3.5)(1)]^1/_{12}}{\ln(4.5/3.5)}$$

$$= 2.1 \text{ ft}^2$$

$$(A_{\text{lm}})_{\text{pipe}} = \frac{[2\pi(3.5)(1) - 2\pi(3.0)(1)]^1/_{12}}{\ln[(3.5)(1)/(3.0)(1)]} = 1.71 \text{ ft}^2$$

$$(A_{\text{lm}})_{\text{glass}} \cong A_{\text{out glass}} \cong A_{\text{in glass}} \cong \frac{2\pi(3.0)}{12} 1$$

$$\cong 1.57 \text{ ft}^2$$

Substituting in equation (a):

$$Q_r =$$

$$\frac{350 - 120}{1/[(12)(0.03)(2.1)] + 0.5/[(12)(26)(1.71)] + 1/[(32)(12)(0.5)(1.57)]}$$

$$= \frac{230}{1.32 + 0.000938 + 0.332}$$

(Notice the relative resistances.)

$\quad Q_r = 140$ Btu/hr for each foot of length

□ □ □

One-dimensional steady-state conduction in spherical coordinates

If we repeat the approach used for the preceding two coordinate frames for conduction in a hollow sphere, we first have from the Fourier equation (see Fig. 8.1-7)

$$\frac{Q_r}{A} = -k\frac{dT}{dr} \tag{8.1-26}$$

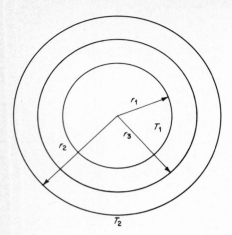

Fig. 8.1-7 Conduction in a hollow sphere.

just as for cylindrical coordinates. However, now the area is a different function of radius:

$$A = 4\pi r^2 \tag{8.1-27}$$

This makes our integration

$$\frac{Q_r}{4\pi}\int_{r_1}^{r_2}\frac{dr}{r^2} = -k\int_{T_1}^{T_2} dT \tag{8.1-28}$$

or

$$\frac{Q_r}{4\pi}\left(-\frac{1}{r_2}+\frac{1}{r_1}\right) = -k(T_2 - T_1) \tag{8.1-29}$$

As before, we multiply by $(r_2 - r_1)$ and get

$$\frac{Q_r(r_2 - r_1)}{4\pi(r_2 - r_1)/(1/r_1 - 1/r_2)} = -k(T_2 - T_1) \tag{8.1-30}$$

but rearranging the denominator of the left-hand side:

$$\frac{4\pi(r_2 - r_1)}{1/r_1 - 1/r_2} = \frac{4\pi(r_2 - r_1)}{(r_2 - r_1)/r_2 r_1} = 4\pi r_1 r_2 \equiv A_{\mathrm{gm}} = \sqrt{A_1 A_2} \tag{8.1-31}$$

and now the appropriate mean area is what we define as the *geometric mean*. Then

$$\frac{Q_r(r_2 - r_1)}{A_{\mathrm{gm}}} = -k(T_2 - T_1) \tag{8.1-32}$$

in *spherical* coordinates, which corresponds to Eqs. (8.1-6) and (8.1-22) in the other systems.

For a *composite* sphere we get by the same procedure as before

$$Q_r = \frac{-1}{\displaystyle\sum_{i=\mathrm{I,II,\dots}} (\Delta r/kA_{gm})_i} (\Delta T)_{\text{overall}} \qquad (8.1\text{-}33)$$

Example 8.1-3 Conduction through shielding You are faced with the problem of the disposal of solid radioactive waste. It has been proposed that the waste be coated with an organic coating containing a shielding material and dumped into the ocean. The coating degrades above 400°F. The ocean may be assumed to keep the *entire outer surface* of the coating at 50°F. The waste may be treated as a sphere of 1 ft radius which gives off 440 Btu/hr. The thermal conductivity of the coating is 0.02 Btu/hr ft °F and is independent of temperature. How thick must the coating be to avoid degradation of the inside surface?

Solution

Let r = outer radius

$$Q_r = - kA_{gm} \frac{\Delta T}{\Delta r}$$

$$440 \text{ Btu/hr} = -\ 0.02\,(\text{Btu/hr ft °F})\ 4\pi r(1)\ \text{ft}^2 \frac{50 - 400}{r - 1}$$

$$352r = 440$$

$$r = 1.25 \text{ ft}$$

The coating is $1.25 - 1 = 0.25$ ft or 3 in. This is not the total answer since the shielding can be no *thicker* than 3 in. The next question to be answered would be whether this is thick enough to allow for the unavoidable degradation of the coating.

□ □ □

Example 8.1-4 Conduction with variable thermal conductivity A firebrick whose thermal conductivity variation with temperature can be described as

$$k = 0.10 + 5 \times 10^{-5}\,T$$

where k = Btu/hr ft °F

$$T = \text{°F}$$

is to be used in a 4-in.-thick furnace liner. The outside temperature of the liner is to be about 100°F and the inside temperature 1500°F. Calculate the heat flux and temperature profile.

Solution   The Fourier equation is

$$q = -k \frac{dT}{dx}$$

Since q will be constant through the firebrick (steady state, rectangular coordinates), we can write

$$q \int_0^L dx = - \int_{T_1}^{T_2} k \, dT$$

Substituting

$$q \int_0^{1/3 \text{ ft}} dx = - \int_{1,500}^{100} (0.10 + 5 \times 10^{-5} T) \, dT$$

$$q = 3 [(0.1)(1,500 - 100) + (5 \times 10^{-5})(1.12 \times 10^6)]$$

$$q = 588 \text{ Btu/hr ft}^2$$

To get the profile

$$q \int_0^x dx = - \int_{1,500}^T (0.10 + 5 \times 10^{-5} T) \, dT$$

$$588 \, x = 0.1 \, (1,500 - T) + \frac{5 \times 10^{-5}}{2} (2.25 \times 10^6 - T^2)$$

$$x = 0.352 - 1.7 \times 10^{-4} T - 4.28 \times 10^{-8} T^2$$

Note that the profile is no longer linear.

□ □ □

8.2   HEAT TRANSFER BY CONVECTION

In this section we discuss heat transfer by convection. Convective heat transfer is due to *fluid motion*. For example, the fluid adjacent to a heated surface is increased in internal energy by conduction from the surface to the fluid. Fluid motion carries the heated fluid away from the wall and the colder fluid in the mainstream is heated by mixing with and conduction from the warmer fluid. Heat is transferred in *laminar* flow mainly by molecular interaction between adjacent fluid "layers." In *turbulent* flow heat is transferred on a macroscopic level due to physical mixing of the fluids. We generally consider the fluid motion, and therefore the resulting heat transfer, as (1) *forced* or (2) *free* convection of the fluid. By *forced convection* we mean that the fluid is moved by an external force such as a pump or the blower on a hot air furnace; by *free convection* we mean the fluid moves because of density differences resulting from temperature variations within the fluid, e.g., the layer of cold air that tumbles down a windowpane in winter.

Convective heat transfer is a confused term because the true convective part of the mechanism only *moves material* (even though it is hot or cold material) and there is no *heat* as we defined earlier associated with this process. All the energy is associated with mass in this step. The ultimate *heat* transfer is always by conduction or radiation when the convected material interacts with the ambient material. A better term would be convective *energy* transfer.

Most problems in convection are too complicated to permit the kind of analytical solution we discussed for conduction, and we adopt an individual heat transfer coefficient approach for many convective problems (see Chap. 9). There are situations where certain models such as the boundary layer can be used to predict convective heat transfer.

The thermal boundary layer[1]

If we consider laminar incompressible flow past a heated flat plate, and make *microscopic* (rather than *macroscopic*) balances of momentum and energy, we are led to differential equations; with suitable assumptions, these become the well-known *boundary layer* equations of Prandtl. In each case, the resulting differential equation in dimensionless form is

$$\frac{d^2\Omega}{d\eta^2} + \frac{Kf(\eta)}{2}\frac{d\Omega}{d\eta} = 0 \tag{8.2-1}$$

with boundary conditions

at $\eta = 0, \Omega = 0$

at $\eta = \infty, \Omega = 1$

In this equation Ω is a dimensionless velocity, temperature, or concentration depending on whether a momentum, energy, or mass balance was

[1] This discussion parallels similar discussions in Secs. 5.6 and 11.2. At the end of Sec. 11.2, we give an example of comparison of the concentration, temperature, and velocity boundary layers.

used. The coefficient K is dimensionless, and η is a dimensionless distance[1] which incorporates both y, the distance from the plate surface, and x, the distance from the nose of the plate (a so-called similarity variable):

$$\eta = y \sqrt{\frac{v_0}{\nu x}} = \frac{y}{x} \sqrt{\text{Re}} \qquad (8.2\text{-}2)$$

where Re is a Reynolds number based on the distance from the front of the plate, x, and the free-stream velocity, v_0. The function $f(\eta)$ is a known function (which cannot be expressed in closed form but can be expressed as a series).

For momentum transfer one obtains

$$\frac{d^2 (v/v_0)}{d\,\eta^2} + \frac{f(\eta)}{2} \frac{d(v/v_0)}{d\eta} = 0 \qquad (8.2\text{-}3)$$

For heat transfer one obtains (where 0 is free stream and s is surface)

$$\frac{d^2 [(T - T_s)/(T_0 - T_s)]}{d\eta^2} + \text{Pr} \frac{f(\eta)}{2} \frac{d[(T - T_s)/(T_0 - T_s)]}{d\eta} = 0$$

$$(8.2\text{-}4)$$

Obviously for momentum transfer, $K = 1$, for heat transfer, $K = \text{Pr}$. The boundary condition at $\eta = 0$ is at the surface of the plate, since $y = 0$ implies $\eta = 0$. Substituting in the appropriate form for Ω we see that this yields $v = 0$, $T = T_s$ as desired. The boundary condition at $\eta = \infty$ is for $x = 0$ or $y = \infty$. Substitution gives the free-stream values $v = v_0$ and $T = T_0$.

We can solve the above equations, and without going into details of the solution method we present a plot of the solution in Fig. 8.2-1. This plot gives *profiles* of velocity or temperature. This can be seen by considering a given fluid at a given distance x from the nose of the plate. This means that everything is fixed in η except y, and so the curves give velocity or temperature as a function of y—the profile in the boundary layer. This profile is plotted slightly different from the usual orientation; the latter is shown in the sketch.

Notice that if $\text{Pr} = 1$, the *dimensionless differential equation* and *boundary conditions* are the *same* for *both cases*. This means the solu-

---

[1] One way to develop the variable η is by the techniques of dimensional analysis we have developed. One writes the variables upon which the velocity depends, and combines these variables to obtain dimensionless groups, one of which is η. For a discussion see A. G. Hansen, "Similarity Analyses of Boundary Value Problems in Engineering," Prentice-Hall, Inc., Englewood Cliffs, N.J., 1964.

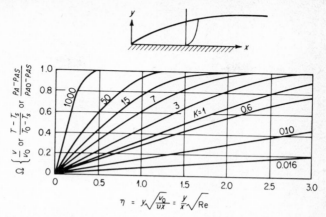

Fig. 8.2-1 Solution of Eq. (8.2-1). *(C. Q. Bennett and J. E. Myers, "Momentum, Heat, and Mass Transfer," McGraw-Hill Book Company, New York, 1962).*

tions are identical and the profiles (and therefore the boundary layers) coincide.

(This example also illustrates the appearance of dimensionless groups as the coefficients in the dimensionless differential equation.)

8.3 HEAT TRANSFER BY RADIATION

Heat transfer by radiation occurs, as was mentioned earlier, in a manner which does not parallel the conventional transfer of mass or momentum. The transfer occurs without an intervening medium (although at one time such a medium, the "ether," was postulated in theoretical attempts to explain the mechanism of radiation). *Thermal* radiation is a relatively small part of the total electromagnetic spectrum as shown in Fig. 8.3-1.

Electromagnetic radiation has both a particulate (e.g., it exerts a force when incident on a surface) and a wavelike (it can be focused, diffracted, etc.) nature. Satisfactory explanation of this dual nature was achieved by quantum theory, first advanced by Max Planck. According to quantum theory, radiant energy is discretized in packets called quanta, whose energy is related to the apparent wavelength as

$$E = h\nu \tag{8.3-1}$$

where E = energy of a quantum

h = Planck's constant = 6.625×10^{-27} erg sec

ν = frequency

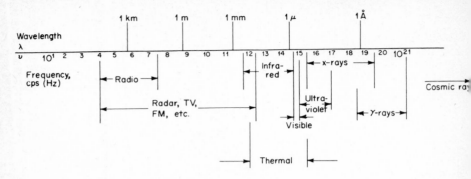

Fig. 8.3-1 The electromagnetic spectrum.

Note that this relation shows quanta of higher frequency to be more energetic. Since

$$c = \lambda \nu \tag{8.3-2}$$

where c = speed of propagation

λ = wavelength

shorter-wavelength quanta are more energetic.

Interaction of radiation and matter

The *interaction* of radiation with matter is accomplished by the transition
of atoms and molecules among various states which change the internal
energy of the matter. These states may be translational, rotational, vibrational, or electronic.

When radiation is incident on matter, several things may happen.
First, it may pass through the substance unaltered; that is, it may be *transmitted*. (We do *not* include absorption followed by reradiation in *transmission*.) Second, it may be *reflected* from the surface of the object.
Reflection can occur in any combination of two extreme cases known as
specular and *diffuse* reflection. For specular reflection the angle of in-
cidence is equal to the angle of reflection; for diffuse reflection the radiation is scattered in all directions (see Fig. 8.3-2). Third, the radiation may
be *absorbed*, that is, converted into internal energy of the material. Absorption takes place near the surface for electrical conductors, but may
take place up to several millimeters from the surface in nonconducting

solids.[1] Liquids, solids, and many gases absorb significant amounts of thermal radiation. For example, monatomic and symmetrical diatomic gases absorb little or no thermal radiation but many polyatomic gases such as ammonia, carbon dioxide, water vapor, sulfur dioxide, long-chain hydrocarbons, etc., do absorb significant amounts of radiation. It should further be pointed out that gases and vapors do not absorb and emit radiation at all wavelengths. This characteristic is shared by liquids to a lesser extent. These materials are selective absorbers. The absorption of photons can result in the direct change of electronic, vibrational, or rotational states of molecules, in the ionization of a molecule, or in the change in translational energy of a free electron.[2]

To treat the relative amounts of energy which enter into each of the above sorts of interaction we define three terms: the *absorptivity* α, the *reflectivity* γ, and the *transmissivity* τ. These represent the *fraction* of the incident radiation which is, respectively, absorbed, reflected, and transmitted. Because the incident quanta must "match" the energy levels available in the absorbing medium for absorption to take place, α, γ, and τ are functions of wavelength. By virtue of the definition

$$\alpha + \gamma + \tau = 1.0 \qquad (8.3\text{-}3)$$

Bodies with a transmissivity of zero are called *opaque*. Most of the problems of interest to engineers, other than those involving absorbing gases, relate to opaque bodies.

Blackbodies   A *blackbody* is one for which

$$\alpha = 1.0 \qquad \text{(all wavelengths)} \qquad (8.3\text{-}4)$$

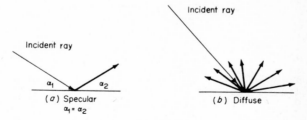

Fig. 8.3-2 Extreme modes of reflection.

[1] S. T. Hsu, "Engineering Heat Transfer," D. Van Nostrand Company, Inc., Princeton, N.J., 1963.
[2] E. M. Sparrow and R. D. Cess, "Radiation Heat Transfer," Brooks/Cole Publishing Company, Belmont, Calif., 1966.

that is, all incident radiation is absorbed. (The reader might like to consider whether one can *see* a blackbody.)

In practice, true blackbodies do not exist, but the concept is valuable in theoretical considerations. Most substances which are thought of as very "black"—carbon black, black plush, black cats, professors' hearts, etc.—are really not very black by the standard of unit absorptivity. To create the best approximation possible to a blackbody in the laboratory one uses an enclosure with a very small opening. Radiation entering this small opening bounces about from wall to wall, with little probability of being reemitted by striking the entrance hole. (See Fig. 8.3-3.) At each reflection some of the radiation is absorbed, and so ultimately virtually all is absorbed.

Emission of radiation   Thus far we have discussed only interaction of radiation and matter via *absorption*. Matter also *emits* radiation by transitions between the same states as listed for absorption. In order to simplify quantitative statements about emission, we define a quantity known as the *emissivity* ε, which is defined to be the ratio of energy emitted at a given wavelength per unit area for the body in question to that of a blackbody at the same temperature.

At this point it should be clear that radiant energy interchange between two bodies is a complex state of affairs. One body emits a series of thermal wavelengths depending on its temperature and composition which individually, upon reaching the second body, may be absorbed, transmitted, or reflected. The absorbed radiation may be *reradiated* by body II in a *different* series of wavelengths (depending on the temperature and composition of body II) which, together with the reflected radiation, may again strike body I and be absorbed, transmitted, or reflected, depending on the wavelength, whereupon some of the absorbed radiation

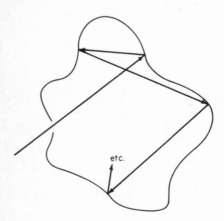

etc.

Fig. 8.3-3 A laboratory blackbody.

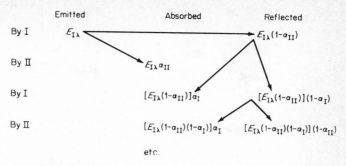

Fig. 8.3-4 History of radiation emitted.

is reradiated and together with the reflected radiation may strike body
II. . . and on and on like an insane game of table tennis played at the
speed of light. The mind recoils at the thought of the introduction of more
than two bodies or an intervening medium which absorbs and scatters.
Obviously, some simplifying assumptions are needed.

**Radiant heat exchange between two opaque bodies with no intervening
medium** Let us begin by considering the case of two bodies:

1. That radiate and absorb in the *same* spectral range (frequency)
2. That are *opaque*
3. That are so situated that *all* the radiation emitted from body I strikes
   body II and vice versa (e. g., infinite parallel planes)

and let us calculate the energy exchanged by radiation between the two
bodies. We can do this by considering the emission of a certain amount of
radiation by body I and subtracting from this amount (1) the portion
which is reflected back and forth and ultimately reabsorbed by body I,
plus (2) the radiation emitted from body II that is absorbed by body I.

Now, *for a given wavelength* λ, let us consider what happens to an
amount of radiation $E_{I\lambda}$ originally emitted by body I (see Fig. 8.3-4). If
we continue this scheme we may write the amount of radiation $E_{I\lambda}$ emitted by body I which is subsequently absorbed by body I as

$$\text{Radiation emitted by I and reabsorbed by I} = E_{I\lambda}(1 + K + K^2 \\ + K^3 + \cdots)(1 - \alpha_{II\lambda})\alpha_{I\lambda} \tag{8.3-5}$$

where

$$K = (1 - \alpha_{II\lambda})(1 - \alpha_{I\lambda}) \tag{8.3-6}$$

but

$$1 + \sum_{i=1}^{n} K^i = \frac{1}{1-K} \qquad \text{(power series expansion)} \qquad (8.3\text{-}7)$$

and so Eq. (8.3-7) reduces to

$$\text{Radiation reabsorbed by } I = \frac{E_{I\lambda}(1-\alpha_{II\lambda})\alpha_{I\lambda}}{1-K} \qquad (8.3\text{-}8)$$

If we follow the same procedure to calculate the emission from body II that is absorbed by body I we get

$$\frac{E_{II\lambda}\alpha_{I\lambda}}{1-K} \qquad (8.3\text{-}9)$$

By subtracting (8.3-8) and (8.3-9) from $E_{I\lambda}$ we get the heat transferred Q

$$Q = E_{I\lambda} - \frac{E_{I\lambda}(1-\alpha_{II\lambda})\alpha_{I\lambda}}{1-K} - \frac{E_{II\lambda}\alpha_{I\lambda}}{1-K} \qquad (8.3\text{-}10)$$

Putting everything over a common denominator and substituting the definition of K

$$Q = \frac{E_{I\lambda}\alpha_{II\lambda} - E_{II\lambda}\alpha_{I\lambda}}{\alpha_{I\lambda} + \alpha_{II\lambda} - \alpha_{I\lambda}\alpha_{II\lambda}} \qquad (8.3\text{-}11)$$

This equation is true at nonequilibrium conditions as well as at equilibrium.

Kirchhoff's law   At *thermal equilibrium*

$$Q = 0 \qquad \text{(net)} \qquad (8.3\text{-}12)$$

Since the denominator of the right-hand side of Eq. (8.3-11) cannot become infinite (the upper limit to absorptivity is 1.0) the numerator must vanish. Therefore

$$E_{I\lambda}\alpha_{II\lambda} - E_{II\lambda}\alpha_{I\lambda} = 0 \qquad (8.3\text{-}13)$$

or

$$\frac{E_{I\lambda}}{\alpha_{I\lambda}} = \frac{E_{II\lambda}}{\alpha_{II\lambda}} \qquad (8.3\text{-}14)$$

This is Kirchhoff's law.

In Eq. (8.3-14) there is nothing to say that one of the bodies cannot be a *blackbody*. Considering the special case for body **II** being black ($\alpha_{II\lambda} = 1.0$):

$$\frac{E_{I\lambda}}{(E_{II\lambda})_b} = \alpha_{I\lambda} \qquad (8.3\text{-}15)$$

But by definition

$$\frac{E_{I\lambda}}{(E_{II\lambda})_b} = \varepsilon_{I\lambda} \qquad (8.3\text{-}16)$$

Therefore, at *thermal* equilibrium

$$\varepsilon_{I\lambda} = \alpha_{I\lambda} \qquad (8.3\text{-}17)$$

This is true for the *total* absorptivity and emissivity (based on the *entire spectrum*) ε and α as well as the *monochromatic* case (for specified wavelength) listed above.

Table 8.3-1 lists some total emissivities.

Blackbody radiation   For blackbody radiation we can show from thermodynamics that

$$q_b = \sigma T^4 \qquad (8.3\text{-}18)$$

This is the *Stefan-Boltzmann law*, and σ, the Stefan-Boltzmann constant, has a value of 0.1714×10^{-8} Btu/hr ft^2 °R^4. Note that the temperature in Eq. (8.3-18) is *absolute*. By using our definition of emissivity we can write for a gray body, for which α is independent of λ (see below),

$$q_g = \varepsilon \sigma T^4 \qquad (8.3\text{-}19)$$

Planck derived from quantum theory an expression for emissive power of a blackbody as a function of wavelength:

$$q = \frac{C_1 \lambda^{-5}}{e^{c_2/\lambda T} - 1} \qquad (8.3\text{-}20)$$

where

$$C_1 = 1.74 \times 10^{-5} \text{ erg cm}^2/\text{sec} \qquad C_2 = 2.59 \text{ cm °R}$$

330   TRANSFER OPERATIONS

Table 8.3-1   Total emissivity of some surfaces[a]

Surface	Temperature, °F	Emissivity
Polished aluminum	73	0.040
Polished copper	242	0.023
Polished iron	800–1800	0.144–0.377
Cast iron newly turned	72	0.435
Oxidized iron	212	0.736
Asbestos board	74	0.96
Red brick	70	0.93
Sixteen different oil paints, all colors	212	0.92–0.96
Water	32–212	0.95–0.963

[a]Courtesy of H. C. Hottel from "Heat Transmission," by W. H. McAdams, 3d ed., McGraw Hill Book Company, New York, 1954.

and the units of $q_{b\lambda}$ are ergs/sec cm^3. The derivative of this distribution with respect to λ can be set equal to zero to find the wavelength of *maximum intensity*; this results in a transcendental equation which can be solved, by trial and error, to yield

$$\lambda_{max} = \frac{0.5216}{T} \qquad (8.3\text{-}21)$$

This is *Wien's displacement law*.

Equation (8.3-20) is plotted in Fig. 8.3-5.[1] The maxima of the curves follow Eq. (8.3-21). Note that as one heats a blackbody the maximum emission in the visible range is first in the red end, and finally becomes more or less distributed uniformly over the whole range at higher temperatures. This is the reason for a bar of metal (which is not too unlike a blackbody) becoming first red hot and then white hot as it is progressively heated.

We now consider the problem of radiant heat exchange between surfaces in light of the above. We will find that we can treat only fairly naïve models, but we fortunately can get fairly accurate results from such models.

Radiant heat exchange between blackbodies   Let us now consider radiant heat exchange between two arbitrarily oriented *black* surfaces as

[1]M. Jacob, "Heat Transfer," vol. I, John Wiley & Sons, Inc., New York, 1950.

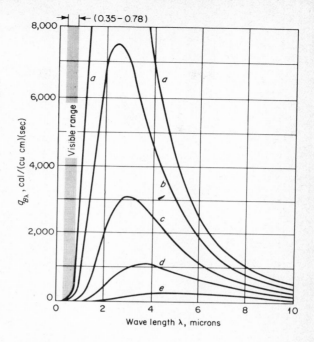

Fig. 8.3-5 Spectral distribution. $(T = (a)$ 1400°K, (b) 1200°K, (c) 1000°K, and (e) 600°K. $(M.$ Jacob, *"Heat Transfer,"* vol. 1, John Wiley & Sons, Inc., New York, 1950.)

shown in Fig. 8.3-6. The radiation going from surface 1 to surface 2 may be written as

$$dQ_{1 \to 2} = \underbrace{I}_{\substack{\text{proportion-}\\\text{ality}\\\text{constant}}} \underbrace{dA_1 \cos \beta}_{\substack{\text{projected}\\\text{area in}\\r \text{ direction}}} \underbrace{\frac{dA_2 \cos\beta_2}{r^2}}_{\substack{\text{solid angle}\\d\omega}} \qquad (8.3\text{-}22)$$

The proportionality constant I is called the *intensity*.

A procedure like the above leads to a similar equation for radiation from surface 2 to surface 1. Taking the difference of the two expressions yields (see Appendix 8A for details of the integration)

$$dQ_{12} = (I_1 - I_2)\, dA_1 \cos \beta_1 \frac{dA_2 \cos \beta_2}{r^2} \qquad (8.3\text{-}23)$$

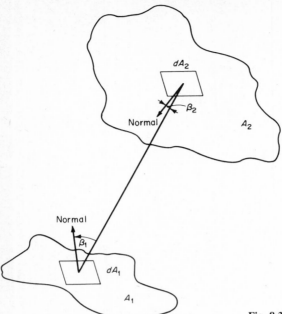

Fig. 8.3-6 Radiation between surfaces.

Equation (8.3-23) must be integrated for each particular geometry. Since this is cumbersome, the results of this integration have been tabulated for common cases and expressed as

$$Q_{21} = \sigma A_1 F_{12}(T_1{}^4 - T_2{}^4) \tag{8.3-24}$$

or

$$Q_{21} = \sigma A_2 F_{21}(T_2{}^4 - T_1{}^4) \tag{8.3-25}$$

where F is the "view factor," and F_{ij} represents the fraction of the total radiation leaving A_i which strikes A_j. Values for F may be found in Fig. 8.3-7.

If the surfaces are connected by a refractory, the absorption and reradiation of the refractory must be accounted for. (See Fig. 8.3-8.) The presence of a refractory leads to increased heat transfer between two surfaces. The appropriate view factor is then $\bar{F}$ from Fig. 8.3-7.

If the surfaces are not black, but can be assumed to be gray—that is, α is independent of λ—the integration can be shown to yield a view factor

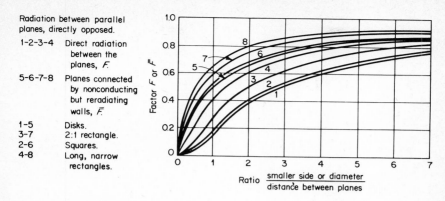

Radiation between parallel planes, directly opposed.

1-2-3-4	Direct radiation between the planes, F.
5-6-7-8	Planes connected by nonconducting but reradiating walls, $\bar{F}$.
1-5	Disks.
3-7	2:1 rectangle.
2-6	Squares.
4-8	Long, narrow rectangles.

Ratio $\dfrac{\text{smaller side or diameter}}{\text{distance between planes}}$

Fig. 8.3-7 View factors. *(W. H. McAdams, "Heat Transmission," 3d ed., McGraw-Hill Book Company, New York, 1954.)*

$$\mathscr{F}_{12} = \frac{1}{1/\bar{F}_{12} + (1/\varepsilon_1 - 1) + (A_1/A_2)(1/\varepsilon_2 - 1)} \tag{8.3-26}$$

where α is assumed equal to ε for each surface singly.

Example 8.3-1 Heat transfer by radiation—blackbody Rather than applying conventional insulation directly to the outside of a furnace, it has been suggested that we use insulation which is somewhat heat sensitive and build a double wall with an air gap as in Fig. 8.3-9 to prevent direct contact of the hot steel wall and the insulation. Given the data shown in Fig. 8.3-9, what will be the inside temperature of the insulation? We neglect convection in the air gap.

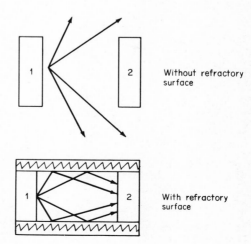

Without refractory surface

With refractory surface

Fig. 8.3-8 Absorption and reradiation in presence of a refractory.

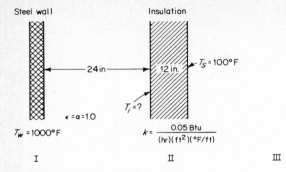

Fig. 8.3-9 System for Example 8.3-1.

Solution The heat transferred passes through I primarily by radiation (neglect natural convection), and through II by conduction. An energy balance shows

$$(a) \quad Q_I = Q_{II}$$

Further:

$$(b) \quad Q_I = \sigma A \, \varepsilon (T_w^4 - T_i^4)$$

(α and ε are evaluated at the temperature of the emitter since the absorption primarily depends on the character of the incident radiation rather than the temperature of the absorber)[1]

where $\sigma = 0.1714 \times 10^{-8}$ Btu/hr ft^2 °R^4

$$T_w = 1460°R$$

$$\varepsilon = 1.0$$

and

$$(c) \quad Q_{II} = kA \, \frac{T_i - T_s}{x_2 - x_1}$$

where $k = 0.05$ Btu/hr ft °F $= 0.05$ Btu/hr ft °R

$x_2 - x_1 = 1$ ft

$$T_s = 560°R$$

[1]G. G. Brown and Associates, "Unit Operations," p. 460, John Wiley & Sons, Inc., New York, 1950.

Substituting equations (b) and (c) in equation (a):

$$\sigma A \varepsilon \, (T_w{}^4 - T_i{}^4) = \frac{kA \, (T_i - T_s)}{x_2 - x_1}$$

Everything is known in the above relation except T_i:

$$0.1714 \times 10^{-8} \, \frac{\text{Btu}}{\text{hr ft}^2 \, {}^\circ\text{R}^4} \, (1.0)(\overline{1460^4} - T_i{}^4) \, {}^\circ\text{R}^4$$

$$= 0.05 \, \frac{\text{Btu}}{\text{hr ft } {}^\circ\text{R}} \, \frac{(T_i - 560) \, {}^\circ\text{R}}{1 \text{ ft}}$$

or, solving by trial,

$$T \cong 1458 \, {}^\circ\text{R} = 998 \, {}^\circ\text{F}$$

Note that the air gap does no good—the high temperature of the insulation is almost the same as the steel wall—in fact, the gap distance does not enter until it is no longer possible to approximate the problem as an infinite plane.

□ □ □

Example 8.3-2 Heat transfer by radiation—gray body A spacecraft which may be approximated as an infinite cylinder is to be used in interplanetary journeys where the surface of the craft "sees" only space: a perfect blackbody with $T = 0{}^\circ\text{R}$. The surface is gray and the emissivity of the surface is 0.1. If the surface of the craft is maintained at 40°F, what is the heat loss per square foot by radiation?

Solution

$$Q_{12} = \sigma A_1 \mathscr{F}_{12}(T_1{}^4 - T_2{}^4)$$

Rearranging:

$$\frac{Q}{A} = \sigma \mathscr{F}_{12}(T_1{}^4 - T_2{}^4) = q$$

But

$$\mathscr{F}_{12} = \frac{1}{1/\overline{F}_{12} + (1/\varepsilon_1 - 1) + A_1/A_2 \, (1/\varepsilon_2 - 1)}$$

and

$$\frac{A_1}{A_2} = 0 \text{ since the space "seen" is essentially infinite in area com-}$$
pared to the surface area of the craft

$\overline{F}_{12} = F_{12}$ since no energy is reradiated by other surfaces

$F_{12} = 1.0$ since the cylinder "sees" only space

Therefore

$$\mathscr{F}_{12} = \frac{1}{1 + (1/\varepsilon_1 - 1) + 0} = \varepsilon_1$$

Substituting

$$\frac{Q}{A} = 0.1714 \times 10^{-8} \text{ Btu/hr ft}^2 \text{ }^\circ\text{R}^4 \text{ } (0.1) \text{ } (500^4 - 0^4) \text{ }^\circ\text{R}^4$$

$$\frac{Q}{A} = 10.7 \text{ Btu/hr ft}^2$$

□ □ □

APPENDIX 8A: INTEGRATION TO OBTAIN VIEW FACTOR

Assume we have radiant heat interchange between two blackbodies without an intervening medium. If all the energy emitted from each of the surfaces is incident upon the other, for example, infinite parallel planes, we can easily calculate the heat interchange. The energy emitted from surface 1 (all of which will be absorbed by surface 2) is

$$Q_1 = A_1 \sigma T_1^4 \tag{8A-1}$$

and that emitted from surface 2 (all of which will be absorbed by surface 1) is

$$Q_2 = A_2 \sigma T_2^4 \tag{8A-2}$$

The net heat transfer, therefore, if $T_2 > T_1$, is

$$Q_{21} = A\sigma(T_2^4 - T_1^4) \tag{8A-3}$$

$$Q_{12} = A\sigma(T_1^4 - T_2^4) \tag{8A-4}$$

where the subscripts are dropped since $A_1 = A_2$. For the case where all radiation emitted by one of the surfaces is not incident on the other surface, the situation is more complex, but we ultimately structure the final equation in the form of Eq. (8A-3), modified by a correction factor, called the view factor, so that

$$Q_{21} = F_{21} A_2 \sigma(T_2{}^4 - T_1{}^4) \tag{8A-5}$$

where F_{21} is based on A_2 or

$$Q_{12} = F_{12} A_1 \sigma(T_1{}^4 - T_2{}^4) \tag{8A-6}$$

Let us consider how one obtains a view factor for different situations. Consider two black surfaces at arbitrary orientation with respect to each other and further consider the elemental areas on each surface, dA_1 and dA_2 (see Fig. 8.3-6). Let β_1 be the angle between the line connecting the differential areas and the normal to dA_2. Designate the length of the line connecting the areas as r. The energy flux emitted from dA_1 is

$$q_{b1} = \sigma T_1{}^4 \tag{8A-7}$$

and from dA_2

$$q_{b2} = \sigma T_2{}^4 \tag{8A-8}$$

The fluxes q_{b1} and q_{b2} in Eqs. (8A-7) and (8A-8) represent the total radiation *emitted* from surfaces 1 and 2. But for an arbitrary orientation this will not necessarily be the amount of radiation *striking* other surfaces. Thus we must determine the amount of radiation emitted in a given direction. To characterize this energy we define the intensity of radiation, which is the radiant energy leaving the emitting surface per unit area per unit solid angle[1] and per unit area of the projection of the surface on a plane perpendicular to the given direction. (See Fig. 8A-1.) For a blackbody, the intensity of radiation is independent of direction, and this is one of the reasons that this concept is useful. Notice that the fact that the intensity is independent of angle *does not* mean that the *amount* of energy emitted in a given direction is independent because the *intensity* contains the *projected* area. Consider the case of a hot plate made of rusty iron, which is very close to a blackbody in its emissive characteristics. If you place your face 1 ft directly above the surface, you certainly feel a larger flux of radiant heat than if you place your face 1 ft away, but at a line of sight of 10° with the surface.

---

[1] A solid angle is the ratio of the surface area subtended on a sphere divided by the square of the radius. It is measured in steradians and if the area subtended is equal to the square of the radius, the solid angle is equal to 1 steradian. It follows if the area subtended is the complete spherical surface (area $= 4\pi r^2$) the solid angle is equal to $4\pi r^2/r^2$ or 4π steradians. Note that this is analogous to radian measure in one dimension.

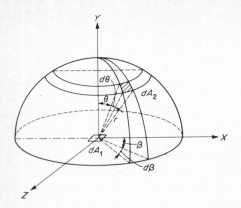

Fig. 8A-1 Emission from radiating surface. *(From "Engineering Heat Transfer" by S. T. Hsu, copyright (c) 1963 by Litton Educational Publishing, Inc., by permission of D. Van Nostrand, Inc.)*

Consider the solid angle shown in Fig. 8A-2. Let the radiant energy per unit time and unit area leaving the surface and *contained within the solid angle shown* be called Δq. By the definition

$$I = \frac{\Delta q}{\cos \beta \, d\omega} \qquad (8A-9)$$

To get the total energy flux passing from the surface into the hemisphere above the surface, we integrate as follows:

$$q = \int_{\omega} I \cos \beta \, d\omega \qquad (8A-10)$$

The solid angle is

$$d\omega = \frac{r \sin \beta \, r \, d\beta \, d\phi}{r^2} \qquad (8A-11)$$

Substituting

$$q = \int_{0}^{2\pi} \int_{0}^{\pi/2} I \cos \beta \sin \beta \, d\beta \, d\phi \qquad (8A-12)$$

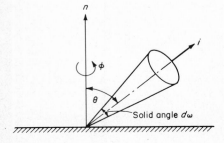

Fig. 8A-2 Intensity of radiation. *(From E. M. Sparrow and R. D. Cess, "Radiation Heat Transfer," copyright (c) 1966 by Wadsworth Publishing Company, Inc., reprinted by permission of the publisher, Brooks/Cole Publishing Company.)*

but if I is independent of β and ϕ, as it is for a blackbody, the expression easily integrates to

$$q = \pi I \qquad \text{(8A-13)}$$

Now, let us consider the case raised just before Eqs. (8A-7) and (8A-8). As in these equations, the energy flux emitted from dA_1 is

$$q_{b1} = \sigma T_1{}^4 \qquad \text{(8A-7)}$$

and from dA_2 is

$$q_{b2} = \sigma T_2{}^4 \qquad \text{(8A-8)}$$

But as shown above, the intensity of radiation from dA_1 is

$$I_1 = \frac{q_{b1}}{\pi} \qquad \text{(8A-14)}$$

Therefore the rate of energy emission from dA_1 per unit solid angle is

$$I_1 \cos \beta_1 \, dA_1 = \frac{\sigma T_1{}^4}{\pi} \cos \beta_1 \, dA_1 \qquad \text{(8A-15)}$$

(The units are Btu/hr unit solid angle.) But the solid angle subtended by surface dA_2 as viewed from dA_1 is $\cos \beta_2 \, dA_2/r^2$. We can write the total energy received by dA_2 from dA_1 as $(\sigma T_1{}^4/\pi) \left[(\cos \beta_1 \cos \beta_2 \, dA_1 \, dA_2)/r^2 \right]$. Since the problem is symmetric, the rate of energy received by dA_1 from dA_2 may be written as $(\sigma T_2{}^4/\pi) \left[(\cos \beta_2 \cos \beta_1 \, dA_2 \, dA_1)/r^2 \right]$. The difference between these two is the net rate of radiant heat interchange:

$$dQ_{12} = \frac{\sigma(T_1{}^4 - T_2{}^4) \cos \beta_1 \cos \beta_2 \, dA_1 \, dA_2}{\pi r^2} \qquad \text{(8A-16)}$$

Now, extending the treatment to finite areas, we obtain

$$Q_{12} = \sigma(T_1{}^4 - T_2{}^4) \left[\int_{A_1} \int_{A2} \frac{\cos \beta_1 \cos \beta_2 \, dA_1 \, dA_2}{\pi \, r^2} \right] \qquad \text{(8A-17)}$$

As noted at the beginning of this appendix, we wish to identify this equation with the form

$$Q_{21} = \sigma F_{21} A_2 (T_2{}^4 - T_1{}^4) \qquad \text{(8A-18)}$$

or

$$Q_{12} = \sigma F_{12} A_1 (T_1^4 - T_2^4) \tag{8A-19}$$

Therefore, it is apparent that the term in brackets in Eq. (8A-17) is $F_{21} A_2$, or $F_{12} A_1$. The factor is commonly called the view factor.

Example 8A-1   View factor for directly opposed "black" parallel disks of unequal area[1]   Determine the view factor for radiant heat transfer between a small disk of area A_1 and a parallel large disk of area A_2. The disks are assumed to be directly opposed; i.e., a line joining their centers is normal to both disks. The large disk has a radius α, and the distance between centers is r_0. This system is shown in Fig. 8A-3.

$$F_{12} = \frac{1}{A_1} \int_{A_2} \int_{A_1} \frac{dA_1 \cos \beta_1 \, dA_2 \cos \beta_2}{\pi r^2}$$

In this problem A_1 is small compared with A_2, and so for the integration over A_1, the other quantities under the integral sign can be considered constant. In addition, the angles β_1 and β_2 are equal; they will be designated simply as β. Therefore the equation for the view factor reduces to

$$F_{12} = \int_{A_2} \frac{\cos^2 \beta \, dA_2}{\pi r^2}$$

The differential element dA_2 is shown in the figure. The area of the element is

$$dA_2 = \rho \, d\psi \, d\rho$$

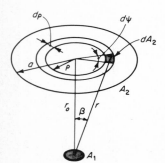

Fig. 8A-3 Two parallel disks. (*C. O. Bennett and J. E. Myers, "Momentum, Heat, and Mass Transfer," McGraw-Hill Book Company, New York 1962.*)

[1]C. O. Bennett and J. E. Myers, "Momentum, Heat, and Mass Transfer," McGraw-Hill Book Company, New York, 1962.

It can also be seen from the construction that $\cos \beta = r_0/r$. Thus the view factor may be written as

$$F_{12} = \int_0^\alpha \int_0^{2\pi} \frac{r_0^2}{\pi r^4} \rho \, d\psi \, d\rho$$

The system is symmetrical, and so the first integration produces the result

$$F_{12} = \int_0^\alpha \frac{2r_0^2}{r^4} \rho \, d\rho$$

The quantity r is a function of ρ as follows:

$$r^2 = \rho^2 + r_0^2$$

Thus we have

$$F_{12} = \int_0^\alpha \frac{2r_0^2 \rho \, d\rho}{(\rho^2 + r_0^2)^2} = \int_0^\alpha \frac{r_0^2 \, d\rho^2}{(\rho^2 + r_0^2)^2}$$

which, upon integration, yields

$$F_{12} = \frac{\alpha^2}{\alpha^2 + r_0^2}$$

□ □ □

PROBLEMS

8.1   A composite furnace wall consists of 9 in. of firebrick ($k = 0.96$ Btu/hr ft °F), 4 1/2 in. of insulating brick ($k = 0.183$ Btu/hr ft °F), and 4 1/2 in. of building brick ($k = 0.40$ Btu/hr ft °F).

   (a) Calculate the heat flux in Btu/hr ft² when the inside surface temperature is 2400°F and the outside is 100°F.

   (b) What are the temperatures at the interfaces of each layer of brick? Is the temperature of the insulating brick below the maximum allowable value of 2000°F?

8.2   A furnace wall consists of two layers of firebrick, $k_1 = 0.8$ Btu/hr ft °F and $k_2 = 0.1$ Btu/hr ft °F, and a steel plate $k_3 = 25$ Btu/hr ft °F as shown below.

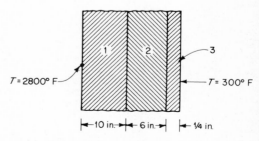

Calculate the heat flux (Btu/hr ft²) and the interface temperatures between materials 1 and 2, 2 and 3.

8.3 A cylindrical conduit carries a gas which keeps the inside surface of the conduit at 600°F. This conduit is insulated with 4 in. of rock wool and the outer surface of this insulation is at a temperature of 100°F. If the conduit is 1/4 in. thick with a 3 1/2-in. ID, find the rate of heat loss per foot length of conduit. The thermal conductivity of the conduit is 0.88 Btu/hr ft °F. The thermal conductivity of the rock wool is given by the equation

$$k = 0.025 + 0.00005T \qquad (T \text{ is in °F and } k \text{ has the same units as the thermal conductivity of the conduit})$$

8.4 A standard 1-in. schedule 40 iron pipe carries saturated steam at 250°F. The pipe is lagged (insulated) with a 2-in. layer of 85 percent magnesia pipe covering, and outside this magnesia there is a 3-in. layer of cork. The inside temperature of the pipe wall is 249°F, and the outside temperature of the cork is 90°F. Calculate (a) the heat loss from 100 ft of pipe, in Btu/hr, and (b) the temperatures at the boundaries between metal and magnesia and between magnesia and cork.

8.5 The thermal insulation on a steam pipe (2 in. OD) consists of magnesia-asbestos (85 percent magnesia) 2 in. thick with a thermal conductivity $k = 0.038$ Btu/hr ft °F. Calculate the temperature at the midpoint of the insulation when the inside surface is at 250°F and the outside surface is at 50°F.

8.6 A 6-in.- nominal-diameter (OD 6.625 in.) pipe carrying Dowtherm vapor at 750°F is covered with a 1.5-in. layer of high-temperature insulation ($k = 0.09$ Btu/hr ft °F) plus a 2-in. layer of 85 percent magnesia ($k = 0.038$ Btu/hr ft °F). If the outside surface of the outer layer has a temperature of 70°F, find (a) the heat loss in Btu/hr per linear foot of pipe, and (b) the temperature at the interface between the two insulations. (*State your assumptions.*)

8.7 If the thermal conductivity of a solid plate varies with temperature as follows:

$$k(T) = 10\left(1 + 0.01\,\frac{T}{°F} + 2.4 \times 10^{-5}\,\frac{T^2}{°F^2}\right) \text{ Btu/hr ft °F}$$

find the heat flow Q through a section of plate 2 by 3 ft and 2 in. thick if the surfaces of the plate are at 300°F and 100°F. Assume 1-D heat transfer.

8.8 Consider the trapezoidal cross-sectioned brick shown below. The brick is surrounded on four sides by insulation. The heat flow is from left to right in the figure and therefore we may consider the conduction of heat as being only in the x direction to a good approximation.

(a) Find the temperature profile in the x direction.
(b) What is the heat flux as a function of x?
(c) In the equation

$$\frac{Q_1}{A_m} = -k(T_2 - T_1)$$

What is the correct mean area?

8.9   The composite body shown below is 10 ft long in the direction perpendicular to the page. Considering that the heat flow is only in the x direction find the temperature at $x = x_1$ and the heat fluxes q_2 and q_3 through the right side of the body.

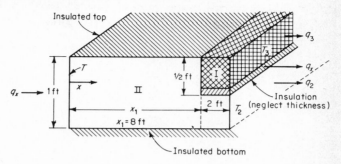

$$k_I = 0.1 \text{ btu/hr ft }°F$$
$$k_{II} = 0.8 \text{ btu/hr ft }°F$$
$$A_1 = 1 \text{ ft} \times 10 \text{ ft} = 10 \text{ ft}^2$$
$$A_2 = A_3 = \tfrac{1}{2} \text{ ft} \times 10 \text{ ft} = 5 \text{ ft}^2$$

8.10

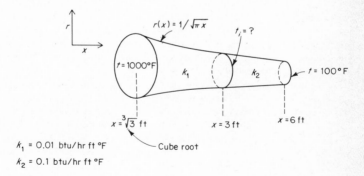

$$k_1 = 0.01 \text{ btu/hr ft }°F$$
$$k_2 = 0.1 \text{ btu/hr ft }°F$$

Shown is a scale model of a missile nose cone. (Dimensions have been chosen to make numerical computations easier.) It has a *circular* cross section of radius r which varies with the axial position x according to the relation

$$r = \frac{1}{\sqrt{\pi x}} \qquad \text{with } r \text{ and } x \text{ in feet (units of 1 are ft}^2)$$

The nose cone is constructed of two materials having thermal conductivities k_1 and k_2. During a recent test of our scale model, the temperature of the large end was measured at 1000°F while that of the small end was 100°F. Assuming heat conduction only in the axial (x) direction, determine for these end temperatures:

    (a) The temperature profile in material 1.
    (b) The temperature t_i where the two materials are joined together.
    (c) The heat flux across the end at the position $x = 6$ ft.
    (d) The heat flow in Btu/hr across the junction at $x = 3$ ft.

8.11 A heat exchanger is cooled at steady state by water flow on the outside surface of its brass wall which has a thermal conductivity of 60 Btu/hr ft °F. If a thermocouple in the brass at point A reads 200°F and point B is 160°F, what is the heat transfer coefficient on the outside surface if the free-stream temperature is 60°F? (Hint: Assume one-dimensional heat transfer.)

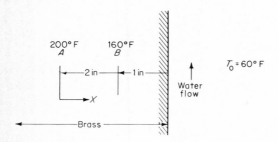

8.12 A window consists of two layers of glass, each 1/4 in. thick, separated by a layer of dry, stagnant air, also 1/4 in. thick. If the temperature drop over the composite system is 30 °F, find the heat loss through such a window, 10 ft long and 4 ft wide. If the window replaces a single glass pane, 3/8 in. thick, at a difference in cost of $120, find the number of days of operation under the specified temperature conditions to pay for the window. Coal with a heating value of 13,200 Btu/lb costs $20 per ton. The furnace efficiency can be assumed to be 50 percent. The thermal conductivity of the window glass is 0.5 Btu/hr ft °F. Is the assumption of stagnant air a good one?

8.13 Calculate the direct radiant heat transfer between two parallel refracting surfaces 12 by 12 ft spaced 12 ft apart, if the surface temperatures are 1300°F and 500°F respectively. What would be the radiant heat transfer if the two plane walls were connected by nonconducting but reradiating walls?

8.14 Two rectangular plates both 6 by 3 in. are placed directly opposite one another with a plate spacing of 2 in. The larger plate is at 240°F, and has an emissivity of 0.5. The other plate is at 110°F, and has an emissivity of 0.33. If there are no other surfaces present, what is the heat flow? If refractory walls now connect the plates, what is the heat flow?

8.15 Calculate the direct radiant heat transfer between two parallel refractory surfaces 10 by 10 ft spaced 10 in. apart. Assume that the temperatures of the surfaces are 600°F and 1200°F respectively. What is the radiant heat transfer if the two planes are connected by nonconducting but reradiating walls?

8.16 Calculate the radiant heat loss from a 3-in. nominal wrought iron oxidized pipe at 200°F which passes through a room at temperature 76°F. What is the heat loss if the pipe is coated with aluminum paint? The emissivities for the oxidized and painted surfaces are 0.8 and 0.3 respectively.

8.17 A large plane, perfectly insulated on one face and maintained at a fixed temperature T_1 on the bare face which has an emissivity of 0.90, loses 200 Btu/hr ft² when exposed to surroundings at absolute zero. A second plane having the same size as the first is also perfectly insulated on one face, but its bare face has an emissivity of 0.45. When the bare face of the second plane is maintained at a fixed temperature T_3 and exposed to surroundings at absolute zero, it loses 100 Btu/hr ft².

Let these planes be brought close together, so that the parallel bare faces are only 1 in. apart, and the heat supply to each be so adjusted that their respective temperatures remain unchanged. What will be the net heat flux between the planes, expressed in Btu/hr ft²?

8.18[1] Air is flowing steadily through a duct whose inner walls are at 800°F. A thermocouple housed in a rusted steel well having a 1-in. OD indicates a temperature of 450°F. The mass velocity is 3600 lb$m$/hr ft^2. Estimate the true temperature of the air.

8.19[1] A thermocouple inserted in a long duct reads 200°F. The duct runs through a furnace that maintains the duct walls at 300°F. Inside the duct a nonabsorbing gas is flowing at such a velocity that the average convective heat transfer coefficient to the thermocouple bulb is 10 Btu/hr ft^2 °F. What is the *gas* temperature at the thermocouple in °F? (You may assume that all surfaces are black and that no heat is conducted along the thermocouple leads. Use T_t for the thermocouple, T_w for the wall, and T_y for the gas.)

[1] For these problems reference must be made to the concept of heat transfer coefficient Chap. 9.

9
Heat Transfer Coefficients

In this chapter we discuss heat transfer coefficients, their correlation and use. The form of the equation adopted for heat transfer is the following: heat flux equals driving force times conductance. This equation is adopted for many types of heat transfer and the "heat transfer coefficient" is calculated from this equation. These heat transfer coefficients, often called convective heat transfer coefficients, are really conductance terms in the flux equation and are adopted for many kinds of mechanisms including radiation. A somewhat analogous relationship was used in defining the friction factor, a momentum transfer coefficient. (However, in the case of the friction factor as defined, f is a resistance rather than a conductance.)

This chapter discusses individual heat transfer coefficients based on the flux equation. We then introduce overall heat transfer coefficients. Finally, there is an extensive discussion of various correlations and correlation equations which are used to estimate heat transfer coefficients in a variety of situations.

9.1 INDIVIDUAL HEAT TRANSFER COEFFICIENTS

The problem of theoretical analysis of convective heat transfer in laminar and turbulent flow is very difficult. Usually one adopts the equation

$$Q = hA(T_S - T) \tag{9.1-1}$$

where T_S is temperature at the surface of the wall. Two cases may be distinguished for the temperature T: one for internal flows and one for external flows. An obvious temperature to use for external flows is the free-stream temperature. For internal flows there are many possibilities, but we usually use the bulk temperature, paralleling the treatment for momentum transfer. The major problem is then the prediction of h. The value of the heat transfer coefficient h for a given situation may be found empirically from measurement of heat transfer rate, temperature drop, and associated heat transfer area. It can also sometimes be predicted from theoretical considerations.

For convective heat transfer from the heated wall of a pipe to the fluid flowing inside whose average temperature (mixing-cup temperature) is $\langle T \rangle$:

$$Q = hA(T_S - \langle T \rangle) \tag{9.1-2}$$

(The temperature difference is conventionally written so as to make Q positive.)

For convection next to a heated surface

$$Q = hA(T_S - T_0) \tag{9.1-3}$$

where T_0 is the temperature "far from the surface" (edge of the thermal boundary layer).

The form of Eq. (9.1-1) is also used for calculating heat transfer in boiling, condensation, and radiation. The boiling coefficient is defined by

$$Q = hA(T_S - T_{sL}) \tag{9.1-4}$$

The primary resistance to heat transfer in boiling is usually the resistance of the vapor film that forms immediately adjacent to the hot surface, and therefore it is reasonable to use the temperature drop across this particular film. The saturated liquid temperature T_{sL} exists on one side of the film, and on the other side, the temperature is that of the surface, T_S.

For condensation, the primary resistance to heat transfer comes from the liquid film that forms next to the surface, and therefore we use

Table 9.1-1   Individual heat transfer coefficients[a]

System	h, $Btu/hr\ ft^2\ °F$
Steam, dropwise condensation	5000–20,000
Steam, film-type condensation	1000–3,000
Water, boiling	300–9,000
Organic vapors, condensing	200–400
Water, heating	50–3,000
Oils, heating or cooling	10–300
Steam, superheating	5–20
Air, heating or cooling	0.2–10

[a] W. H. McAdams, "Heat Transmission," 3d ed., McGraw-Hill Book Company, New York, 1954.

the temperature drop across this film (assuming condensation of a pure vapor). The condensing coefficient is defined by

$$Q = hA(T_{sV} - T_S) \tag{9.1-5}$$

where T_{sV} is the saturated vapor temperature.

One can also define a radiation coefficient

$$Q = h_r A(T_1 - T_2) \tag{9.1-6}$$

where T_1 and T_2 are the temperatures of the two surfaces between which the radiation is transferred. (The radiation coefficient depends on a power of the absolute temperature, and therefore it is not as useful since it is very sensitive to the temperature at which the process occurs.)

Table 9.1-1 gives typical values of individual heat transfer coefficients in various situations. This table is to give the reader an idea of the relative magnitude of h under various conditions; it is not for use in calculation.

9.2   OVERALL HEAT TRANSFER COEFFICIENTS

In Chap. 8 we considered the problem of a series of resistances in conductive heat transfer. Consider now adding convective resistances to this path, such as for the insulated pipe in Fig. 9.2-1. The inside heat transfer coefficient is denoted as h_i, and the outside heat transfer coefficient as h_0. We would like to develop an equation where Q is equal to an *overall* conductance times an appropriate area times the *overall* temperature drop. We normally base the *overall* coefficient either on the inside heat transfer area or on the outside heat transfer area. For example, for a pipe, we would use either the inside area of the pipe or the outside area of the pipe. It is important to recognize that every overall heat transfer coefficient has associated with it an appropriately defined area. And further, these two quantities (the overall heat transfer coefficient and the associated area) must be used together.

At steady state the heat flow through the pipe wall is given by any of the following equations

$$Q = h_i A_i (T_1 - T_2) \qquad (9.2\text{-}1)$$

$$Q = \frac{k_p A_{p\mathrm{lm}} (T_2 - T_3)}{\Delta r_p} \qquad (9.2\text{-}2)$$

$$Q = \frac{k_I A_{I\mathrm{lm}} (T_3 - T_4)}{\Delta r_I} \qquad (9.2\text{-}3)$$

$$Q = h_0 A_0 (T_4 - T_5) \qquad (9.2\text{-}4)$$

Again, we could use any *one* of the above equations to calculate the heat flow, but the required temperature drop is usually not available. Only the *overall* temperature drop is usually at our disposal. To develop an equation in terms of overall temperature drop, we use exactly the same development as we did for a series of purely conductive resistances; that is, we add up all the individual temperature drops to get the overall temperature drop, and we substitute in terms of the heat flux equation to determine the overall resistance. Writing Eqs. (9.2-1) to (9.2-4) in terms of the appropriate temperature differences:

$$T_1 - T_2 = \frac{Q}{h_i A_i} \qquad (9.2\text{-}5)$$

$$T_2 - T_3 = \frac{Q \, \Delta r_p}{k_p A_{p\mathrm{lm}}} \qquad (9.2\text{-}6)$$

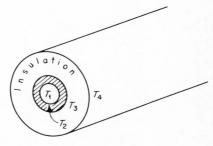

Fig. 9.2-1 Insulated pipe.

$$T_3 - T_4 = \frac{Q \, \Delta r_I}{k_I A_{I\text{lm}}} \tag{9.2-7}$$

$$T_4 - T_5 = \frac{Q}{h_0 A_0} \tag{9.2-8}$$

Adding Eqs. (9.2-5) to (9.2-8) gives

$$Q = \frac{T_1 - T_5}{1/h_i A_i + \Delta r_p / k_p A_{p\text{lm}} + \Delta r_I / k_I A_{I\text{lm}} + 1/h_0 A_0} = \frac{(\Delta T)_{\text{overall}}}{\Sigma R} \tag{9.2-9}$$

where ΣR represents the sum of the thermal resistances.

First, let us develop the average coefficient associated with the inside area. If we multiply Eq. (9.2-9) by the ratio of the inside area of the pipe to itself A_i / A_i (we are multiplying by 1)

$$Q = A_i \, (\Delta T)_{\text{overall}} \left[\frac{1}{1/h_i + \Delta r_p A_i / k_p A_{p\text{lm}} + \Delta r_I A_i / k_I A_{I\text{lm}} + A_i / h_0 A_0} \right] \tag{9.2-10}$$

The term in the brackets in Eq. (9.2-10) is called the overall heat transfer coefficient based on the *inside* area U_i and

$$Q = U_i A_i \, (\Delta T)_{\text{overall}} \tag{9.2-11}$$

Now, let us develop the overall coefficient associated with the *outside* area. If we multiply Eq. (9.2-9) by the ratio of outside area of the insulation to itself A_0 / A_0

$$Q = A_0 \, (\Delta T)_{\text{overall}} \left[\frac{1}{A_0 / h_i A_i + \Delta r_p A_0 / k_p A_{p\text{lm}} + \Delta r_I A_0 / k_I A_{I\text{lm}} + 1/h_0} \right] \tag{9.2-12}$$

The term in the brackets is the overall heat transfer coefficient based on the *outside* area, U_0, and

$$Q = U_0 A_0 \, (\Delta T)_{\text{overall}} \tag{9.2-13}$$

Typical values of U are shown in Table 9.2-1; as with Table 9.1-1, they are illustrative values and are not meant for calculation purposes.

Many fluids form deposits on heat transfer surfaces, causing an additional thermal resistance. This resistance can become large enough that the exchanger must be periodically cleaned. This may be done chemically, or, for deposits inside the tubes, by removing the exchanger head and running a rod with a scraper attached which mechanically abrades the deposits.

Obviously, when designing an exchanger, to facilitate cleaning one must consider where the deposits will form. It is customary to assign a heat transfer coefficient to account for the resistance of these deposits. This coefficient is called the fouling coefficient h_f. (Another term, the "fouling factor" $= 1000/h_f$, is sometimes used.) The area associated with the fouling coefficient is taken as the area of the fouled surface on which the material is deposited. The overall heat transfer coefficient based on outside area for an insulated pipe with dirt on the inside of the pipe is

$$U_0 = \frac{1}{A_0/h_i A_i + A_0/h_f A_i + \Delta r_p A_0/k_p A_{p\text{lm}} + \Delta r_I A_0/k_I A_{I\text{lm}} + 1/h_0}$$
(9.2-14)

where h_f is the fouling coefficient. Fouling coefficients vary tremendously depending on the particular fluids; typical values are shown in Table 9.2-2.

The procedure used above can be applied to a general problem with any number of conductive and convective resistances in series to yield the same sort of result; that is, the overall heat transfer coefficient is the reciprocal of the sum of thermal resistances, which can be written for conduction as

$$\frac{\Delta x_l A_j}{k_l A_{\text{mean}}}$$
(9.2-15)

and for convection as

$$\frac{A_j}{h_l A_l}$$
(9.2-16)

where Δx is the appropriate thickness coordinate, A_j is the area upon which U_j is based, and A_{mean} is the appropriate mean area for the conduction path, e. g., logarithmic for cylindrical geometry. In equation form, for m conductive and n convective resistances:

$$U_j = \frac{1}{\sum_{l=1}^{m} (\Delta x_l A_j/k_l A_{\text{mean}}) + \sum_{l=1}^{n} (A_j/h_l A_l)}$$
(9.2-17)

This equation is for *series* resistances, which are by far the most common. The above general approach can also be easily applied for parallel or series-parallel resistances.

Table 9.2-1   Overall heat transfer coefficients

Table 9.2-1a   **Miscellaneous: a range of values of miscellaneous overall coefficients[a] U, expressed in Btu/hr ft² °F as found in practice. Under special conditions higher or lower values may be realized.**

Type of heat exchanger	State of controlling resistance		Typical fluid	Typical apparatus
	Free convection, U	Forced convection, U		
Liquid to liquid	25–60	50–300	Water	Liquid to liquid heat exchangers
Liquid to liquid	5–10	20–50	Oil	
Liquid to gas (atm. pressure)	1–3	2–10		Hot-water radiators
Liquid to boiling liquid	20–60	50–150	Water	Brine coolers
Liquid to boiling liquid	5–20	25–60	Oil	
Gas (atm. pressure) to liquid	1–3	2–10		Air coolers, economizers
Gas (atm. pressure) to gas	0.6–2	2–6		Steam superheaters
Gas (atm. pressure) to boiling liquid	1–3	2–10		Steam boilers
Condensing vapor to liquid	50–200	150–800	Steam-water	Liquid heaters and condensers
Condensing vapor to liquid	10–30	20–60	Steam-oil	
Condensing vapor to liquid	40–80	60–300	Organic vapor-water	
Condensing vapor to liquid		15–300	Steam-gas mixture	
Condensing vapor to gas (atm. pressure)	1–3	6–16		Steam pipes in air Air heaters
Condensing vapor to boiling liquid	40–100			Scale-forming evaporators
Condensing vapor to boiling liquid	300–800		Steam-water	
Condensing vapor to boiling liquid	50–150		Steam-oil	
Condensing vapor to boiling liquid		50–400	Steam-organic liquid	Steam-jacketed tubes

[a]J. H. Perry (ed.), "Chemical Engineers' Handbook," 3d ed., McGraw-Hill Book Company, New York, 1956.

Table 9.2-1b   Coils immersed in liquids. Overall coefficients U, expressed as Btu/hr ft² °F

Substance inside coil	Substance outside coil	Coil material	Agitation	U	Reference
Steam	Water	Lead	Agitated	70	a
Steam	Sugar and molasses solutions	Copper	None	50–240	b
Steam	Boiling aqueous solution			600	c
Cold water	Dilute organic dye intermediate	Lead	Turboagitator at 95 rpm	300	c
Cold water	Warm water	Wrought iron	Air bubbled into water surrounding coil	150–300	d
Cold water	Hot water	Lead	0.40 rpm paddle stirrer	90–360	e
Brine	Amino acids		30 rpm	100	c
Cold water	25% oleum at 60°C	Wrought iron	Agitated	20	f
Water	Aqueous solution	Lead	500 rpm, sleeve propeller	250	a
Water	8% NaOH		22 rpm	155	c
Steam	Fatty acid	Copper (pancake)	None	96–100	g
Milk	Water		Agitation	300	h
Cold water	Hot water	Copper	None	105–180	i
60°F water	50% aqueous sugar solution	Lead	Mild	50–60	j
Steam and hydrogen at 1,500 psi	60°F	Steel		100–165	j

NOTE. Chilton, Drew, and Jebens, *Ind. Eng. Chem.*, **36**: 510 (1944) give film coefficients for heating and cooling agitated fluids using a coil in a jacketed vessel.

a Read, private communication.
b Stose and Whittemore, thesis, Massachusetts Institute of Technology, 1922.
c Chambers and Steves, private communication.
d Chilton and Colburn, private communication.
e Pierce and Terry, *Chem., & Met. Eng.*, **30**:872 (1924).
f Boertlein, private communication.
g Mills and Daniels, *Ind. Eng. Chem.*, **26**:248–250 (1934).
h Feldmeier, *Adv. paper, ASME Meeting*, Dec. 4, 1934; published in "Heat Transfer," 69–74, ASME, New York, 1936.
i Storrow, *J. Soc. Chem. Ind.*, **64**:322 (1945).
j Private communication.

Table 9.2-1c Miscellaneous: special equipment and materials

Type of equipment	Hot material	Cold material	U	Remarks	Reference
High-pressure boiler	Molten salt	Boiling water	100–150		a
Tubular exchanger	Molten salt	Oil	52–80		a
Steam superheater	Molten salt	Steam	70		a
Air heater	Molten salt	Air	6		a
Catalyst case	Gas	Molten salt	6	Fins on outside of tube	a
Double-pipe Karbate exchanger	Water	Water	300–500		b
Karbate trombone cooler	20°Bé. HCl	Water	300	Water $\Gamma_1' = 1750$	c
Karbate tube reboiler	Steam	20% HCl	136	Vertical thermosiphon reboiler	j
	Steam	35% HCl	472–575		j
Double-pipe pyrex glass exchanger using heat exchanger tubing	Air–water vapor	Water	25–75	Cooling water in annulus	d
	Water	Water	80–110		d
	Condensing steam	Water	100–125		d
Glass trombone cooler	50% sugar solution	60°F water	50–60	Sugar solution inside pipe	i
Glass pipe in trough	20°Bé. HCl	Water	25		c
Votator	Water	Water	520–1120	Rotor velocity = 300 —1900 rpm	e
Pebble heater	Solid pebbles	Air	4	Heating gases to 1900°F using $1/2$-in. pebbles	f
		Methane	9		f
		Hydrogen	22		f
Long-tube vertical evaporator	Condensing steam	Water	300–1200	Water $\Gamma_2' = 400$–21,000 inside tubes	g
Falling-film condenser	Condensing steam	Water	574–2300		h
Stainless-steel conveyor belt	Molten TNT	50°F air	5–7	Air blowing under and over belt	i
Partial condenser	Hydrocarbons and chlorinated hydrocarbons	Boiling propane	55–76	Refrigerated condenser	j
Shell and tube reboiler	Hot water	Hydrocarbons	42–88	Hot water in tubes	j
Reboiler	Steam	Chlorinated hydrocarbons	67	Clean reboiler, $\Delta t = 12°F$	j
			20	Same reboiler after several months service, $\Delta t = 96°F$	j

U = Btu/hr ft² °F.

Γ_1' = lb/hr ft of pipe length for each side of pipe.

Γ_2' = lb/hr ft of periphery.

a Newton and Shimp, *Trans. A.I.Ch.E.*, **41**: 197 (1945).

b Werking, *Trans. A.I.Ch.E.*, **35**: 489 (1939).

c Lippman, *Chem. & Met. Eng.*, **52** (3): 112 (1945).

d Thompson and Foust, *Chem. & Met. Eng.*, **47**: 410 (1940).

e Houlton, *Ind. Eng. Chem.*, **36**: 522–528 (1944).

f Norton, *Chem. & Met. Eng.*, **53** (7):116(1946).

g Cessna, Lientz, and Badger, *Trans. A.I.Ch.E.*, **36**: 759 (1940).

h McAdams, Drew, and Bays, *Trans. ASME*, **62**: 627(1940).

i Private communication.

j Breidenbach and O'Connell, *Trans. A.I.Ch.E.*, **42**: 761(1946).

354

Table 9.2-1d **Overall coefficients for heat exchangers in petroleum service**

Service of exchanger	Fluids		Velocity in tubes, ft/sec	Δt_m °F	Overall coefficient U, Btu/hr ft² °F
	Tubes	Shell			
Stabilizer reflux condensers	Water	Condensing vapors + residual gas	3.0	13.5	94
			5.0	22	145
		108 – 118° API	0.3–0.6		55–67[a]
			0.7		98–125[a]
Partial condensers	39° API crude	58° API gasoline	2.4	147	24
	55° API crude	62° API naphtha	4.4	80	37
	55° API crude	62° API naphtha	6.8	90	48
Stabilizer reboiler	Steam	58° API oil		33.5	42
Absorber reboiler	Steam	37° API oil		41.4	45
Stabilizer reboiler	Steam	67–74° API			43–183[a]
Oil preheater	42° API oil	Steam	1.4	32	108
Exchangers	60° API	58° API	1.4	65	74
	63° API	57° API	4.6	69	139
	70–82° API	67–74° API	0.3–0.7		18–37[a]
	70–82° API	67–74° API	0.3–0.7		35–45[a]
	43° API	37° API	1.6	59	33
	39° API crude	13° API residue	3.9	262	19
Coolers	Water	57° API		40	52
	Water	44° API		97	40
	Water	67–74° API	0.2		20[a]
	Water	67–74° API	0.4–0.7		51–53[a]

[a] McGiffin, *Trans. A.I.Ch.E.*, **38**:761 (1942). All other data in Table 9.2-1d are from Higgins, "Heat Transfer," p. 56, a special publication of the ASME, New York, 1936.

Table 9.2-1e Jacketed vessels. Overall coefficients

U, expressed in Btu/hr ft^2 °F

Fluid inside jacket	Fluid in vessel	Wall material	Agitation	U	Reference
Steam	Water	Enameled cast iron	0–400 rpm	96–120	a
Steam	Milk	Enameled cast iron	None	200	b
Steam	Milk	Enameled cast iron	Stirring	300	b
Steam	Milk boiling	Enameled cast iron	None	500	b
Steam	Milk	Enameled cast iron	200 rpm	86	a
Steam	Fruit slurry	Enameled cast iron	None	33–90	a
Steam	Fruit slurry	Enameled cast iron	Stirring	154	a
Steam	Water	Cast iron and loose lead lining	Agitated	4–9	c
Steam	Water	Cast iron and loose lead lining	None	3	c
Steam	Boiling SO$_2$	Steel	None	60	c
Steam	Boiling water	Steel	None	187	c
Hot water	Warm water	Enameled cast iron	None	70	a
Cold water	Cold water	Enameled cast iron	None	43	a
Ice water	Cold water	Stoneware	Agitated	7	c
Ice water	Cold water	Stoneware	None	5	c
Brine, low velocity	Nitration slurry		35–58 rpm	32–60	d
Water	Sodium alcoholate solution	"Frederking" (cast-in-coil)	Agitated, baffled	80	d
Steam	Evaporating water	Copper		381	e
Steam	Evaporating water	Enamelware		36.7	e
Steam	Water	Copper	None	148	f
Steam	Water	Copper	Simple stirring	244	f
Steam	Boiling water	Copper	None	250	f
Steam	Paraffin wax	Copper	None	27.4	g
Steam	Paraffin wax	Cast iron	Scraper	107	g
Water	Paraffin wax	Copper	None	24.4	g
Water	Paraffin wax	Cast iron	Scraper	72.3	g
Steam	Solution	Cast iron	Double scrapers	175–210	h
Steam	Slurry	Cast iron	Double scrapers	160–175	h
Steam	Paste	Cast iron	Double scrapers	125–140	h
Steam	Lumpy mass	Cast iron	Double scrapers	75–96	h
Steam	Powder (5% moisture)	Cast iron	Double scrapers	41–51	h

References:
a Poste, *Ind. Eng. Chem.*, **16**:469 (1924).
b Bowen, *Agr. Eng.*, **11**:27 (1930).
c Read, private communication.
d Chambers and Steves, private communication.
e Robson, *Australian Chem. Ind. J. & Proc.*, **3**:47–54 (1936).
f Chemical Engineering Charts No. 4. *Ind. Chemist*, **82**:374 (1931).
g Huggins, *Ind. Eng. Chem.*, **23**:749–753 (1931).
h Laughlin, *Trans. A.I.Ch.E.*, **36**:345 (1940).

Table 9.2-1f   Values of U for ammonia condensers

Type of condenser	Water rate		$\Delta t_m =$ 1.5°F	$\Delta t_m =$ 3.5°F	$\Delta t_m =$ 7°F	$\Delta t_m =$ 2—6°F
			Btu/hr ft² (°F overall Δt)			
Vertical tube and shell	$\Gamma_v =$	400	220	170	150	
		800	275	225	215	
		1,200	310	270	260	
		1,600	350	315	300	
		2,000		390	340	
		2,400		430	370	
Horizontal drip	$\Gamma_h =$	400				250
		800				330
		1,200				400
Double pipe	$V =$	4	350	270	230	
		6	410	320	280	
		8	470	390	350	

$\Gamma_v =$ lb of water/hr/ft of periphery.

$\Gamma_h =$ lb of water/hr/ft of tube length for each side of tube.

$V =$ ft/sec.

Example 9.2-1   Heat transfer resistance in series   A composite flat furnace wall consists of 9 in. of firebrick ($k = 0.96$ Btu/hr ft °F), $4^1/_2$ in. of insulating brick ($k = 0.183$ Btu/hr ft °F), and $4^1/_2$ in. of building brick ($k = 0.40$ Btu/hr ft °F). The individual heat transfer coefficients at the inside and outside walls are 10 Btu/hr ft² °F and 1 Btu/hr ft² °F respectively. The temperature of the gas within the furnace is 2400°F and the ambient air external to the furnace is 100°F. (a) Determine the overall heat transfer coefficient. (b) What is the heat transfer rate per square foot of wall area? (c) What is the temperature at the surface of each layer of brick? (d) Is the temperature of the insulating brick below the maximum allowable value of 2000°F?

Solution

(a) $Q = U_0 A_0 (\Delta T)_{\text{overall}} = U_i A_i (\Delta T)_{\text{overall}}$

and since $A_0 = A_i = A_a = A_b = A_c$:

$$U_0 = U_i = \frac{1}{1/h_i + \Delta x_a/k_a + \Delta x_b/k_b + \Delta x_c/k_c + 1/h_o}$$

where the subscripts refer to regions as shown in Fig. 9.2-2.

Table 9.2-2 Fouling coefficients[a]

Water				
Temperature of heating medium	Up to 240°F		240–400°F	
Temperature of water	125°F or less		Above 125°F	
Water velocity, ft/sec	3 and less	Over 3	3 and less	Over 3
Distilled	2,000	2,000	2,000	2,000
Seawater	2,000	2,000	1,000	1,000
Treated boiler feed water	1,000	2,000	1,000	1,000
Treated makeup for cooling tower	1,000	1,000	500	500
City, well, Great Lakes	1,000	1,000	500	500
Brackish, clean river water	500	1,000	330	500
River water, muddy, silty	330	500	250	330
Hard (over 15 g/gal)	330	330	200	200
Chicago Sanitary Canal	125	170	100	125

Chemicals	
Inorganic:	
Gases (oil-bearing or dirty)	500
Liquids (heating or vaporization)	500
Refrigerant brines	1,000
Organic:	
Gases	
Process	1,000
Utility (oil-bearing, refrigerant, etc.)	500
Condensing vapors (condensers)	1,000
Liquids	
Process	1,000
Vaporizing liquids (reboilers)	500
Heat transfer media	1,000
Refrigerant liquids	1,000
Polymer-forming liquids	200
Oils (vegetable and heavy gas oil)	330
Asphalt and residuum	100

[a] R. H. Perry, C. H. Chilton, and S. D. Kirkpatrick (eds.), "Chemical Engineers Handbook," 4th ed., McGraw-Hill Book Company, New York, 1963.

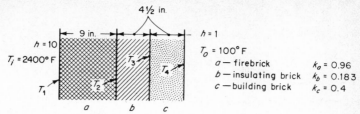

Fig. 9.2-2 Sketch for Example 9.2-1.

Evaluating terms individually:

$$\frac{1}{h_i} = (1 \text{ hr ft}^2 \text{ °F})/(10 \text{ Btu}) = 0.1 \text{ hr ft}^2 \text{ °F/Btu}$$

$$\frac{1}{h_0} = (1 \text{ hr ft}^2\text{°F})/(1 \text{ Btu}) = 1 \text{ hr ft}^2 \text{ °F/Btu}$$

$$\frac{\Delta x_a}{k_a} = \frac{9/12 \text{ ft}}{0.91 \text{ Btu/hr ft °F}} = 0.78 \text{ hr ft}^2 \text{ °F/Btu}$$

$$\frac{\Delta x_b}{k_b} = \frac{(4.5/12) \text{ ft}}{0.183 \text{ Btu/hr ft °F}} = 2.06 \text{ hr ft}^2 \text{ °F/Btu}$$

$$\frac{\Delta x_c}{k_c} = \frac{(4.5/12) \text{ ft}}{0.4 \text{ Btu/hr ft °F}} = 0.94 \text{ hr ft}^2 \text{ °F/Btu}$$

Then

$$U_0 = U_i = \frac{1}{0.1 + 0.78 + 2.06 + 0.94 + 1} = 0.205 \text{ Btu/hr ft}^2 \text{ °F}$$

(b) $\dfrac{Q}{A_0} = U_0 \, \Delta T = 0.205 \text{ (Btu/hr ft}^2 \text{ °F) } (2400 - 100) \text{ °F}$

$$= 472 \text{ Btu/hr ft}^2$$

(c) The interface temperatures are found by applying Eq. (9.2-17) four times, each time considering a different ΔT:

$$2400 - T_1 = \frac{Q}{Ah_i} = \frac{472 \text{ Btu/hr ft}^2 \text{ °F}}{\text{hr ft}^2 \text{ (10) Btu}}$$

$$= 47.2 \text{ °F}$$

$$T_1 = 2353 \text{ °F}$$

$$2353 - T_2 = \frac{Q \, (\Delta x_a/k_a)}{A} = (472)(0.78) = 368\text{°F}$$

$$T_2 = 1985°F$$

$$1985 - T_3 = (472)(2.06) = 972°F$$

$$T_3 = 1063°F$$

$$1063 - T_4 = (472)(0.94) = 443°F$$

$$T_4 = 570°F$$

The profile is sketched in Fig. 9.2-3.

(d) From the sketch, it can be seen that the insulating brick remains below the maximum temperature of 2000°F permitted.

□ □ □

Obviously if one of the thermal resistances in the denominator of the right-hand side of Eq. (9.2-17) becomes large with respect to the other resistances, this resistance "controls" the process; that is, the transmission of heat is much more sensitive to a change in this resistance than a *proportionate* change in the others.

Example 9.2-2 Controlling resistance A pilot-plant distillation column is made of sections of 10-in.-OD glass $^3/_8$ in. thick. It has been proposed to replace one section with an aluminum section with 0.37-in. wall thickness so that machining may be easily done to adapt certain instruments to the column for an experiment. The objection has been raised that the higher thermal conductivity of the aluminum will cause a great

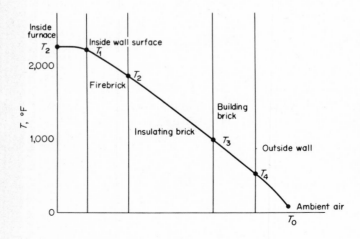

Fig. 9.2-3 Sketch for Example 9.2-1.

deal more heat to be transferred, generating more liquid in the column (internal reflux) and upsetting operations. The inside heat transfer coefficient is estimated to be about 1000 Btu/hr ft² °F and the outside coefficient about 1.0 Btu/hr ft² °F. The thermal conductivity of the glass is 7.5 Btu/hr ft² °F/in. and of aluminum is 1000 Btu/hr ft² °F/in. Discuss.

Solution

$$\frac{1}{U_0} = \frac{1}{h_0} + \frac{\Delta r A_0}{k A_{\mathrm{lm}}} + \frac{A_0}{h_i A_i}$$

For all practical purposes $A_0 = A_i = A_{lm}$, and so

$$\frac{1}{U_0} = \frac{1}{h_0} + \frac{\Delta r}{k} + \frac{1}{h_i}$$

where

$$\frac{1}{h_0} = \frac{1}{1.0} = 1.0 \text{ hr ft}^2 \text{ °F/Btu}$$

and

$$\frac{1}{h_i} = \frac{1}{1000} = 0.001 \text{ hr ft}^2 \text{ °F/Btu}$$

For glass

$$\frac{\Delta r}{k} = \frac{0.375 \text{ in.}}{7.5 \text{ Btu/(hr ft}^2 \text{ °F/in.)}} = 0.05 \text{ hr ft}^2 \text{ °F/Btu}$$

For aluminum

$$\frac{\Delta r}{k} = \frac{0.37 \text{ in.}}{1000 \text{ Btu/(hr ft}^2 \text{ °F/in.)}} = 0.00037 \text{ hr ft}^2 \text{ °F/Btu}$$

Therefore for glass

$$U_0 = \frac{1}{1.0 + 0.05 + 0.001} = 0.95 \text{ Btu/hr ft}^2 \text{ °F}$$

and for aluminum

$$U_0 = \frac{1}{1.0 + 0.00037 + 0.001} = 0.999 \text{ Btu/hr ft}^2 \text{ °F}$$

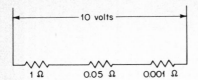

Fig. 9.2-4 Electrical analogy.

The change is only about 5 percent. Since the temperature drop will be about the same, about 5 percent more heat will transfer with the aluminum section. Whether this will be important to the project will rest on other considerations; here, the important thing to note is that a more than hundredfold decrease in the conduction resistance gave only a 5 percent increase in U. The reason is that obviously *the outside resistance is controlling*. Changes in the other resistances in series matter much less than changes in the outside coefficient.

The situation is analogous to an electrical circuit where we have a 1-, a 0.05-, and a 0.001-ohm resistor hooked in series as shown. (See Fig. 9.2-4.) We get little more current even by removing the 0.001-ohm resistor entirely.

There is an important lesson to be learned here. *The engineer should spend his time looking for and working on the controlling step of the process.* When such a step can be identified, vast amounts of work can frequently be saved by neglecting the calculation of less important steps. This is true of fluid flow, kinetics, mass transfer, and many other problems, as well as combinations of these problems. For example, mass transfer can often control a mass transfer/chemical kinetics problem so that the kinetics can be neglected and the reaction assumed to be at equilibrium.

These comments obviously hold for *series* processes, and the decision as to how large a resistance term might be neglected depends on the particular situation. It is exceedingly unwise, however, to place oneself in the position of recommending installation of a larger pipe instead of simply recommending opening the valve.

□ □ □

Example 9.2-3 Overall heat transfer coefficient with fouling A multipass heat exchanger has two passes on the shell side and four passes on the tube side. The tubes are 1-in. OD, 16 BWG. Water on the shell side is used to cool oil inside the tubes. The individual heat transfer coefficients (Btu/hr ft² °F) are $h_{oil} = 48$, $h_{water} = 170$, $h_{scale} = 500$. Determine the overall heat transfer coefficient based on the outside area neglecting the resistance of the metal wall.

Solution Heat exchanger tubes, like pipe, come in standard sizes. The wall thickness is usually specified in BWG (Birmingham Wire Gage) units. A table of tubing characteristics is shown as Table 9.2-3. As can be seen in Table 9.2-3, for 1-in., 16 BWG tubes,

$$\frac{A_0}{A_i} = \frac{0.2618 \ \text{ft}^2/\text{ft}}{0.2278 \ \text{ft}^2/\text{ft}} = 1.15$$

The scale will appear on the water side (outside) of the tubes, and so h_{scale} is associated with A_0. We assume that the inside remains unfouled. The expression for U is then

$$U_0 = \frac{1}{1/h_0 + 1/h_f + A_0/A_i h_i} = \frac{1}{1/170 + 1/500 + 1.15/48} \ \frac{\text{Btu}}{\text{hr ft}^2 \ ^\circ\text{F}}$$
$$= 31.2 \ \text{Btu/hr ft}^2 \ ^\circ\text{F}$$

□ □ □

9.3 CORRELATIONS FOR PREDICTION OF HEAT TRANSFER COEFFICIENTS

Heat transfer coefficients can be evaluated based on models at several levels of sophistication. At perhaps the lowest level of sophistication, dimensional analysis may be applied to selected variables for a given process, a set of dimensionless products found that (hopefully) completely describe the process, and the dimensionless groups empirically correlated.

If one has more insight into mechanism, the differential equation which describes the process may be derived and made dimensionless. In this way the fundamental variables are not arbitrarily selected, but enter in a natural way because of their appearance in the differential equations. The differential equations frequently cannot be solved; however, the dimensionless groups which are obtained as coefficients in the differential equations can be correlated with more confidence because the fundamental variables in these groups were obtained by a more rigorous procedure than arbitrarily selecting the fundamental variables. (Note that when we speak of correlating the dimensionless groups, we mean a series of experiments which will give us the dependence of these groups on one another. This is a matter of solving a *physical* analog of the system rather than a *mathematical* analog.)

At the highest level of sophistication, the proper mechanism is used in the derivation of the differential equation, the differential equation can be solved, and the solution to the differential equation is the correlating

Table 9.2-3 Tubing characteristics[a]

OD of tubing	BWG gage	Thickness in.	Internal area, in.²	ft² external surface per ft length	ft² internal surface per ft length	Weight per ft length steel, lb[b]	ID tubing, in.	Moment of inertia, in.⁴	Section modulus, in.²	Radius of gyration, in.	Constant C[c]	OD/ID	Metal area (transverse metal area), in.²
1/4	22	0.028	0.0295	0.0655	0.0508	0.066	0.194	0.00012	0.00098	0.0792	46	1.289	0.0195
1/4	24	0.022	0.0333	0.0655	0.0539	0.054	0.206	0.00011	0.00083	0.0810	52	1.214	0.0159
1/4	26	0.018	0.0360	0.0655	0.0560	0.045	0.214	0.00009	0.00071	0.0824	56	1.168	0.0131
3/8	18	0.049	0.0603	0.0982	0.0725	0.171	0.277	0.00068	0.0036	0.1164	94	1.354	0.0502
3/8	20	0.035	0.0731	0.0982	0.0798	0.127	0.305	0.00055	0.0029	0.1213	114	1.233	0.0374
3/8	22	0.028	0.0799	0.0982	0.0835	0.104	0.319	0.00046	0.0025	0.1227	125	1.176	0.0305
3/8	24	0.022	0.0860	0.0982	0.0867	0.083	0.331	0.00038	0.0020	0.1248	134	1.133	0.0244
1/2	16	0.065	0.1075	0.1309	0.0969	0.302	0.370	0.0022	0.0086	0.1556	168	1.351	0.0888
1/2	18	0.049	0.1269	0.1309	0.1052	0.236	0.402	0.0018	0.0072	0.1606	198	1.244	0.0694
1/2	20	0.035	0.1452	0.1309	0.1126	0.174	0.430	0.0014	0.0056	0.1649	227	1.163	0.0511
1/2	22	0.028	0.1548	0.1309	0.1162	0.141	0.444	0.0012	0.0046	0.1671	241	1.126	0.0415
5/8	12	0.109	0.1301	0.1636	0.1066	0.602	0.407	0.0061	0.0197	0.1864	203	1.536	0.177
5/8	13	0.095	0.1486	0.1636	0.1139	0.537	0.435	0.0057	0.0183	0.1903	232	1.437	0.158
5/8	14	0.083	0.1655	0.1636	0.1202	0.479	0.459	0.0053	0.0170	0.1938	258	1.362	0.141
5/8	15	0.072	0.1817	0.1636	0.1259	0.425	0.481	0.0049	0.0156	0.1971	283	1.299	0.125
5/8	16	0.065	0.1924	0.1636	0.1296	0.388	0.495	0.0045	0.0145	0.1993	300	1.263	0.114
5/8	17	0.058	0.2035	0.1636	0.1333	0.350	0.509	0.0042	0.0134	0.2016	317	1.228	0.103
5/8	18	0.049	0.2181	0.1636	0.1380	0.303	0.527	0.0037	0.0118	0.2043	340	1.186	0.089
5/8	19	0.042	0.2298	0.1636	0.1416	0.262	0.541	0.0033	0.0105	0.2068	358	1.155	0.077
5/8	20	0.035	0.2419	0.1636	0.1453	0.221	0.555	0.0028	0.0091	0.2089	377	1.126	0.065

$3/4$	10	0.134	0.1825	0.1963	0.1262	0.884	0.482	0.0129	0.0344	0.2229	285	1.556	0.260
$3/4$	11	0.120	0.2043	0.1963	0.1335	0.809	0.510	0.0122	0.0326	0.2267	319	1.471	0.238
$3/4$	12	0.109	0.2223	0.1963	0.1393	0.748	0.532	0.0116	0.0309	0.2299	347	1.410	0.220
$3/4$	13	0.095	0.2463	0.1963	0.1466	0.666	0.560	0.0107	0.0285	0.2340	384	1.339	0.196
$3/4$	14	0.083	0.2679	0.1963	0.1529	0.592	0.584	0.0098	0.0262	0.2376	418	1.284	0.174
$3/4$	15	0.072	0.2884	0.1963	0.1587	0.520	0.606	0.0089	0.0238	0.2410	450	1.238	0.153
$3/4$	16	0.065	0.3019	0.1963	0.1623	0.476	0.620	0.0083	0.0221	0.2433	471	1.210	0.140
$3/4$	17	0.058	0.3157	0.1963	0.1660	0.428	0.634	0.0076	0.0203	0.2455	492	1.183	0.126
$3/4$	18	0.049	0.3339	0.1963	0.1707	0.367	0.652	0.0067	0.0178	0.2484	521	1.150	0.108
$3/4$	20	0.035	0.3632	0.1963	0.1780	0.269	0.680	0.0050	0.0134	0.2532	567	1.103	0.079
$7/8$	10	0.134	0.2892	0.2291	0.1589	1.061	0.607	0.0221	0.0505	0.2662	451	1.441	0.312
$7/8$	11	0.120	0.3166	0.2291	0.1662	0.969	0.635	0.0208	0.0475	0.2703	494	1.378	0.285
$7/8$	12	0.109	0.3390	0.2291	0.1720	0.891	0.657	0.0196	0.0449	0.2736	529	1.332	0.262
$7/8$	13	0.095	0.3685	0.2291	0.1793	0.792	0.685	0.0180	0.0411	0.2778	575	1.277	0.233
$7/8$	14	0.083	0.3948	0.2291	0.1856	0.704	0.709	0.0164	0.0374	0.2815	616	1.234	0.207
$7/8$	16	0.065	0.4359	0.2291	0.1950	0.561	0.745	0.0137	0.0312	0.2873	680	1.174	0.165
$7/8$	18	0.049	0.4742	0.2291	0.2034	0.432	0.777	0.0109	0.0249	0.2925	740	1.126	0.127
$7/8$	20	0.035	0.5090	0.2291	0.2107	0.313	0.805	0.0082	0.0187	0.2972	794	1.087	0.092
1	8	0.165	0.3526	0.2618	0.1754	1.462	0.670	0.0392	0.0784	0.3009	550	1.493	0.430
1	10	0.134	0.4208	0.2618	0.1916	1.237	0.732	0.0350	0.0700	0.3098	656	1.366	0.364
1	11	0.120	0.4536	0.2618	0.1990	1.129	0.760	0.0327	0.0654	0.3140	708	1.316	0.332
1	12	0.109	0.4803	0.2618	0.2047	1.037	0.782	0.0307	0.0615	0.3174	749	1.279	0.305
1	13	0.095	0.5133	0.2618	0.2121	0.918	0.810	0.0280	0.0559	0.3217	804	1.235	0.270
1	14	0.083	0.5463	0.2618	0.2183	0.813	0.834	0.0253	0.0507	0.3255	852	1.199	0.239
1	15	0.072	0.5755	0.2618	0.2241	0.714	0.856	0.0227	0.0455	0.3291	898	1.167	0.210
1	16	0.065	0.5945	0.2618	0.2278	0.649	0.870	0.0210	0.0419	0.3314	927	1.149	0.191
1	18	0.049	0.6390	0.2618	0.2361	0.496	0.902	0.0166	0.0332	0.3366	997	1.109	0.146
1	20	0.035	0.6793	0.2618	0.2435	0.360	0.930	0.0124	0.0247	0.3414	1060	1.075	0.106

Table 9.2-3 Tubing characteristics[a]

OD of tubing	BWG gage	Thickness, in.	Internal area, in.²	ft² external surface per ft length	ft² internal surface per ft length	Weight per ft length steel, lb[b]	ID tubing in.	Moment of inertia, in.⁴	Section modulus, in.²	Radius of gyration, in.	Constant C[c]	OD/ID	Metal area (transverse metal area), in.²
1¼	7	0.180	0.6221	0.3272	0.2330	2.057	0.890	0.0890	0.1425	0.3836	970	1.404	0.605
1¼	8	0.165	0.6648	0.3272	0.2409	1.921	0.920	0.0847	0.1355	0.3880	1037	1.359	0.565
1¼	10	0.134	0.7574	0.3272	0.2571	1.598	0.982	0.0741	0.1186	0.3974	1182	1.273	0.470
1¼	11	0.120	0.8012	0.3272	0.2644	1.448	1.010	0.0688	0.1100	0.4018	1250	1.238	0.426
1¼	12	0.109	0.8365	0.3272	0.2702	1.329	1.032	0.0642	0.1027	0.4052	1305	1.211	0.391
1¼	13	0.095	0.8825	0.3272	0.2775	1.173	1.060	0.0579	0.0926	0.4097	1377	1.179	0.345
1¼	14	0.083	0.9229	0.3272	0.2838	1.033	1.084	0.0521	0.0833	0.4136	1440	1.153	0.304
1¼	16	0.065	0.9852	0.3272	0.2932	0.823	1.120	0.0426	0.0682	0.4196	1537	1.116	0.242
1¼	18	0.049	1.042	0.3272	0.3016	0.629	1.152	0.0334	0.0534	0.4250	1626	1.085	0.185
1¼	20	0.035	1.094	0.3272	0.3089	0.456	1.180	0.0247	0.0395	0.4297	1707	1.059	0.134
1½	10	0.134	1.192	0.3927	0.3225	1.955	1.232	0.1354	0.1806	0.4853	1860	1.218	0.575
1½	12	0.109	1.291	0.3927	0.3356	1.618	1.282	0.1159	0.1546	0.4933	2014	1.170	0.476
1½	14	0.083	1.398	0.3927	0.3492	1.258	1.334	0.0931	0.1241	0.5018	2181	1.124	0.370
1½	16	0.065	1.474	0.3927	0.3587	0.996	1.370	0.0756	0.1008	0.5079	2299	1.095	0.293
2	11	0.120	2.433	0.5236	0.4608	2.410	1.760	0.3144	0.3144	0.6660	3795	1.136	0.709
2	13	0.095	2.573	0.5236	0.4739	1.934	1.810	0.2586	0.2586	0.6744	4014	1.105	0.569
2½	9	0.148	3.815	0.6540	0.5770	3.719	2.204	0.7592	0.6074	0.8332	5951	1.134	1.094

^{a} Standards of Tubular Exchanger Manufacturers Association, 4th ed., 1960.

^{b} Weights are based on low-carbon steel with a density of 0.2833 lb/in.^{3}. For other metals multiply by the following factors:

Aluminum	0.35
AISI 400 series stainless steels	0.99
AISI 300 series stainless steels	1.02
Aluminum bronze	1.04
Aluminum brass	1.06
Nickel-chrome-iron	1.07
Admiralty	1.09
Nickel and nickel-copper	1.13
Copper and cupronickels	1.14

^{c} Liquid velocity $= \dfrac{\text{lb per tube per hr}}{C \times \text{sp. gr. of liquid}}$ in ft/sec (sp. gr. of water at 60°F $= 1.0$).

function for the variables. As we saw in Chap. 3, each of these levels of sophistication has its own particular advantages and disadvantages depending on the situation to be treated. Usually it is best to attempt analysis of the functional relation of the variables at all these levels before reaching a decision on how to correlate the data.

Notice that since we have shown there is an analogy between heat, mass, and momentum transfer, we may use this analogy to solve a heat transfer analog of a momentum transfer problem, or a mass transfer analog of a heat transfer problem. The utility of this approach depends on the mechanism of the two processes (for example, mass transfer and heat transfer) being the same, so that the analog is described by the same basic differential equation as the process. In this book, we are concerned with application of the correlating equations obtained from one of the above approaches. We will not emphasize *solution* of the differential equation, but rather emphasize *using* the solutions in application to practical problems.

To make the forms of our correlating equations plausible (but certainly not to justify them in general) we consider an application involving the lowest level of sophistication above, that is, the selection of a set of variables, and dimensional analysis of this set to form dimensionless groups.

For the moment let us accept the fact that for heat transfer to a fluid in turbulent flow in a pipe, the pertinent variables are the pipe diameter D, the thermal conductivity of the fluid k, the velocity of the fluid v, the density of the fluid ρ, the viscosity of the fluid μ, the specific heat of the fluid $\hat{C}_p$, and the heat transfer coefficient h. Functionally

$$f(D,k,\mu,\rho,v,\hat{C}_p,h) = 0 \tag{9.3-1}$$

Applying the methods of Chap. 3 in the mass, length, temperature, time system gives the following dimensionless groups (omitting the intermediate algebra):

$$\frac{hD}{k} = F\left(\frac{Dv\rho}{\mu}, \frac{\hat{C}_p\mu}{k}\right) \tag{9.3-2}$$

$$\text{Nu} = F(\text{Re,Pr}) \tag{9.3-3}$$

where Nu is the Nusselt number, Re is the Reynolds number, and Pr is the Prandtl number (the ratio of momentum diffusivity to thermal diffusivity).

Similarly, for a free (natural) convection problem the variables are

$$f(D,k,\mu,\rho,v,\hat{C}_p,g,\beta,\Delta T,L,h) = 0 \tag{9.3-4}$$

where the additional variables beyond those in Eq. (9.3-1) are the acceleration due to gravity g, the coefficient of thermal expansion β, the temperature difference between the heated surface and the fluid ΔT, and the length of the heated section L. Dimensional analysis of the variables in Eq. (9.3-4) gives as a complete set of dimensionless products for these variables

$$\frac{hD}{k} = F\left(\frac{Dv\rho}{\mu}, \frac{\hat{C}_p\mu}{k}, \frac{L}{D}, \frac{L^3\rho^2 g \,\Delta T\, \beta}{\mu^2}, \frac{L^3\rho^2 g}{\mu^2}, \frac{L^2\rho^2 k\, \Delta T}{\mu^3}\right)$$

$$\tag{9.3-5}$$

For problems we will treat, the last two groups can be neglected. For example, for a heated flat plate in the absence of forced convection, $D = \infty$ and $v \cong 0$, and so

$$\frac{hD}{k} = G\left(\frac{L^3\rho^2 g \,\Delta T\, \beta}{\mu^2}, \frac{\hat{C}_p\mu}{k}\right) \tag{9.3-6}$$

or

$$\text{Nu} = G(\text{Gr,Pr}) \tag{9.3-7}$$

where Gr is the Grashof number (the ratio of buoyant to viscous forces).

We emphasize again that the design equations we will consider have been developed in a variety of ways, but all incorporate the groups from the dimensional analysis above. To develop each of the equations we present would be a long process indeed, and so we shall simply point to the dimensional analysis as a rationale which should make the form of the equations *plausible* to the reader, although it does not justify the specific form *strictly*.

We can consider a simple example which shows in slightly more detail how the form of certain design equations originates. Many design equations have been developed by extension of a postulated analogy of momentum, heat, and mass transfer called the *Reynolds* analogy, developed by the same Reynolds of the Reynolds number.

The Reynolds analogy assumes that momentum, heat, and mass are transferred by the *same* process. By this we mean that the transfer in each case is described by the same dimensionless mathematical model, or, in other words, that the dimensionless profiles of velocity, temperature, and concentration coincide. (See Example 11.2-1.) As we shall see, this analogy is subject to some severe limitations.

We can arbitrarily extend our models of transport by molecular processes to include convection processes by writing a so-called eddy coefficient to account for the convective part of the transport.[1]

$$\tau = -\left(\mu^t + \mu\right)\frac{dv}{dy} \tag{9.3-8}$$

$$q = -\left(k^t + k\right)\frac{dT}{dy} \tag{9.3-9}$$

$$J_A^* = -\left(\mathscr{D}_{AB}{}^t + \mathscr{D}_{AB}\right)\frac{dc_A}{dy} \tag{9.3-10}$$

The coefficients with the superscript t (for turbulence) are the eddy viscosity, the eddy thermal conductivity, and the eddy diffusivity, respectively. Remember that the original coefficients were *properties of only the fluid*. The eddy coefficients, by their very nature, are *properties of the flow field* as well.

We now can take the ratio of any two of these equations. Since we are interested in heat transfer at the moment, let us divide Eq. (9.3-8) by Eq. (9.3-9):

$$\frac{\tau}{q} = \frac{(\mu^t + \mu)\,dv/dy}{(k^t + k)\,dT/dy} \tag{9.3-11}$$

Since we are interested in developing design equations, we consider Eq. (9.3-11) as applied at a solid surface, e. g., the wall of a tube or the surface of a plate. We replace the flux at the wall in terms of our design coefficients h and f (or C_D), and for simplicity we restrict the discussion to fluids with constant density and heat capacity:

$$\frac{{}^1\!/_2\,\rho\langle v\rangle^2 f}{h(\langle T\rangle - T_w)} = \frac{\left[\,(\mu^t + \mu)\,dv/dy\,\right]_{y=0}}{\left[\,(k^t + k)\,dT/dy\,\right]_{y=0}} \tag{9.3-12}$$

(Here we have used f, $\langle v\rangle$, and $\langle T\rangle$ as we would for flow in a tube—C_D, v_o, and T_0 would be used for external flows.)

The Reynolds assumption says that the *dimensionless* temperature and velocity profiles will coincide. (We could dedimensionalize y, the dis-

---

[1] The mass transfer equation, Eq. (9.3-10), is discussed in Sec. 11.1.

tance from the surface, but since it is the same in each derivative we would accomplish nothing). And so, dedimensionalizing, we take the difference between the quantity and its value at the wall divided by the difference between its bulk value and the wall value:

$$\frac{dv}{dy} = \frac{(\langle v \rangle - 0)\, d\,[(v - 0)/(\langle v \rangle - 0)]}{dy} \qquad (9.3\text{-}13)$$

$$\frac{dT}{dy} = (\langle T \rangle - T_w)\,\frac{d\,[(T - T_w)/(\langle T \rangle - T_w)]}{dy} \qquad (9.3\text{-}14)$$

Substituting

$$\frac{{}^{1}/_{2}\rho\langle v\rangle^{2} f}{h(\langle T \rangle - T_w)} = \frac{(\mu^{t} + \mu)}{(k^{t} + k)}\; \frac{\langle v \rangle\, d(v/\langle v \rangle)/dy\,\big|_{y=0}}{(\langle T \rangle - T_w)d[(T - T_w)/(\langle T \rangle - T_w)]/dy\,\big|_{y=0}} \qquad (9.3\text{-}15)$$

or

$$\frac{{}^{1}/_{2}\rho\langle v\rangle f}{h} = \frac{(\mu^{t} + \mu)}{(k^{t} + k)}\; \frac{d(v/\langle v\rangle/dy)\,\big|_{y=0}}{d[(T - T_w)/(\langle T \rangle - T_w)]/dy\,\big|_{y=0}} \qquad (9.3\text{-}16)$$

By the Reynolds assumption, the dimensionless velocity and temperature profiles coincide; therefore it follows that the slopes at the wall (and everywhere) must be the same, so that

$$\frac{d(v/\langle v\rangle)/dy\,\big|_{y=0}}{d[(T - T_w)/(\langle T \rangle - T_w)]/dy\,\big|_{y=0}} = 1.0 \qquad (9.3\text{-}17)$$

$$\frac{{}^{1}/_{2}\rho\langle v\rangle f}{h} = \frac{\mu^{t} + \mu}{k^{t} + k} \qquad (9.3\text{-}18)$$

We now examine the right-hand side of Eq. (9.3-18) to see what conclusions we might draw. Our dimensional analysis has shown that the dimensionless group involving the ratio of viscosity and thermal conductivity is the Prandtl number:

$$\text{Pr} = \frac{\mu \hat{C}_p}{k} = \frac{\mu/\rho}{k/\rho\hat{C}_p} = \frac{\nu}{\alpha} \qquad (9.3\text{-}19)$$

where v is the momentum diffusivity (kinematic viscosity) and α is the thermal diffusivity, so named because they represent the coefficient in the differential equation whose solution gives respectively the velocity profile and temperature profile as a function of time and therefore describes the "diffusion" of momentum and thermal energy.[1] We rewrite the right-hand side of Eq. (9.3-18) as follows:

$$\frac{\tfrac{1}{2}\rho\langle v\rangle f}{h} = \frac{1}{\hat{C}_p}\frac{\mu^t/\rho + \mu/\rho}{k^t/\rho\hat{C}_p + k/\rho\hat{C}_p} = \frac{1}{\hat{C}_p}\frac{v^t + v}{\alpha^t + \alpha} \tag{9.3-20}$$

However, if momentum and thermal energy are transported in the same flow field by the same mechanism, it is reasonable to expect v^t to equal α^t. (This would *not* be reasonable if there were, for example, form drag, which has no parallel in thermal energy transfer.) If *in addition* $v = \alpha$ (Pr = 1.0), then Eq. (9.3-20) becomes

$$\frac{\tfrac{1}{2}\rho\langle v\rangle f}{h} = \frac{1}{\hat{C}_p} \tag{9.3-21}$$

or

$$\frac{h}{\hat{C}_p\rho\langle v\rangle} = \frac{f}{2} \tag{9.3-22}$$

We can define $h/\hat{C}_p\rho\langle v\rangle$ as the Stanton number (St); it is apparent that

$$\text{St} = \frac{\text{Nu}}{\text{Re Pr}} \tag{9.3-23}$$

Note that this equation is greatly restricted by the assumptions.

Equation (9.3-22) is the Reynolds analogy. This is interesting not only because it suggests a way of correlating data but also because it shows that under certain circumstances the heat transfer coefficient can be found from friction factor measurements (and vice versa).

We can also take the ratio of momentum transfer to mass transfer or heat transfer to mass transfer and arrive at analogous equations to Eq.

---

[1] It is plausible that v should measure the diffusion of momentum; for example, in a transient situation the more viscous a fluid is, the faster momentum transfers, but since the momentum involves both mass and velocity, a denser fluid will "soak up" more momentum rather than transferring it. A crude analogy is the diffusion of water in steel wool and a sponge: the sponge has a large capacity to absorb water and so it diffuses more slowly. The same comments hold for thermal energy: the lower the thermal conductivity the slower the diffusion; the denser and the higher the heat capacity, the more capacity for thermal energy.

(9.3-22). We emphasize again that the transport mechanism must be the same for the above procedure to be valid.

In the above paragraphs we showed how functional relations involving the variables associated with heat transfer can be developed for correlation of heat transfer data. Many useful design equations are determined by applying dimensional analysis and the Reynolds analogy to experimental data. Some of the equations we will discuss have also evolved from analytical solutions of the differential equations of change (microscopic forms of our macroscopic balances) or from integral solutions of the boundary layer equations. Our emphasis here is on *use* of the design equations to motivate subsequent study of the *development* of these equations.

Discussion of average driving forces

Before going further in discussion of design equations, we must digress briefly to discuss driving forces. The driving force for momentum transfer during flow of incompressible fluids in a pipe remains constant as we go down the pipe, because the velocity profile for fully developed flow does not change. But in heat transfer, since the fluid changes in energy, the *temperature* profile *does* change, and therefore the driving force changes. We must develop an appropriate average temperature driving force for flow in a pipe or closed circuit.

We would like to be able to describe heat transfer in a conduit by an equation of the form

$$Q = UA(\Delta T)_{av} \tag{9.3-24}$$

where Q is the total rate of heat transfer, U is the overall heat transfer coefficient based as always on A, a selected area, and $(\Delta T)_{av}$ is a driving force calculated in such a manner as to make Eq. (9.3-24) correct. We restrict ourselves to cases where U may be assumed to be constant along the pipe (frequently an excellent assumption) to avoid having to define *both* an average ΔT and an average U. (We would thereby be confronted with an *embarras du choix* in which we can choose *any* definition for one of the means and use the other to adjust accordingly.) This problem is very similar to the one considered earlier of defining an appropriate mean area to use with an overall ΔT in multilayer heat conduction problems.

Our approach is basically the same as it was in the heat conduction problem—we calculate the left-hand side of Eq. (9.3-24), divide both sides by UA, and write what remains in terms of temperature—this, by definition, must be the required average ΔT.

We begin by writing Q in terms of the sum of amounts of heat trans-

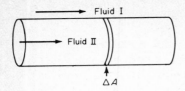

Fig. 9.3-1 System for derivation of average driving force.

ferred across elemental areas dA (see Fig. 9.3-1). The relative direction of flow of the fluids is unimportant to the final solution as we will develop it.

$$Q = \lim_{\Delta A \to 0} \sum_A \delta Q \qquad (9.3\text{-}25)$$

but

$$\delta Q = U \, \Delta T \, \Delta A \qquad (9.3\text{-}26)$$

and so

$$Q = \lim_{\Delta A \to 0} \sum_A U \, \Delta T \, \Delta A \qquad (9.3\text{-}27)$$

but this is

$$Q = \int_A U \, \Delta T \, dA \qquad (9.3\text{-}28)$$

We now must evaluate the integral, but we face the perplexing problem of how ΔT varies with A (U is assumed constant and so causes no difficulty). We attack this problem by changing the dependent variable in the integral, because we know that

1. We can write Q as an exact differential (there is no shaft work, and the difference between heat and work is an exact differential):

$$dQ = U \, \Delta T \, dA \qquad (9.3\text{-}29)$$

$$\frac{dA}{dQ} = \frac{1}{U \, \Delta T} \qquad (9.3\text{-}30)$$

This equation relates Q and A.

2. We also can relate Q and ΔT by observing that

$$dQ = w_I \, \hat{C}_{p\,I} \, dT \tag{9.3-31}$$

$$dQ = -w_{II} \, \hat{C}_{p\,II} \, dT_{II} \tag{9.3-32}$$

and

$$d \, \Delta T = d(T_I - T_{II}) = dT_I - dT_{II} \tag{9.3-33}$$

or, substituting,

$$d \, \Delta T = \frac{dQ}{w_I \hat{C}_{p\,I}} + \frac{dQ}{w_{II} \hat{C}_{p\,II}} = dQ \left(\frac{1}{w_I \hat{C}_{p\,I}} + \frac{1}{w_{II} \hat{C}_{p\,II}} \right) \tag{9.3-34}$$

$$\frac{dQ}{d \, \Delta T} = \frac{1}{1/w_I \hat{C}_{p\,I} + 1/w_{II} \hat{C}_{p\,II}} \tag{9.3-35}$$

which relates Q and ΔT.

We change the independent variable by writing

$$Q = U \int_A \Delta T \, dA = U \int \Delta T \, \frac{dA}{dQ} \, \frac{dQ}{d \, \Delta T} \, d \, \Delta T \tag{9.3-36}$$

$$= U \int_{\Delta T_1}^{\Delta T_2} \Delta T \, \frac{1}{U \, \Delta T} \, \frac{1}{1/w_I \hat{C}_{p\,I} + 1/w_{II} \hat{C}_{p\,II}} \, d \, \Delta T \tag{9.3-37}$$

$$= \frac{1}{1/w_I \hat{C}_{p\,I} + 1/w_{II} \hat{C}_{p\,II}} \int_{\Delta T_1}^{\Delta T_2} d \, \Delta T \tag{9.3-38}$$

$$Q = \frac{\Delta T_2 - \Delta T_1}{1/w_I \hat{C}_{p\,I} + 1/w_{II} \hat{C}_{p\,II}} \tag{9.3-39}$$

where ΔT_1 and ΔT_2 are the temperature differences at the ends of the tube. We have assumed $\hat{C}_p$ to be independent of T, and we have treated the $w\hat{C}_p$ terms as constants since they do not change with ΔT as we go along the tube.

We now substitute for Q in Eq. (9.3-24) using Eq. (9.3-39), solve for $(\Delta T)_{av}$, and obtain

$$(\Delta T)_{av} = \frac{\Delta T_2 - \Delta T_1}{UA \, (1/w_I \hat{C}_{p\,I} + 1/w_{II} C_{p\,II})} \tag{9.3-40}$$

This is not yet what we want because we have much on the right-hand side that is not temperature difference. Accordingly, we look for a way to eliminate those things in terms of temperature difference. We observe that no single equation permits this, but a combination of Eqs. (9.3-29) and (9.3-34) to eliminate dQ will yield

$$d \, \Delta T = U \, \Delta T \, dA \left(\frac{1}{w_I \hat{C}_{p\,I}} + \frac{1}{w_{II} \hat{C}_{p\,II}} \right) \tag{9.3-41}$$

or

$$\int_{\Delta T_1}^{\Delta T_2} \frac{d \, \Delta T}{\Delta T} = U \left(\frac{1}{w_I \hat{C}_{p\,I}} + \frac{1}{w_{II} \hat{C}_{p\,II}} \right) \int_0^A dA \tag{9.3-42}$$

(where we indicate the area integration rather than writing it explicitly in radius and length). This yields

$$\ln \frac{\Delta T_2}{\Delta T_1} = UA \left(\frac{1}{w_I \hat{C}_{p\,I}} + \frac{1}{w_{II} \hat{C}_{p\,II}} \right) \tag{9.3-43}$$

and now we can substitute in Eq. (9.3-40) to obtain

$$(\Delta T)_{av} = \frac{\Delta T_2 - \Delta T_1}{\ln (\Delta T_2 / \Delta T_1)} \tag{9.3-44}$$

In other words, for heat exchange in tubes with fully developed flow, fluids with constant properties, and constant U, Eq. (9.3-24) is correct if used with the logarithmic mean temperature difference.

If $\Delta T_1 = \Delta T_2$, the logarithmic mean then becomes indeterminate. A simple application of L'Hospital's rule rescues the situation, however:

$$\lim_{\Delta T_2 / \Delta T_1 \to 1} \frac{\Delta T_2 - \Delta T_1}{\ln (\Delta T_2 / \Delta T_1)} = \lim \frac{(\Delta T_2 / \Delta T_1 - 1) \Delta T_1}{\ln (\Delta T_2 / \Delta T_1)} \tag{9.3-45}$$

$$= \lim_{\Delta T_2 / \Delta T_1 \to 1} \frac{d/d(\Delta T_2 / \Delta T_1) [(\Delta T_2 / \Delta T_1 - 1) \Delta T_1]}{d/d(\Delta T_2 / \Delta T_1) [\ln (\Delta T_2 / \Delta T_1)]} \tag{9.3-46}$$

$$= \lim_{\Delta T_2 / \Delta T_1 \to 1} \frac{\Delta T_1}{\Delta T_1 / \Delta T_2} = \Delta T_2 \tag{9.3-47}$$

but since $\Delta T_1 = \Delta T_2$ we can write for this case:

$$(\Delta T)_{av} = \Delta T \tag{9.3-48}$$

Frequently the arithmetic mean is used in place of the logarithmic mean with sufficient accuracy. A discussion of the accuracy of this process is included in Appendix 9A.

In the above development we assume the heat transfer coefficient to be constant. If the heat transfer coefficient is not constant, there is an additional degree of freedom in the problem in that we can *arbitrarily* define $(\Delta T)_{av}$ and define the average coefficient accordingly. Immediately below we consider the design procedure for using average heat transfer coefficients based on a variety of driving forces.

Design equations

To this point we have illustrated the utility of using heat transfer coefficients without concerning ourselves greatly with how a heat transfer coefficient may be predicted. In this section we will present some typical design equations.

Ideally, one would like for a design equation to permit calculation of a single average coefficient which could be used to predict the heat transfer rate, much as a single friction factor could frequently be used to predict pressure drop in pipes. In heat transfer, however, several considerations intrude:

1. *Fluid properties*

   When thermal gradients are imposed on a liquid flow field, the assumption of constant physical properties frequently breaks down because viscosity is typically a fairly sensitive function of temperature. This in turn affects the velocity field. Usually less important than the viscosity change, but critical for certain problems, is the effect of temperature on heat capacity, thermal conductivity, and density (plus the associated natural convection).

2. *Driving forces*

   As opposed to momentum transfer in a pipe where an unchanging velocity profile is frequently obtained for incompressible fluids, in the case of heat transfer the fluid temperature is continually increasing or decreasing.

3. *Entrance effects*

   Above we discussed an example in which the momentum and thermal boundary layers were assumed to coincide. This may or may not be the case, and also since the lengths of pipe or tube used for heat transfer are frequently short, entrance effects may become very important. Fortunately, however, neglect of entrance effects usually gives a heat transfer coefficient that is smaller than the correct one, and so such an assumption tends to be conservative.

These considerations make it far more difficult to obtain design equations for heat transfer with a wide applicability than was possible for momentum transfer.

Laminar flow

For forced convection in *laminar* flow in tubes it is possible to construct mathematical models (in this case, partial differential equations) which can be solved for certain boundary conditions that give results useful in predicting heat transfer in real systems. These models usually use an assumed, constant velocity profile, constant physical properties, and a wall condition of either constant wall temperature or constant heat flux through the wall. All these features of the model may or may not correspond to the true situations for a given real world problem.

The solutions of the various differential equations by which the process is modeled are usually in the form of somewhat complex infinite series which, when evaluated, furnish a temperature profile in the tube at any given point. The details of the models and their solution lie outside our purpose here, which is to introduce the reader to the use of existing design equations in the treatment of real problems.

The mathematical models may be thought of as "black boxes" which take as input the conditions of the problem and yield as output temperature profiles, just as Newton's second law takes an initial position, a force, and a mass, and upon integration yields position versus time (a "position profile" in time rather than space).

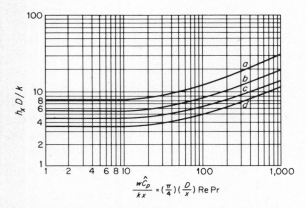

Fig. 9.3-2 Local Nusselt number for developed laminar flow in a pipe. (*a*) Rodlike flow, uniform heat flux; (*b*) rodlike flow, uniform wall temperature; (*c*) parabolic flow, uniform heat flux; (*d*) parabolic flow, uniform wall temperature. (*C. O. Bennett and J. E. Myers, "Momentum, Heat, and Mass Transfer," McGraw-Hill Book Company, New York, 1962.*)

Obviously, if one has a temperature profile from a model this can easily be converted to a heat transfer coefficient. Using the physical fact that the fluid sticks to the wall leads one to conclude that heat transfer in the fluid just at the wall is by conduction only. All the heat transferred, however, must pass through this layer. This means that the heat flux calculated for conduction just at the wall must equal that calculated using a heat transfer coefficient:

$$h(\langle T \rangle - T_w) = q = -k \left. \frac{dT}{dr} \right|_{r=R} \tag{9.3-49}$$

or

$$h = \frac{-k}{\langle T \rangle - T_w} \left. \frac{dT}{dr} \right|_{r=R} \tag{9.3-50}$$

but the bulk temperature, the wall temperature, and the slope of the temperature profile at the wall can all be obtained from the temperature profile, which is a function of k, and so a knowledge of the temperature profile is equivalent to a knowledge of h. The results are plotted in dimensionless form as the Nusselt number.

Results of solving such a mathematical model are shown in Fig. 9.3-2. This figure is based on a fluid with constant physical properties, flowing with a constant velocity profile as indicated. Two different wall boundary conditions are given. Note that this figure gives the *local* coefficient.

The *average* coefficient is easiest to use for many problems, and this is shown in Fig. 9.3-3 for the constant wall-temperature case. Average coefficients are given for use with both arithmetic mean and logarithmic mean temperature differences (that is, mean difference in fluid temperature and wall temperature for the end conditions of the tube). The above solution neglects axial conduction.[1]

The viscosity frequently varies significantly, making the assumption of constant velocity profile a poor one. Sieder and Tate[2] empirically developed an equation for the constant wall-temperature case to account for changing viscosity:

$$\mathrm{Nu} = 1.86 \, \mathrm{Re}^{1/3} \, \mathrm{Pr}^{1/3} \left(\frac{D}{L} \right)^{1/3} \left(\frac{\mu}{\mu_0} \right)^{0.14} \tag{9.3-51}$$

---

[1] The work from which Figs. 9.3-2 and 9.3-3 were derived was started by Graetz in about 1885. For a more extensive discussion the reader is referred to S. T. Hsu, "Engineering Heat Transfer," D. Van Nostrand Company, Inc., Princeton, N.J., 1963.

[2] E. N. Sieder and G. E. Tate, *Ind. Eng. Chem.*, **28**:1429 (1936).

In this equation all properties are evaluated at the arithmetic mean bulk temperature except μ_0, which is evaluated at the wall temperature, and the Nusselt number contains as h the *average* coefficient for the tube of length L.

Coefficients in short tubes may be a couple of orders of magnitude higher than predicted by Eq. (9.3-51) because of entrance effects. Fortunately the prediction is conservative.

The entrance effect may be treated using Fig. 9.3-4, which gives the Nusselt number as a function of the entrance length L_e. Note that the implication of this chart is that the temperature profile is fully established at about

$$\frac{L_e}{D} = 0.05 \text{ Re Pr} \qquad\qquad (9.3\text{-}52)$$

since this is where h (and the temperature profile) becomes constant.

Turbulent flow

The results of dimensional analysis of the variables for heat transfer to a fluid in *turbulent* flow have been used by different investigators to correlate heat transfer data. The resulting equations usually bear the name of

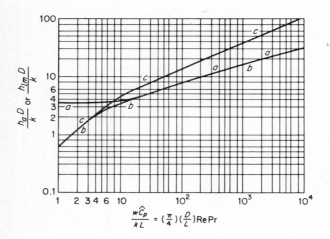

Fig. 9.3-3 Average Nusselt number for developed laminar flow in a pipe with uniform wall temperature. (*a*) Parabolic flow, $h_{1m}D/k$; (*b*) parabolic flow, h_aD/k; (*c*) rodlike flow, h_aD/k. (*C. O. Bennett, and J. E. Myers, "Momentum, Heat, and Mass Transfer," McGraw-Hill Book Company, New York, 1962.*)

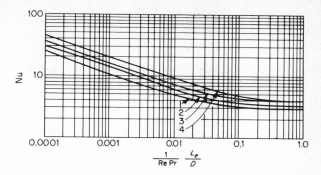

Fig. 9.3-4 Entrance length for flow inside tubes. (1) Nusselt;
(2) Kays: constant tube wall temperature; (3) Kays: constant
heat flux; (4) Kays: linear temperature variation along axis.
(*From "Engineering Heat Transfer by S. T. Hsu, copyright (c)
1963 by Litton Educational Publishing, Inc., by permission of
D. Van Nostrand, Inc.*)

the investigator. In general, all the equations discussed are of the form of
Eq. (9.3-3), and are power-function representations of the function f of
Eq. (9.3-3). The differences in the equations usually are related to the
temperature at which the fluid properties are evaluated and whether
change in viscosity with temperature is important.

The Dittus-Boelter[1] equation for turbulent flow in tubes uses fluid
properties evaluated at the arithmetic mean bulk temperature of the fluid,
with $a = 0.4$ when the fluid is heated and $a = 0.3$ when the fluid is cooled.
(For most gases $\mathrm{Pr} \cong 1$ and this variation of the exponent is of little
consequence.)

$$\frac{hD}{k} = 0.023 \left(\frac{D\langle v\rangle\rho}{\mu}\right)^{0.8} \left(\frac{\widehat{C}_p\mu}{k}\right)^{a} \tag{9.3-53}$$

The Dittus-Boelter equation resulted from an attempt to fit the data
shown in Fig. 9.3-5 with a set of two curves. The data presented include
both liquids and gases.

The Colburn equation[2] results from correlating data using fluid
properties evaluated at the wall temperature, except for $\widehat{C}_p$ which is
evaluated at the arithmetic mean bulk fluid temperature

$$\frac{h}{\langle v\rangle\rho\widehat{C}_p} \left(\frac{\widehat{C}_p\mu}{k}\right)^{2/3} = 0.023 \left(\frac{D\langle v\rangle\rho}{\mu}\right)^{-0.2} \tag{9.3-54}$$

[1] F. W. Dittus and L. M. K. Boelter, *Univ. of Calif. Publ. in Eng.*, **2**:443 (1930).
[2] A. P. Colburn, *Purdue Univ. Eng. Bull.*, **26**:1(1942).

To show the parallel to the Dittus-Boelter equation and to our dimensional analysis results this may be rearranged as

$$\frac{hD}{k} = 0.023 \left(\frac{D\langle v\rangle\rho}{\mu}\right)^{0.8} \left(\frac{\hat{C}_p\mu}{k}\right)^{1/3} \tag{9.3-55}$$

or

$$\text{Nu} = 0.023\, \text{Re}^{0.8}\, \text{Pr}^{1/3} \tag{9.3-56}$$

The Sieder-Tate equation for turbulent flow takes into account possible variation in the viscosity of the fluid near the wall because of thermal gradients. This equation is valid up to $\text{Pr} = 10^4$. The properties of the fluids are evaluated at the arithmetic average bulk temperature of the fluid, except for μ_0, which is evaluated at the average wall temperature.

$$\frac{h}{\langle v\rangle\rho\hat{C}_p} \left(\frac{\hat{C}_p\mu}{k}\right)^{2/3} \left(\frac{\mu_0}{\mu}\right)^{0.14} = 0.023 \left(\frac{D\langle v\rangle\rho}{\mu}\right)^{-0.2} \tag{9.3-57}$$

or, rearranged,

$$\text{Nu} = 0.023\, \text{Re}^{0.8}\, \text{Pr}^{1/3} \left(\frac{\mu}{\mu_0}\right)^{0.14} \tag{9.3-58}$$

Example 9.3-1 Comparison of the Dittus-Boelter, Colburn, and Sieder-Tate equations Determine the heat transfer coefficient for water flowing in a 1-in.-OD, 16 BWG tube at a velocity of 10 ft/sec. The temperature of the tube is 180°F and the water enters at 70°F and leaves at 130°F. Physical properties:

Average fluid temperature $= \dfrac{70 + 130}{2} = 100°\text{F}$

$\rho = 62.0$ lb$m$/ft^3 at 100°F

$\mu = 0.684$ cp at 100°F

$\mu = 0.346$ cp at 180°F

$\hat{C}_p = 0.999$ Btu/lb$m$°F at 100°F

$k = 0.363$ Btu/hr ft °F at 100°F

ID of tubing $= 0.870$ in.

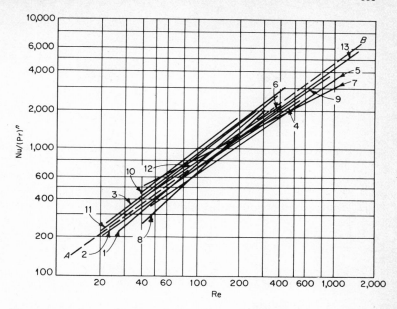

Fig. 9.3-5 Plot of Dittus-Boelter equation. *(From "Engineering Heat Transfer" by S. T. Hsu, copyright (c) 1963 by Litton Educational Publishing, Inc., by permission of D. Van Nostrand, Inc.)*

1 Morris and Whitman, *Ind. Eng. Chem.*, **20**:234 (1928).

2 Josse, *Mitt. Masch. Lab. Kgl. Tech. Hochsch. Berlin*, Heft V, 15 (1913).

2 Hilliger, *Ver. Deut. Ingr. Z.*, **60**:881 (1916).

3 Schulze, *Mitt. Wärmest. ver. Deut. Eisenhüttenlte.* (117): (1928).

4 *Ibid.*

5 *Ibid.*

6 *Ibid.*

7 *Ibid.*

8 Bray and Saylor, taken from McAdams and Frost, *Refrig. Eng.*, **10**:323 (1924).

9 Webster, *ibid.*

10 Stanton, *ibid.*

11 *Ibid.*

12 *Ibid.*

13 Nusselt, *Hab. Schrift Kgl. Sach. Tech. Hochsch. Dresden*, February 1909.

$n = 0.3$ for cooling.

$n = 0.4$ for heating.

Solution   The Reynolds number is

$$\text{Re} = \frac{D\langle v\rangle\rho}{\mu} = \frac{(0.870 \text{ ft}/12)(10 \text{ ft/sec})(62.0 \text{ lb}m/\text{ft}^3)}{(0.684)\,(6.72 \times 10^{-4})\,\text{lb}m/\text{ft sec}} = 97,792$$

The Prandtl number is

$$\text{Pr} = \frac{\hat{C}_p\mu}{k} =$$

$$\frac{(0.999)\,\text{Btu/lb}m\,^{\circ}\text{F}\,[\,(0.684)(6.72 \times 10^{-4})\,\text{lb}m/\text{ft sec}\,](3,600 \text{ sec/hr})}{0.363 \text{ Btu/hr ft }^{\circ}\text{F}}$$

$$= 4.554$$

The Dittus-Boelter equation yields

$$h = (0.023)\,\text{Re}^{0.8}\,\text{Pr}^{0.4}\,\frac{k}{D}$$

$$= \left[(0.023)(97,792)^{0.8}\,(4.554)^{0.4}\,\frac{(0.363)\,\text{Btu}}{\text{hr ft }^{\circ}\text{F}}\right]\frac{12}{0.870 \text{ ft}}$$

$$= 2075 \text{ Btu/hr ft}^2\,^{\circ}\text{F}$$

The Colburn equation gives

$$h = 0.023\,\text{Re}^{0.8}\,\text{Pr}^{1/3}\,\frac{k}{D}$$

$$= [(0.023)\,(97,792)^{0.8}\,(4.554)^{1/3}\,(0.363 \text{ Btu/hr ft }^{\circ}\text{F})]\,\frac{12}{0.870 \text{ ft}}$$

$$= 1876 \text{ Btu/hr ft}^2\,^{\circ}\text{F}$$

From the Sieder-Tate equation:

$$h = 0.023\,\text{Re}^{0.8}\,\text{Pr}^{0.3}\left(\frac{\mu}{\mu_o}\right)^{0.14}\frac{k}{D}$$

$$= (0.023)\,(97,792)^{0.8}\,(4.554)^{1/3}\left(\frac{0.684}{0.346}\right)^{0.14}\frac{0.363 \text{ Btu}}{\text{hr ft }^{\circ}\text{F}(0.870/12 \text{ ft})}$$

$$h = 2063 \text{ Btu/hr ft}^2\,^{\circ}\text{F}$$

If the above differences are significant for the given overall situation, one should refer to the original literature to determine the best procedure. This is true for all design equations as well as for the ones above.

□ □ □

Example 9.3-2 Average heat transfer coefficient from local coefficients In order to calculate heat transfer for a length of pipe it is more convenient to use average heat transfer coefficients rather than local coefficients. Determine the average heat transfer coefficient for a pipe with constant wall temperature based on the local coefficient and average temperature difference given by

$$\frac{(T_S - \langle T_{in} \rangle) + (T_S - \langle T_{out} \rangle)}{2}$$

where T_S is the temperature of the wall and $\langle T_{in} \rangle$ and $\langle T_{out} \rangle$ are bulk temperatures.

Solution

$$Q = h_a A \frac{(T_S - \langle T_{in} \rangle) + (T_S - \langle T_{out} \rangle)}{2} = h_a A \ (T_S - \langle T \rangle)_{av}$$

We can write the heat transferred over a differential length of pipe as

$$h_x (T_S - \langle T \rangle) \ 2\pi R \ dx$$

where h_x is the *local* coefficient. Integrating, we obtain the total heat transfer as

$$Q = \int_0^L h_x (T_S - \langle T \rangle) \ 2\pi R \ dx$$

and so we have

$$Q = 2\pi R \int_0^L h_x (T_S - \langle T \rangle) dx = h_a \ (T_S - \langle T \rangle)_{av} \ 2\pi R L$$

or

$$h_a = \frac{\int_0^L h_x (T_S - \langle T \rangle) dx}{(T_S - \langle T \rangle)_{av}}$$

To obtain a numerical value for h_a we must know h_x and $T_S - \langle T \rangle$ as a function of x. In laminar flow, these functions can be determined from the analytical solution which gives T as a function of r and x as an infinite series, as outlined above. This is how Fig. 9.3-3 was constructed. For turbulent flow, the functions would have to be determined by experimental measurement and the expression integrated numerically.

□ □ □

For flowing fluids heated in noncircular conduits or in annuli, the equivalent diameter (four times the hydraulic radius) is commonly used in place of the tube diameter in the equations above for estimating heat transfer coefficients. For annular flow where D_2 is the larger diameter,

$$D_{eq} = D_2 - D_1 \qquad (9.3\text{-}59)$$

An equation developed specifically for the outer wall of the inside pipe of an annulus is[1]

$$Nu = 0.02 \ Re^{0.8} \ Pr^{1/3} (D_2/D_1)^{0.45} \qquad (9.3\text{-}60)$$

for $12,000 < Re < 220,000$. Fluid properties are evaluated at the arithmetic mean bulk temperature. For the inner wall the heat transfer coefficient may be found from the Sieder-Tate equation, Eq. (9.3-32).

Heat transfer coefficients for fluid flow *normal* to a cylinder (pipe) depend on the configuration of the boundary layer. The formation of a large wake changes the heat transfer characteristics. For *forced* convection

$$\frac{hD}{k} = b \left(\frac{D\langle v \rangle \rho}{\mu} \right)^n \qquad (9.3\text{-}61)$$

where values of b and n depend on the Reynolds number as shown in Table 9.3-1. The equation retains the functional form derived from dimensional analysis earlier.

For *free* convection with flow normal to a horizontal cylinder when $10^4 < Pr \ Gr < 10^9$

$$\frac{hD}{k} = 0.53 \left(\frac{D^3 \rho^2 g \beta \ \Delta T}{\mu^2} \right)^{1/4} \left(\frac{\hat{C}_p \mu}{k} \right)^{1/4} \qquad (9.3\text{-}62)$$

This equation can be presented as a nomograph for $Gr \ Pr > 10^9$ as shown in Fig. 9.3-6. This is a common way of presenting design equations.

[1] C. C. Monrad and J. F. Pelton, *Trans. ASME*, **38**:593 (1942).

Table 9.3-1   Values of b and n for Eq. (9.3-61)[a]

$D<v>\rho/\mu$	n	b
1–4	0.330	0.891
4–4 × 10	0.385	0.821
4 × 10–4 × 10³	0.466	0.615
4 × 10³–4 × 10⁴	0.618	0.174
4 × 10⁴–2.5 × 10⁵	0.805	0.0239

[a] R. Hilpert, *Forsch. Gebiete Ingenieurw.*, **4**:215 (1933).

For natural convection outside vertical tubes or plates Fig. 9.3-7 may be used for Gr Pr $> 10^9$. As noted, it also can be used for gases or liquids flowing upward at low velocity *in* vertical tubes.

For convection normal to the outside of pipes or tubes arranged in successive rows, we need a modified equation since the heat transfer to a tube is affected by the wake of tubes upstream:

$$\frac{hD}{k} = b\left(\frac{DG_{\max}}{\mu}\right)^n \tag{9.3-63}$$

where $G_{\max} = \rho v'$ and where v' is the velocity at the point of minimum flow cross-sectional area. The values of b and n depend on the tube arrangement. For a square grid with center-to-center spacing of the tubes equal to twice the tube diameter, $b = 0.299$ and $n = 0.632$. For a pattern of equilateral triangles with center-to-center spacing equal to twice the tube diameter, $b = 0.482$ and $n = 0.556$. The equation above was developed for gases. To use Eq. (9.3-63) for liquids multiply it by 1.1 $Pr^{1/3}$.

Example 9.3-3   Heat transfer with flow normal to pipes   Water is being heated by flowing it across a bank of rows of 1-in., 16 BWG gage copper tubes 10 ft long. There are five tubes per row. The tubes are spaced on a square grid with center-to-center spacing of twice the tube diameter. The inside temperature of the tube walls is 230°F (neglect conduction resistance). If water flowing at 80 gal/min enters the exchanger at 60°F and leaves at 140°F find the mean heat transfer coefficient for the water. Assume the space between the outside tube and the containing shell is one-half tube diameter.

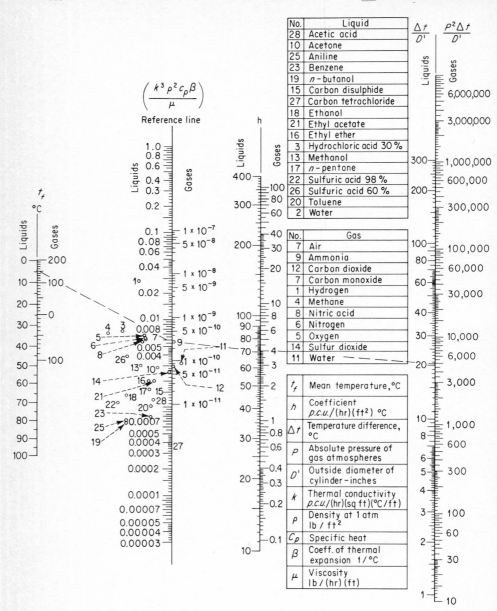

Fig. 9.3-6 Convection outside horizontal cylinders. (*R. H. Perry, C. H. Chilton, and S. D. Kirkpatrick* (eds.), *"Chemical Engineers' Handbook,"* 4th ed., McGraw-Hill Book Company, New York, 1963.)

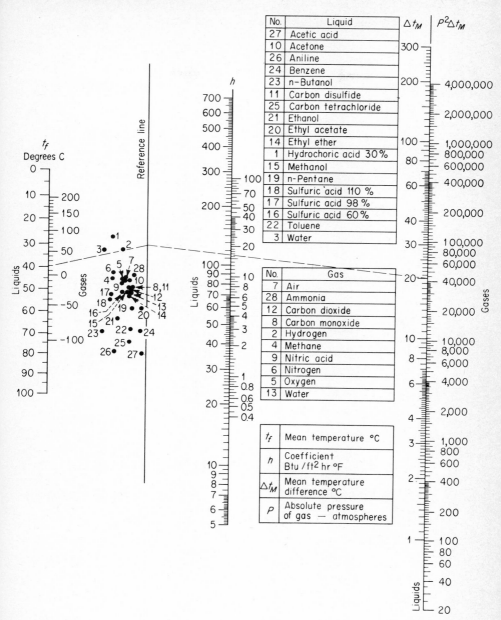

No.	Liquid
27	Acetic acid
10	Acetone
26	Aniline
24	Benzene
23	n-Butanol
11	Carbon disulfide
25	Carbon tetrachloride
21	Ethanol
20	Ethyl acetate
14	Ethyl ether
1	Hydrochoric acid 30%
15	Methanol
19	n-Pentane
18	Sulfuric acid 110 %
17	Sulfuric acid 98 %
16	Sulfuric acid 60%
22	Toluene
3	Water

No.	Gas
7	Air
28	Ammonia
12	Carbon dioxide
8	Carbon monoxide
2	Hydrogen
4	Methane
9	Nitric acid
6	Nitrogen
5	Oxygen
13	Water

t_f	Mean temperature °C
h	Coefficient Btu /ft^2 hr °F
Δt_M	Mean temperature difference °C
P	Absolute pressure of gas — atmospheres

g. 9.3-7 Convection inside or outside vertical tubes. (*R. H. Perry, C. H. Chilton, and S. D. Kirkpa-ck (eds.), "Chemical Engineers' Handbook," McGraw-Hill Book Company, New York, 1963.*)

Solution   The minimum flow cross-sectional area is

$$\text{Area} = 4 \text{ openings} \frac{[(1 \text{ ft})/12](10 \text{ ft})}{\text{opening}}$$

$$+ 2 \text{ openings} \frac{[(1/2 \text{ ft})/12](10 \text{ ft})}{\text{opening}}$$

$$= (5) \left(\frac{10}{12} \right) = 0.417 \text{ ft}^2$$

Note that we are neglecting end effects. The Reynolds number is

$$\text{Re} = \frac{DG_{\max}}{\mu} = \left(\frac{1}{12} \text{ ft} \right) \left[\frac{\text{ft sec}}{(0.684)(6.72 \times 10^{-4}) \text{ lb}m} \right] \times$$

$$\left(80 \frac{\text{gal}}{\text{min}} \right) \left(\frac{\text{ft}^3}{7.48 \text{ gal}} \right) \left(\frac{62 \text{ lb}m}{\text{ft}^3} \right) \left(\frac{1}{0.417 \text{ ft}^2} \right) \left(\frac{\text{min}}{60 \text{ sec}} \right)$$

$$= 4,805$$

$$\text{Pr} = \frac{\mu \hat{C}_p}{k} = \left[(0.684)(6.72 \times 10^{-4}) \frac{\text{lb}m}{\text{ft sec}} \right] \times$$

$$\left(\frac{0.999 \text{ Btu}}{\text{lb}m \text{ °F}} \right) \left(\frac{3,600 \text{ sec}}{\text{hr}} \right) \left(\frac{\text{hr ft °F}}{0.363 \text{ Btu}} \right)$$

$$= 4.554$$

$$h = 1.1b\text{Re}^n \text{Pr}^{1/3} \frac{k}{D}$$

$$= [(1.1)(0.229)(4,805)^{0.632}(4.554)^{1/3} \, 0.363 \text{ Btu/hr ft °F} \frac{12}{1 \text{ ft}}$$

$$= 386 \text{ Btu/hr ft}^2 \text{ °F}$$

□ □ □

A useful design equation for air flowing outside and at right angles to a bank of finned tubes is[1]

$$h = 0.17 \frac{v^{0.6}}{D^{0.4}} \left(\frac{L}{L - D} \right)^{0.6} \tag{9.3-64}$$

[1] See R. H. Perry, C. H. Chilton, and S. D. Kirkpatrick (eds.), "Chemical Engineers Handbook," 4th ed., pp. 10-24, McGraw-Hill Book Company, New York, 1963.

where v is the freestream velocity of the air in feet per minute, D is the outside diameter in inches of the bare tube (root diameter of fin), and L is the center-to-center spacing in inches of the tubes in a row. This equation is, unfortunately, dimensional, but the extreme complexity of the situation does not permit other than rough estimation from an overall design equation. Obviously to obtain more precise answers fin geometry must be considered, which further complicates the situation.

The correlation of Froessling is used for determining the heat transfer coefficient from spheres. In this case, to the functional form of Eq. (9.3-3) a constant is added such that

$$\frac{hD}{k} = 2 + 0.6 \left(\frac{\hat{C}_p \mu}{k}\right)^{1/3} \left(\frac{Dv_\infty \rho}{\mu}\right)^{1/2} \tag{9.3-65}$$

The Froessling correlation is also used to correlate data for porous media, because one approach to modeling flow in a porous medium is to consider flow over a collection of submerged objects. For such a case, one replaces v_∞ with $10.73\ v'_\infty$ where v'_∞ is the velocity of approach in the porous media (see Chap. 5).

The constant term in Eq. (9.3-65) accounts for the fact that heat transfer continues (by conduction) even when the Reynolds number goes to zero (for example, zero flow velocity).

9.4 HEAT TRANSFER WITH PHASE CHANGE

The transfer of heat when phase change occurs, such as in boiling and condensation, is usually described by a heat transfer coefficient as defined in Eqs. (9.1-4) and (9.1-5). Heat transfer which accompanies phase change usually occurs at a high rate. As we have seen in Table 9.1-1 coefficients for these cases can be extremely high.

Boiling—mechanism

Boiling may occur by more than one mechanism. The two main classes which are usually distinguished are (1) nucleate boiling and (2) film boiling. The type of boiling one observes in heating water for a cup of tea or coffee, or in boiling an egg in a pan, usually is nucleate boiling. In this class of boiling, the vapor is evolved as a multitude of bubbles. As one increases heat flux, however, these bubbles grow large enough that they coalesce and form a continuous film of vapor at the surface, creating the regime known as film boiling.

An interesting illustration of film boiling is the Leidenfrost phenomenon. If one pours small amounts of water on the surface of a hot plate, the drops skitter about the surface and appear to bounce up and down.

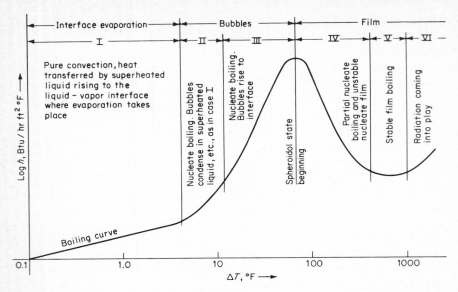

Fig. 9.4-1 Boiling curves. [*E. F. Fauber, and R. L. Scorah, Trans. ASME*, **70**:369 (1948).]

Leidenfrost explained this behavior by the fact that as the drop touches the hot surface, a film of vapor is formed between the drop and the surface which accelerates the drop in the upward direction. Since the film of vapor forms a relatively good insulator between the drop and the surface, the heat transfer then begins to decrease and the film thins until the drop again touches the surface, more water is evaporated, the drop shoots upward once more, etc., until all the water in the drop has been evaporated.

It is relatively obvious that for heat transfer purposes, nucleate boiling is far preferable to film boiling. Much work has been done on the formation of bubbles during boiling, and explanation of nucleation phenomena is far from trivial. If one considers the problem of nucleating extremely small bubbles of vapor, the surface tension forces become very important since the ratio of surface area to volume of a sphere increases as the radius decreases. For this reason, it is easier to nucleate a bubble from a cavity which initially contains some trapped gas, so that the radius of curvature when the bubble begins to grow is relatively large, and less pressure is required to overcome the surface tension force. There is an excellent motion picture available which illustrates this type of bubble nucleation.[1]

We can see an illustration of the common mechanisms of boiling by

[1]Educational Service Inc., "Surface Tension in Fluid Mechanics."

considering the case of a platinum wire (immersed in water) through which an electric current is passed. If we conduct such an experiment, we will obtain results such as are shown in Fig. 9.4-1, in which we have plotted the heat flux versus the temperature of the wire. The heat flux as a function of wire temperature is readily obtained because (1) we know the surface area of the wire, (2) we can measure the electrical power input, and (3) we know the temperature of the wire because we know its resistance characteristics as a function of temperature.

The mechanism observed in the temperature range of 212 to 300°F is nucleate boiling. Bubbles of vapor form at discrete sites on the heated surface and then detach almost immediately. As the temperature of the surface increases, bubbles form at more and more sites, until eventually the bubbles coalesce and the surface becomes covered by a continuous film of vapor. This vapor film prevents the liquid from reaching the surface and the heat supplied to evaporate the liquid must all pass through the vapor film. The mechanism then is film boiling.

Between the nucleate and film boiling regimes there is a transition region where heat flux drops as the temperature rises, as shown in the figure. In film boiling vaporization takes place at the vapor-liquid interface.

Condensation—mechanism

Condensation, like boiling, occurs by two distinct mechanisms. The mechanisms are analogous to those for boiling and they are called dropwise condensation and film condensation. For vertical surfaces, if the condensing liquid does not wet the surface readily, drops will form, and as soon as the drop is large enough that its weight overcomes the surface tension forces holding it to the surface, the drop runs rapidly from the surface, clearing the surface for more condensation.

Large drops that run from the surface coalesce with small drops in their path and sweep them from the surface also. In this dropwise case, most of the heat transfer takes place directly between the vapor and the solid surface, and therefore the heat transfer coefficient is quite high.

If the condensing liquid wets the surface readily, however, one will obtain a continuous film of liquid over the heat transfer surface and heat transfer then is limited by conduction through this film of liquid. There has been a great deal of work done on additives or surface coatings which will promote dropwise condensation, but, in fact, in process equipment one usually works in the film condensing regime.

A third type of heat transfer with phase change is associated with freezing (or crystallization, adduction, inclusion, clathrate formation, etc.) processes. We normally distinguish only one mechanism for these processes (although there are probably many) because we are forming a solid phase directly from either a liquid phase or a gas phase and this simplistic view describes many processes adequately.

Boiling and condensation are both extremely interesting and extremely important processes in heat transfer; however, to discuss the theory necessary to develop these areas in detail would require a text in itself. Below, therefore, we will simply present some of the more common design equations to illustrate their general form. In the solution of specific practical problems, one should always refer to the literature for data and theory, or to experiments which one develops oneself to get the appropriate correlation.

Boiling

For film boiling around a horizontal tube, one can predict a heat transfer coefficient by using the equation[1]

$$h = h_c + h_r \left(\frac{3}{4} + \frac{1}{4} \frac{h_r}{h_c} \frac{1}{2.62 + h_r/h_c} \right)^{1/4} \tag{9.4-1}$$

where the convective film coefficient and the radiative coefficient are given by the equations

$$h_c = 0.62 \left[\frac{g\rho_V(\rho_L - \rho_V)\lambda k_V{}^3}{D\mu_V(T_S - T_L)} \right]^{1/4} \tag{9.4-2}$$

and

$$h_r = \frac{\varepsilon\sigma(T_S{}^4 - T_L{}^4)}{T_S - T_L} \tag{9.4-3}$$

For film boiling around a vertical tube, one modifies the above correlation by calculating a Reynolds number defined by

$$\mathrm{Re} = \frac{4w}{\pi D\mu_V} \tag{9.4-4}$$

and calculating the heat transfer coefficient from the equation[2]

$$h \left[\frac{\mu^2{}_V}{g\rho_V(\rho_L - \rho_V)k_V{}^3} \right]^{1/3} = 0.0020 \, \mathrm{Re}^{0.6} \tag{9.4-5}$$

[1] L. A. Bromley, *Chem. Eng. Progr.*, **46**:221 (1950).
[2] S. T. Hsu, "Engineering Heat Transfer," p. 426, D. Van Nostrand Company, Inc., Princeton, N.J., 1963.

Table 9.4-1   Values of C for Eq. (9.4-6)[a]

Combination	C
n-Pentane—chromium	0.015
Water—platinum	0.013
Water—copper	0.013
Carbon tetrachloride—copper	0.013
Benzene—chromium	0.010
Water—brass	0.006
35% K_2CO_3	0.0054
n-Butyl alcohol—copper	0.00305
50% K_2CO_3-copper	0.00275
Ethyl alcohol—chromium	0.0027
Isopropyl—copper	0.00225

[a]W. M. Rohsenow, *Trans. ASME*, **74**:969 (1952).

For nucleate boiling one usually uses an equation not written explicitly in terms of a heat transfer coefficient which is as follows:[1]

$$\frac{\hat{C}_{pL}(T_S - T_V)}{\lambda} = C\left[\frac{q}{\mu_L \lambda} \sqrt{\frac{4.17 \times 10^8 \sigma}{g(\rho_L - \rho_V)}}\right]^{0.33} Pr_L^{1.7} \tag{9.4-6}$$

See Table 9.4-1 for values of C.

Condensation

For condensation on vertical tubes, for laminar flow the mean heat transfer coefficient is given by

$$h_a \left(\frac{\nu^2}{k^3 g}\right)^{1/3} = 1.47 \, Re_x^{-1/3} \tag{9.4-7}$$

where

$$Re_x = \frac{4x_e \langle v \rangle \rho}{\mu} \tag{9.4-8}$$

and x_e is the film thickness. For turbulent flow, $Re_x > 2,000$

$$h_a \left(\frac{\nu^2}{k^3 g}\right)^{1/3} = 0.0077 \, Re_x^{0.4} \tag{9.4-9}$$

[1]W. M. Rohsenow, *Trans. ASME*, **74**:969 (1952).

For condensation on a row of horizontal tubes stacked vertically where condensate flow is laminar, the mean coefficient for the entire bank of tubes is

$$h_a = 0.725 \left[\frac{k^3 \rho g \lambda}{\nu N D (T_{sv} - T_S)} \right]^{1/4} \tag{9.4-10}$$

where N is the number of tubes in a row.

The reader should be cautioned that condensing coefficients can be affected substantially by the presence of noncondensable gases. For example, if one has a heat exchanger in which ethanol condenses in the presence of nitrogen, a layer of nitrogen builds up at the condensing surface, and changes the problem from one in heat transfer to one in mass transfer, since the limiting step now is the diffusion of the ethanol across the stagnant nitrogen layer next to the tube. This is an exaggeration—all the gas will not remain in a layer next to the tube wall but bubbles will rise to the top of the exchanger. For this reason, most condensers on distillation columns are provided with some type of vent for noncondensable gases.

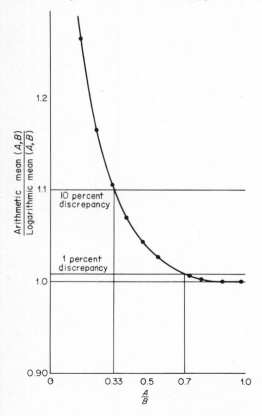

Fig. 9A-1 Discrepancy in using arithmetic mean for logarithmic mean.

APPENDIX 9A: APPROXIMATION OF LOGARITHMIC MEAN BY ARITHMETIC MEAN

It is frequently desirable for ease of calculation to replace the logarithmic mean of two quantities A and B by the arithmetic mean. We can investigate the accuracy of this approximation by considering the function

$$\frac{\text{Arithmetic mean } (A, B)}{\text{Logarithmic mean } (A, B)} = \frac{(A + B)/2}{(A - B)/\ln (A/B)} \tag{9A-1}$$

$$= \frac{A/B + 1}{2 (A/B - 1)} \ln \frac{A}{B} \tag{9A-2}$$

A plot of this function is shown in Fig. 9A-1. As can be seen, a discrepancy of less than 1 percent is obtained by replacing the logarithmic mean by the arithmetic mean for ratios of A to B of from 0.7 to 1.0. Surprisingly, the discrepancy is less than 10 percent down to ratios as low as 0.33. The error grows rapidly outside this range, however.

PROBLEMS

9.1 Calculate the heat loss per foot from a 3-in. schedule 40 pipe (ID $= 3.07$ in., OD $= 3.500$ in., where $k = 25$ Btu/hr ft °F) covered with a $1\frac{1}{4}$-in. thickness of asbestos insulation ($k = 0.1$ Btu/hr ft °F). The pipe transports a fluid at a bulk mixing-cup temperature of 350°F with an inside heat transfer coefficient h_i, 30 Btu/hr ft² °F, and is exposed to air at a bulk mixing temperature of 80°F with an outer surface heat transfer coefficient of 2 Btu/hr ft² °F.

   (*a*) Calculate the heat loss per foot.
   (*b*) Calculate T, T_1, T_2.

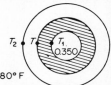

9.2 Oil at a bulk temperature of 200°F is flowing in a 2-in. schedule 40 pipeline (ID $= 2.067$, OD $= 2.375$, $k = 26$ Btu/hr ft °F) at a bulk velocity of 4 ft/sec. If the outside of the pipe is covered with insulation $\frac{1}{2}$-in. thick ($k = 0.1$ Btu/hr ft °F) and the outside surface of the insulation is 90°F, find

   (*a*) The inside heat transfer coefficient evaluated at 200°F.
   (*b*) The heat loss from the pipe per foot.
   (*c*) The inside pipe temperature.
At 200°F for oil

$k \;\; = \;\; 0.074 \text{ Btu/hr ft °F}$

$\text{Pr} = 62$

$\hat{C}_p = 0.51 \text{ Btu/lb}m \text{ °F}$

$\mu \;\; = \;\; 250 \times 10^{-5} \text{ lb}m/\text{ft sec}$

$\rho \;\; = \;\; 54 \text{ lb}m/\text{ft}^3$

9.3 If you were standing in a house where the inside air temperature was 80°F, in which case would the inside wall temperature be the coldest: (*a*) when the outside air temperature is at 0°F and there is a 4-mph wind blowing parallel to the wall, or (*b*) when the outside air temperature is at 10°F and there is a 45-mph wind blowing parallel to the wall. Justify your answer with calculations. Data:

$$\text{Thermal resistance of wall} = \frac{1.50}{A} \text{ hr ft}^2 \text{ °F/Btu}$$

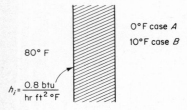

0°F case *A*

10°F case *B*

80° F

$h_i = \dfrac{0.8 \text{ btu}}{\text{hr ft}^2 \text{ °F}}$

Physical properties of air $0° \le T \le 10°$

$Pr = 0.73$

$k = 0.0133$ Btu/hr ft °F

$\mu = 1.11 \times 10^{-5}$ lbm/ft sec

$\rho = 0.086$ lbm/ft³

$Nu_L = 0.086 \, Pr^{1/3} \, Re_L^{0.8}$

9.4 Consider the wall shown below.

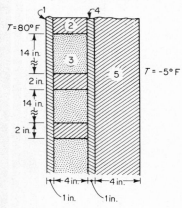

Materials

1. Wallboard
   $k_1 = 0.40$ Btu/hr ft °F

2. Wood
   $k_2 = 0.14$ Btu/hr ft °F

3. Insulation
   $k_3 = 0.016$ Btu/hr ft °F

4. Beaver board
   $k_4 = 0.35$ Btu/hr ft °F

5. Brick
   $k_5 = 0.38$ Btu/hr ft °F

If $h_{in} = 2$ Btu/hr ft² °F and $h_{out} = 12$ Btu/hr ft² °F, find Q/A through the wall and the inside wall temperature.

9.5 (*a*) A single pane of glass ($k = 0.50$ Btu/hr ft °F) which is $1/4$ in. thick separates the air in a room which is at bulk temperature of 70°F from the outdoors where the bulk atmospheric temperature is 0°F. If a thermocouple is attached to the surface of the glass inside the room it reads some temperature between 0 and 70°F. What is the temperature? Assume that the heat transfer coefficients h_i and h_o are both constant at 1 Btu/hr ft² °F.

(*b*) It has been suggested that the outside bulk atmospheric temperatures could be determined approximately by fastening a thermocouple to the inner surface of the window as in part (*a*) and covering it with a thick layer of insulation ($k = 0.06$ Btu/hr ft °F). What thickness of insulation would be required to give a reading of the thermocouple within 5° of the bulk outdoors temperature? Assume that the coefficients are the same as in part (a).

9.6 A jacketed kettle is used for concentrating 1,000 lbm/hr of an aqueous salt solution by means of steam at 250°F. The solution boils at 220°F and has a latent heat of 965 Btu/lbm. The kettle is of steel ($k = 35$ Btu/hr ft °F) and is $1/4$-in. thick. Determine the increase in capacity that might be expected if the kettle were replaced by one of copper ($k = 220$ Btu/hr ft °F).

9.7 Two thermometers are placed in an airstream flowing through a circular duct. One of these thermometers which has been silvered on the outside reads 100°F. The other is a plain mercury-in-glass thermometer and reads 150°F. Assuming that the readings are correct, calculate the temperature of the duct walls and that of the gas. The coefficient for the air on both thermometers is 2 Btu/hr ft² °F. The emissivity for the silvered surface is 0.02 and the glass surface, 0.94.

9.8 A hot fluid is to be cooled in a double-pipe heat exchanger from 245 to 225°F. Compare the advantage of counterflow over parallel flow. The cold fluid is heated from 135 to 220°F.

9.9 Calculate the heat loss per lineal foot from a 3-in. schedule 40 steel pipe (3.07-in. ID, 3.500-in. OD, $k = 25$ Btu/hr ft °F) covered with 1-in. thickness of asbestos insulation ($k = 0.11$ Btu/hr ft °F). The pipe transports a fluid at 300°F with an inner unit surface conductance of 40 Btu/hr ft² °F (h_{in}) and is exposed to ambient air at 80°F with an outer unit surface conductance of 3 Btu/hr ft² °F (h_{out}).

9.10 Hot light oil is to be cooled by water in a double-pipe heat exchanger from 250 to 170°F. The water is heated from 90 to 150°F. If the oil flow rate is 2.200 lbm/hr, $\hat{C}_{p\,oil} = 0.56$ Btu/lbm °F, find the water flow rate. If $U_0 = 45$ Btu/hr ft² °F, compare counter current and cocurrent heat exchangers. Assuming a clean heat exchanger, which resistance—water side, metal, or oil side—would you expect to be the controlling resistance?

9.11 A hypothetical countercurrent heat exchanger having an *infinite* heat transfer area is used to cool 10,000 lbm/hr of water entering at 150°F. The coolant is 5,000 lbm/hr of water entering at 100°F.

    a) What are the exit temperatures of the streams?

    b) What is the rate of heat transferred?

9.12 A fluid is heated in a heat exchanger by condensing steam. Derive the log mean ΔT to use when the heat capacity $\hat{C}_p$ is equal to $a + bT$.

9.13 Determine the heat loss from a bare, horizontal steam pipe, 150-ft long, 6-in. OD, at 275°F. This pipe is located indoors where the temperature is 70°F.

    Calculate the loss in the same room after this pipe is covered with a 3-in. layer of insulation ($k = 0.11$ Btu/hr ft °F).

9.14 A composite wall is placed around a chemical reactor vessel. It is made up of concrete with steel beams centrally located for reinforcement and an outer wall of stainless steel. A typical cross section is shown below.

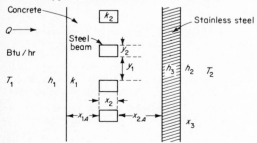

Assuming one-dimensional heat transfer throughout (i.e., vertical temperature uniformity), derive the following expression for Q:

$$Q = \frac{(T_1 - T_2)(y_1 + y_2)}{1/h_1 + x_1/k_1 + x_3/k_3 + [x_2/(k_1 y_1 + k_2 y_2)]/(y_1 + y_2) + 1/h_2}$$

where $x_1 = x_{1A} + x_{2A}$. Take a typical repeating section 1-ft side (perpendicular to the paper) and $(y_1 + y_2)$ high as your area. Then calculate Q based on this area and for the following case:

No.	T, °F	h, Btu/hr ft² °F	k, Btu/hr ft °F	x, in.	y, in.
1	100	5	0.5	10	5
2	400	2	25	2	3
3	...	...	10	$1/2$	...

9.15 Insulation placed on a small-diameter pipe or electrical wire to prevent heat loss may actually increase the heat loss instead. This occurs because as the insulation radius r_2 increases, the insulation thermal resistance increases but the outside surface area is increased at the same time. The result is that the heat loss Q will increase with increasing r_2, until a maximum Q is reached and then further increase in r_2 lowers Q. If the inside surface temperature of the insulation T_1 is known and the surroundings are at T_2, find the value of r_2 for maximum Q. Assume that T_1, T_2, k, h_2, and r_1 are constant.

9.16 Light oil enters the tube side of a heat exchanger (ID = 1.00 in.) at 150°F and exits at 300°F.

Property	150°F	300°F
ρ, lbm/ft³	54.3	51.8
μ, lbm/ft sec	530×10^{-5}	83×10^{-5}
k, Btu/hr ft °F	0.075	0.073
Pr	122	22

The light oil flows through the tube-side heat exchanger at 1,200 gal/min. There are 60 tubes in the heat exchanger. Calculate the tube-side heat transfer coefficient.

9.17 Water is flowing in the shell side of a heat exchanger. The water is flowing parallel to the tubes. The water enters at 80°F and leaves at 150°F. The inside diameter of the shell is 12 in., the outside tube diameter is 1.125 in., and there are 50 tubes in the heat exchanger. The water flow rate through the heat exchanger is 800 gal/min.

Property	80°F	150°F
ρ, lbm/ft³	62.2	61.2
μ, lbm/ft sec	0.578×10^{-3}	0.292×10^{-3}
k, lbm/hr ft °F	0.353	0.384
Pr	5.89	2.74

(a) Calculate the equivalent diameter for the flow on the shell side of the heat exchanger.

(b) Calculate the average shell-side heat transfer coefficient.

9.18 Water is flowing in a thin-walled copper tube under conditions such that the coefficient of convective heat transfer from water to the tube is given by

$$h_i = 25 \langle v \rangle^{1/3}$$

where $\langle v \rangle$ is in ft/sec and h_i is in Btu/hr ft² °F. Heat from the water flows through the tube wall to the surrounding atmosphere at a rate such that the convection coefficient from the outer surface of the tube to the atmosphere is given by

$$h_0 = 0.5(\Delta T)^{1/4}$$

where h_0 is in Btu/hr ft² °F and ΔT is the temperature drop between the tube wall and the atmosphere in °F. Assume that the heat transfer resistance of the tube wall is negligible. Find the minimum water velocity necessary to keep ice from forming on the pipe wall when the bulk water temperature is 35°F and the atmosphere is at −25°F.

9.19 The coefficient on the shell side of a shell and tube exchanger having 1.0-in.-OD tubes is 200 Btu/hr ft² °F for water in turbulent flow. Estimate the coefficient for a second similar exchanger constructed of ³/₄-in.-OD tubes under identical service conditions.

9.20 If you were designing a condenser or feed water heater that would heat incoming water in tubes by condensing steam on the outside of the tubes, which orientation (either vertical or horizontal) of the tubes would give better heat transfer to the water? Assume saturated steam at 7 psia, a 1-in.-OD 6-ft-long tube with an average wall temperature of 125°F. Explain fully your choice of tube orientation.

9.21 A ¹/₂-in.-OD 5-ft-long tube is to be used to condense steam at 6 psia. Estimate the mean heat transfer coefficient for this tube in the (a) horizontal and (b) vertical positions. Assume that the average tube-wall temperature is 180°F.

9.22 Water flowing turbulently is being heated in a concentric pipe exchanger by steam condensing in the annulus at 220°F. The water enters at 60°F and leaves at 80°F. If the water rate is doubled, estimate the new exit temperature.

9.23 In the production of salt from seawater, the seawater is run into a large pond 100 ft long, 20 ft wide, and 2 ft deep. There the water is evaporated by exposure to the sun and dry air, leaving solid salt on the bottom of the pond.

At a certain point in this cycle of operations, the level of brine in the pond is 1 ft deep. On the evening of a clear day a workman finds that the sun has heated the brine to 90°F During the following 10-hr period of the night, the average air temperature is 70°F, and the average effective blackbody temperature of the sky is − 100°F. Neglecting any vaporization or crystallization effects and assuming negligible heat transfer between the brine and the earth, estimate the temperature of the brine at the end of the 10-hr period. The properties of the brine may be assumed to be approximately those of water, and the brine temperature remains uniform from top to bottom.

The coefficient of heat transfer from the surface of the pond to air by natural convection may be taken as

$$h_c = 0.38(\Delta T_S)^{0.25}$$

where T_S is the temperature of the surface minus the temperature of the ambient air in °F, and h_c is heat transfer coefficient in Btu/hr ft² °F.

9.24 A boiler is producing methanol vapor at 1 atm using hot oil flowing through a single 1.0-in.-OD × 0.134-in.-wall stainless-steel 304 tube as a source of heat. From one test of the equipment the following data are available:

Oil temperature—210°F entering, 196°F leaving; inside coefficient—100 Btu/hr ft² °F. Determine the heat transfer coefficient for boiling and the approximate heat flux.

10
Heat Transfer Applications

This chapter, like the parallel chapters on momentum (Chap. 7) and mass transfer (Chap. 13), is divided into two sections. The first section is a relatively detailed presentation of the determination of heat transfer coefficients and a comparison of these experimentally determined coefficients with the design equations presented earlier. In addition the Wilson method of determining heat transfer coefficients for heat exchangers is discussed.

In the second section we discuss heat transfer applications with reference to the natural gasoline plant as introduced in Chap. 1. Several different parts of this plant provide opportunity for heat transfer calculations associated with heat exchangers, which are the usual application of the design equations given in Chap. 9 and data such as collected in the first section.

As with the parallel chapters a series of plates is included at the end of the chapter—in this case to provide a look at heat exchanger equipment.

10.1 EXPERIMENTAL DETERMINATION OF HEAT TRANSFER COEFFICIENTS

To this point, we have simply presented design equations for various types of heat transfer coefficients without explaining anything about how one obtains data to develop such relationships. In this section we wish to illustrate a method of obtaining such data. To determine a single heat transfer coefficient, it is usually necessary to do one of two things: either to make all other heat transfer coefficients in the system so large that the overall heat transfer coefficient becomes identical with the coefficient we are attempting to develop, or somehow to account for the behavior of the other coefficients by a design calculation. Another problem also faces us. To obtain the most generally useful correlation we should determine *local* coefficients which can then subsequently be integrated to obtain *average* coefficients; however, this is a difficult problem experimentally. The illustration will be for determination of a particular average coefficient rather than for a local coefficient.

Suppose we wish to determine the heat transfer coefficient for fluids inside tubes. The experimental apparatus is shown in Fig. 10.1-1. Light oil flows in a long copper tube which has thermocouples embedded in grooves in the surface. The tube is long enough before entering the heated section that the velocity profile is fully developed. In the heated section steam condenses on the outer wall of the tube and transfers heat

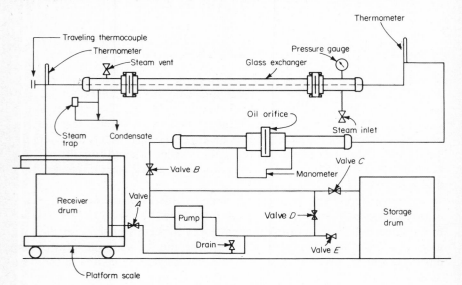

Fig. 10.1-1 Equipment schematic for determining heat transfer coefficients. (*Chemical Engineering Department, University of Wisconsin.*)

through the tube to the oil. We measure the temperature of the oil at two points relatively close together and far enough inside the heated section that thermal entrance effects are absent. We want to calculate the average coefficient and compare it with the value given by various mathematical models. The data for the experiment are given in Table 10.1-1.

Table 10.1-1 Experimental heat transfer data and summary calculations

Run no.	1	2	3	4	5	6	7
1 w, lbm oil per hr	1,430	1,200	841	2,040	2,920	3,730	2,680
2 G', lbm/hr ft², $(w/0.0016)(10^{-4})$	89.4	75.0	52.6	120.8	183	234	167
3 T_0 average oil temperature °F	136.7	142.6	146.2	148.2	151.5	154.9	157.2
4 $\hat{C}_p$, average heat capacity	0.504	0.509	0.511	0.512	0.514	0.515	0.517
5 $(T_5 - T_1)_{oil}$, increase in oil temperature °F	18.7	18.7	18.2	14.3	12.5	10.2	11.2
6 Q, Btu/hr, $w\hat{C}_p(T_5 - T_1)$ oil	13,460	11,450	7840	14,970	18,780	19,600	15,500
7 T_{ms}, average temperature of outside surface	227.3	228.4	230.3	226.5	224.8	224.0	225.7
8 ΔT_s, temperature drop across steam film $T_s - T_{ms}$	6.9	5.2	3.3	7.5	9.5	10.6	8.5
9 h_s, steam film $Q_s/A_s\Delta T_s$	1,970	2,220	2,400	2,020	2,000	1,870	1,845
10 ΔT_m, temperature drop across metal, $(QL_m/k_m A_m)\,(0.000245)$	3.3	2.8	1.9	4.9	4.6	4.8	3.8
11 T_{mo}, metal temperature at oil surface $T_{ms} - \Delta T_m$	224	225.6	228.4	221.6	220.2	219.2	221.9
12 ΔT_0, temperature drop across oil film $T_{mo} - T_0$ av	87.3	83.0	82.0	73.4	68.7	64.3	64.2
13 h_0, oil film $Q/A_0\Delta T_0$	241	216	149	318	426	476	377
14 ΔT, overall temperature drop $T_s - T_0$ av	97.5	91.0	87.4	85.8	82.8	79.7	77.0
15 U_0, overall $Q/A_0\Delta T$	216	197	140	272	354	384	337
16 Mass velocity, $^1 10^{-4} G$	89.4	75.0	52.6	120.8	183	234	167
17 T_0, average oil temperature	136.7	142.6	146.2	148.2	151.5	154.9	157.2
18 T_{mo}, average oil surface temperature	224	225.6	228.4	221.6	220.2	219.2	221.9
19 T_f, average oil film temperature	180.4	184.1	187.3	184.9	185.9	187	189.3
20 μ, viscosity at temperature T_0	9.44	8.78	8.45	8.27	8.00	7.74	7.57
21 μ_s, viscosity at temperature T_{mo}	4.66	4.63	4.58	4.69	4.72	4.73	4.69
22 μ_f, viscosity at temperature T_f	6.16	6.03	5.83	5.94	5.90	5.85	5.74
23 $\hat{C}_p$ average heat capacity	0.504	0.509	0.511	0.572	0.514	0.515	0.517
24 k, thermal conductivity	0.081	0.0810	0.0811	0.0808	0.0808	0.0807	0.0807
25 Prandtl no. $\hat{C}_p\mu/k$	58.5	59.1	53.1	52.4	50.9	49.3	48.6
26 $(\hat{C}_p\mu/k)^{-0.4}$	0.196	0.202	0.204	0.205	0.207	0.210	0.211
27 $(\hat{C}_p\mu/k)^{2/3}$	15.4	14.7	14.3	14.2	13.9	13.6	13.4
28 $(\hat{C}_p\mu_f/k)^{2/3}$	11.5	11.4	11.2	11.4	11.4	11.4	11.2
29 Reynolds no. $(DG'/\mu) \times 10^{-3}$	4.27	3.86	2.81	6.99	10.32	13.66	9.96
30 $(DG'/\mu)^{0.8}$	810	749	580	1,200	1,645	2,060	1,600
31 $(DG'/\mu)^{-0.2}$	.187	.191	.203	.170	.158	.149	.158
32 $(DG'/\mu_f)^{-0.2}$	1.089	1.123	1.197	1.017	0.935	0.888	0.948
33 Nusselt no., hD/k	134	120	83	178	238	266	211
34 Stanton no., $(h/\hat{C}_p G') \times 10^{-4}$	5.32	5.64	5.56	5.17	4.65	3.95	4.36
35 μ/μ_s	2.02	1.89	1.84	1.76	1.69	1.64	1.62
36 $(\mu/\mu_s)^{-0.14}$	.906	.915	.918	.924	.929	.933	0.933
37 $(hD/k)/(\hat{C}_p\mu/k)^{-0.4}$	26.3	24.2	16.9	36.5	49.3	55.8	44.5
38 $(h/\hat{C}_p G')(\hat{C}_p\mu/k)^{2/3}(\mu/\mu_s)^{-0.14} \times 10^{-4}$	74.2	75.6	73.0	67.9	60.0	50.2	54.5
39 $(h/\hat{C}_p G')(\hat{C}_p\mu_f/k)^{2/3} \times 10^{-4}$	61.0	64.2	67.3	59.1	53.2	45.0	49.7

The physical properties of the oil are given in Table 10.1-2. The data on the tube are given in Table 10.1-3.

Sample calculations associated with Table 10.1-1 are carried out below. All calculations are for run 1.

1. Weight of oil flow lb$m$/hr

$$w = (100 \text{ lb}m)\frac{3,600}{252} = 1,430 \text{ lb}m/\text{hr}$$

6. Heat transferred, Btu/hr

$$Q = w\hat{C}_p \ (T_s - T_1)$$
$$= (1,430)(0.504)(18.7) = 13,460 \text{ Btu/hr}$$

9. Steam, heat transfer coefficient

$$h_s = \frac{Q_s}{A_s \Delta T_s}$$
$$= \frac{13,460}{0.99 \times 6.9} = 1970 \text{ Btu/hr ft}^2 \text{ °F}$$

Table 10.1-2   Physical properties of heat transfer oil

T, °F	Density ρ, lbm/ft³	Viscosity μ, lbm/hr ft	Spec. heat $\hat{C}_p$, Btu/ lb °F	Thermal conductivity k, Btu/hr ft °F	$\dfrac{\hat{C}_p \mu}{k}$	$\left(\dfrac{\hat{C}_p \mu}{k}\right)^{2/3}$	$\left(\dfrac{\hat{C}_p \mu}{k}\right)^{0.4}$	$\left(\dfrac{\hat{C}_p \mu}{k}\right)^{1/3}$
80	51.7	20.10	0.476	0.0822	75.45	17.90	5.64	4.23
90	51.6	17.10	0.481	0.0820	66.30	16.45	5.35	4.05
100	51.4	14.80	0.486	0.0818	58.80	15.20	5.11	3.89
110	51.2	14.80	0.492	0.0816	52.45	14.05	4.88	3.74
120	51.0	11.80	0.498	0.0814	46.90	13.05	4.66	3.61
130	50.7	10.10	0.502	0.0812	41.90	12.10	4.45	3.47
140	50.5	9.01	0.508	0.0810	37.93	11.32	4.28	3.36
150	50.2	9.11	0.513	0.0808	34.42	10.60	4.12	3.25
160	49.9	7.36	0.518	0.0806	31.55	10.05	3.98	3.16
170	49.7	6.75	0.525	0.0804	29.23	9.50	3.86	3.08
180	49.4	6.19	0.530	0.0802	27.00	9.00	3.74	3.00
190	49.2	5.71	0.535	0.0800	24.87	8.62	3.61	2.94
200	48.8	5.25	0.540	0.0798	23.43	8.18	3.52	2.66
210	48.6	4.96	0.545	0.0796	22.05	7.88	3.45	2.81
220	48.4	4.76	0.550	0.0794	21.00	7.63	3.38	2.76

Table 10.1-3 Section areas and thermal resistances of tube sections

Section no.	0	1	2	3	4	5
A_s	0.165	0.275	0.220	0.220	0.275	0.165
A_m	0.138	0.227	0.184	0.184	0.227	0.138
A_0	0.1065	0.178	0.142	0.142	0.178	0.1065
$\dfrac{L}{kA_m}$	0.00146	0.00089	0.0011	0.0011	0.00089	0.00146

10. Temperature drop across metal

$$\Delta T_m = \frac{QL_m}{k_m A_m}$$

$$= \frac{13{,}460 \times 0.0125}{62 \times 0.822} = 3.3°\text{F}$$

13. Oil, heat transfer coefficient

$$h_0 = \frac{Q}{A_0 \Delta T_0}$$

$$= \frac{13{,}460}{0.64 \times 87.3} = 241 \text{ Btu/hr ft}^2\,°\text{F}$$

15. Overall heat transfer coefficient

$$U_0 = \frac{Q}{A_0 \Delta T}$$

$$U_0 = \frac{13{,}460}{0.64 \times 87.5} = 216 \text{ Btu/hr ft}^2\,°\text{F}$$

Check calculation of U_0

$$\frac{1}{U_0} = \frac{1}{h_0} + \frac{L_m A_0}{kA_m} + \frac{A_0}{h_s A_s}$$

$$= \frac{1}{241} + \frac{(0.0125)(0.64)}{(62)(0.822)} + \frac{0.64}{(1{,}970)(0.99)}$$

$$U_0 = 216 \text{ Btu/hr ft}^2\,°\text{F}$$

16. Mass velocity of oil

$$G' = \frac{1,430}{0.0016} = 894,000 \text{ lb}m/\text{hr ft}^2$$

25. Prandtl number

$$\frac{\hat{C}_p \mu}{k} = \frac{(0.504)(9.44)}{0.0811} = 58.5$$

29. Reynolds number

$$\frac{DG'}{\mu} = \frac{(0.0814)(89.4 \times 10^4)}{9.44} = 4,270$$

33. Nusselt number

$$\frac{hD}{k} = \frac{(241)(0.0184)}{0.0811} = 134$$

34. Stanton number

$$\frac{h}{\hat{C}_p G'} = \frac{241}{(0.504)(89.4 \times 10^4)} = 5.32 \times 10^{-4}$$

h_0, by Dittus-Boelter equation

$$\left(\frac{hD}{k}\right)\left(\frac{\hat{C}_p \mu}{k}\right)^{-0.4} = 0.023 \left(\frac{DG'}{\mu}\right)^{0.8}$$

$$h(1.77)(5.10) = 0.023 \, (4,270)^{0.8}$$

$$h_0 = 163 \text{ Btu/hr ft}^2\,^\circ\text{F}$$

h_0, Sieder-Tate equation

$$\frac{h}{\hat{C}_p G'}\left(\frac{\hat{C}_p \mu}{k}\right)^{2/3}\left(\frac{\mu}{\mu_s}\right)^{-0.14} = 0.023 \left(\frac{DG'}{\mu}\right)^{-0.2}$$

$$h_0 \frac{1}{45 \times 10^4} (15.1)(0.906) = 0.023 \, (4,270)^{0.8}$$

$$h_0 = 133 \text{ Btu/hr ft}^2\,^\circ\text{F}$$

h_0, Colburn equation

$$\frac{h}{\widehat{C}_p G'} \left(\frac{\widehat{C}_p \mu}{k}\right)^{2/3} = 0.023 \left(\frac{DG'}{\mu_f}\right)^{-0.2}$$

$$h \left(\frac{1}{45 \times 10^4}\right) (15.1) = .023 \, (6550)^{-0.2}$$

$h_0 = 198$ Btu/hr ft^2°F

h_s, Wilson method (see below)
From sum of least squares

$b = 2.57 \times 10^{-4}$

$$= \frac{A_0 L_m}{k_m A_m} + \frac{A_0}{h_s A_s}$$

$$0.000257 = \frac{(0.64)(0.0124)}{(0.822)(62)} + \frac{0.64}{0.99 h_s}$$

$h_s = 6400$ Btu/hr ft^2°F

h_s, Nusselt equation (Eq. 9.4-9)

$$h_s = 0.725 \left[\frac{k_f{}^3 \rho_f{}^2 \lambda g}{D \mu_f \Delta T}\right]^{1/4}$$

$$= 0.725 \left[\frac{(0.074)(3,540)(4.18 \times 10^8)(970)}{(87.3)(0.070)(6.72 \times 10^{-4})(0.2838)(3,600)}\right]^{1/4}$$

$$= 3170 \text{ Btu/hr ft}^2°F$$

The experimental results given in Table 10.1-4 are compared with the prediction of the Dittus-Boelter equation on Fig. 10.1-2. In this case the experimental results are higher but they do have about the same Reynolds number dependence as predicted by the equation.

Table 10.1-4   Data for Fig. 10.1-2[a,b]

DG'/μ	$[hD/k][\widehat{C}_p \mu/k]^{-0.4}$
2,810	16.9
3,860	24.2
4,270	26.3
6,990	36.5
9,960	44.5
10,320	49.3
13,660	55.8

[a] Slope   0.77
[b] Intercept   0.0328

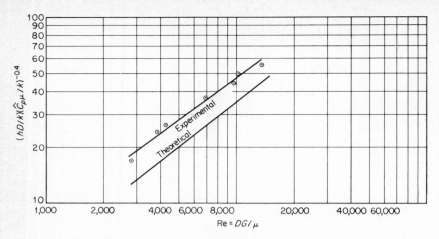

Fig. 10.1-2 Comparison of oil coefficients to Dittus-Boelter equations.

The experimental results are calculated in the form of the Sieder-Tate equation in Table 10.1-5, and they are compared with the prediction of the Sieder-Tate equation on Fig. 10.1-3. The experimental data are higher but the Reynolds dependence is about correct.

The experimental results are calculated in the form of the Colburn equation in Table 10.1-6 and compared with this equation on Fig. 10.1-4. The experimental data are higher but again the slope is the same.

Table 10.1-7 gives a summary of the experimental results for the oil side heat transfer coefficient as a function of mean velocity and a comparison with the three equations.[1]

[1] These data are from actual laboratory measurements. In each case (Figs. 10.2-1, 2, and 3), the theoretical and experimental slopes agree, but the experimental intercept is higher than the theoretical. This shows the presence of a systematic error in the experiment.

Table 10.1-5   Data for Fig. 10.1-3[a, b]

DG'/μ	$[h/\widehat{C}_p G'][\widehat{C}_p \mu/k]^{2/3}[\mu/\mu_s]^{-0.14}$
2,810	73.0
3,860	75.6
4,270	74.2
6,990	67.9
9,960	54.5
10,320	60.0
13,660	50.2

[a] Slope −0.229
[b] Intercept   0.388

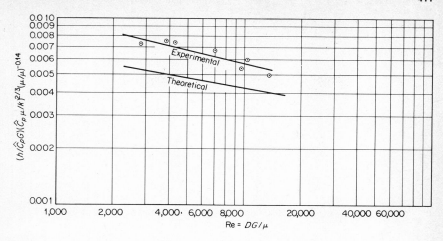

Fig. 10.1-3 Comparison of oil coefficients to Sieder-Tate equation.

Wilson method

A useful graphical method for evaluating individual heat transfer coefficients from overall heat transfer coefficients was proposed by Wilson.[1] Inspection of the Dittus-Boelter equation shows that if the temperature of a liquid in passing through a given heat exchange apparatus does not change markedly, the liquid heat transfer coefficient is proportional to $G'^{0.8}$, where G' is the mass velocity of the liquid in the tubes of the exchanger. Thus,

$$h_0 = aG'^{0.8} \qquad\qquad (10.1\text{-}1)$$

[1] E. E. Wilson, *Trans. ASME*, **37**:47 (1915).

Table 10.1-6   Data for Fig. 10.1-4[a, b]

DG'/μ_f	$[h/\widehat{C}_p G'][\widehat{C}_p \mu_f/k]^{1/3}$
4,100	62.3
5,610	64.2
6,550	61.0
9,700	59.1
12,750	49.7
14,000	59.1
18,000	53.2

[a] Slope -0.208
[b] Intercept 0.0346

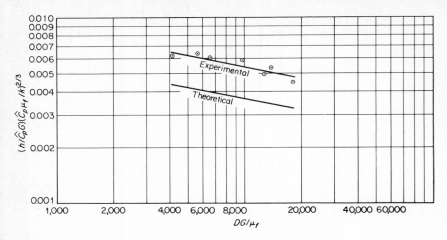

Fig. 10.1-4 Comparison of oil coefficients to Colburn equation.

where a = some constant. The overall heat transfer coefficient, based on the liquid surface area, may be written as

$$\frac{1}{U_0} = \frac{1}{h_0} + \frac{A_0 L_m}{k_m A_m} + \frac{A_0}{h_s A_s} \qquad (10.1\text{-}2)$$

If the heat exchange apparatus is a condenser, i. e., liquid on one side of the tubes and a condensing vapor on the other, the value of h_s does not change markedly throughout the length of the apparatus, provided the metal wall temperature change is not too large. As an approximation, h_s may be assumed constant. Also term $A_0 L_m/k_m A_m$ in Eq. (10.1-2) will be constant. Thus Eq. (10.1-2) may be approximated as

$$\frac{1}{U_0} = \frac{1}{aG'^{0.8}} + b \qquad (10.1\text{-}3)$$

where

$$b = \frac{A_0 L_m}{k_m A_m} + \frac{A_0}{h_s A_s}$$

Equation (10.1-3) will plot as a straight line, provided that the assumptions are correct, if $1/U_0$ is plotted as ordinate against $1/G'^{0.8}$ as abscissa. The slope of this line will be $1/a$ and the intercept will be equal to b. From the slope, values of h_0 corresponding to various G' values may be found,

Table 10.1-7 Summary data and results

Oil coefficient

G', lbm/hr ft²	Reynolds no.	h_0, Btu/hr ft² °F			
		Dittus-Boelter	*Sieder-Tate*	*Colburn*	*Experimental*
52.6 × 10⁴	2,810	109	93	107	149
75.0 × 10⁴	3,860	143	120	137	216
89.4 × 10⁴	4,270	163	133	156	241
120.8 × 10⁴	6,990	206	185	198	318
167.0 × 10⁴	9,960	284	258	265	377
183.0 × 10⁴	10,320	294	268	282	426
234.0 × 10⁴	13,660	386	332	348	476

Comparison of slopes and constants

Equation	Experimental		Theoretical	
	Slope	Constant	Slope	Constant
Dittus-Boelter	0.77	0.0328	0.8	0.0225
Sieder-Tate	−0.229	0.0388	−0.2	0.0270
Colburn	−0.208	0.0346	−0.2	0.0230

and from the intercept, a single average value of h_s may be evaluated. Thus, the individual resistances may be estimated readily from measurements of the overall heat transfer coefficient.

The steam-side coefficient was determined in the previous experiment using the Wilson method.

Table 10.1-8 presents the experimental data from the experiment.

Table10.1-8 Data for Wilson plot (Fig. 10.1-5)

G'	h_0	$(1/U_0)$	$(1/G'^{0.8})$
526,000	149	7.7 × 10⁻⁶	0.00260
750,000	216	9.8 × 10⁻⁶	0.00283
894,000	241	10.52 × 10⁻⁶	0.00297
120,800	318	13.66 × 10⁻⁶	0.00368
167,000	377	17.35 × 10⁻⁶	0.00463
183,000	426	20.0 × 10⁻⁶	0.00507
234,000	476	26.4 × 10⁻⁶	0.00715

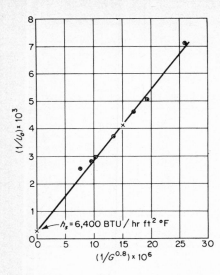

Fig. 10.1-5 Wilson correlation.

The data are plotted on Fig. 10.1-5. The steam-side heat transfer coefficient calculated from the intercept is given in the figures.

The Wilson method is often used to determine the fouling coefficient of an exchanger by obtaining the data for a Wilson plot at various times and determining the intercept value.

10.2 SELECTED APPLICATIONS

Heat transfer applications come in many forms; however, most design time by far is spent in the design of heat exchangers. Many of these units, fortunately, embody common features. Since the natural gasoline plant introduced in Chap. 1 contains several typical heat exchange situations we will use this system for examples of preliminary heat exchanger design. We begin with the simplest example of a heat exchanger; that is, a double-pipe heat exchanger with liquid on both sides. From this we go to the much more commonly used shell and tube exchanger, and from there to the economic analysis of this case.

In the era of thermal pollution, it becomes increasingly important to find a different heat sink from rivers, streams, and lakes to use for the disposal of heat from a process plant. A very popular form of exchanger is the "Fin-Fan" exchanger in which heat is exchanged with atmospheric air. (This, of course, merely transfers the thermal pollution problem from the water to the air, but at the moment the air has a much larger capacity to absorb this pollutant than do existing water sources.) We also consider an example of such an exchanger below.

Thermal driving forces for heat exchangers

Heat exchangers are used to transfer heat from one fluid to another. The simplest form of heat exchanger is a container where a hot and cold fluid can be mixed. In such an exchanger the amount of heat transferred is found directly from an energy balance. In most commercial applications we must transfer heat from one fluid to another while keeping the fluids separate. Heat exchangers to accomplish heat transfer with separated fluids come in many shapes and forms.

In a simple double-pipe exchanger, basically two fluids flow past each other countercurrently, cocurrently, or crosscurrently, with one fluid in the inner pipe and one in the annulus; or they flow through complex multipass exchangers. The principal design objective in every case is to determine the required heat transfer area for a given situation.

In general

$$dQ = U_0 \, (\Delta T)_{\text{overall}} \, dA_0 \tag{10.2-1}$$

where U_0 is the overall heat transfer coefficient based on the outside area and $\Delta T_{\text{overall}}$ is the difference between the mixing-cup temperature of the hot and cold fluids. The temperature difference for the double-pipe exchanger sketched in Fig. 10.2-1 is shown in Fig. 10.2-2. An energy balance gives

$$dQ = w_c \hat{C}_{pc} dT_c = w_h \hat{C}_{ph} dT_h = U_0 (T_h - T_c) \, dA_0 \tag{10.2-2}$$

where the subscripts c and h refer to the cold and hot fluid respectively. From Fig. 10.2-2[1]

[1] See Sec. 9.3 for a more precise derivation of LMTD.

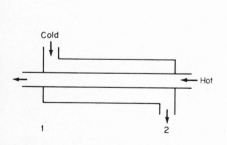

Fig. 10.2-1 Double-pipe heat exchanger.

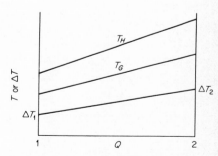

Fig. 10.2-2 Temperature differences in a double-pipe heat exchanger.

$$\frac{d \, \Delta T}{dQ} = \frac{\Delta T_2 - \Delta T_1}{Q} \tag{10.2-3}$$

or

$$\frac{d \, \Delta T}{U_0 \, \Delta T \, dA_0} = \frac{\Delta T_2 - \Delta T_1}{Q} \tag{10.2-4}$$

Integrating between 1 and 2

$$\int_{\Delta T_1}^{\Delta T_2} \frac{d\Delta T}{U_0 \, \Delta T} = \frac{\Delta T_2 - \Delta T_1}{Q} \int_0^{A_0} dA_0 \tag{10.2-5}$$

and

$$Q = U_0 A_0 \left[\frac{\Delta T_2 - \Delta T_1}{\ln \, (\Delta T_2/\Delta T_1)} \right] \tag{10.2-6}$$

where the term in brackets is the logarithmic mean temperature difference (LMTD). If U_0 is a function of temperature it must be taken inside the integral. If the flow in the exchanger in Fig. 10.2-1 is cocurrent instead of countercurrent as indicated, we would obtain the same expression for the LMTD.

Figure 10.2-3 is a sketch of a shell and tube heat exchanger. Usually the hot fluid is run through the tubes and the cold fluid through the shell. Final decision on where to put the hot or cold fluid depends on fouling, corrosion, and pressure-drop considerations. To find heat transfer area for more complex exchangers Eq. (10.2-6) is modified to give

$$Q = U_0 A_0 Y \, \text{LMTD} \tag{10.2-7}$$

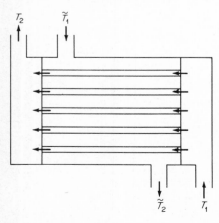

Fig. 10.2-3 Shell and tube exchanger.

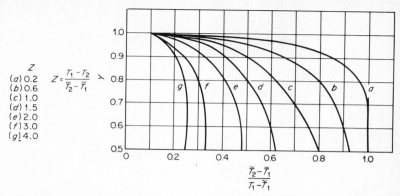

$$Z = \frac{T_1 - T_2}{\tilde{T}_2 - \tilde{T}_1}$$

Z
(a) 0.2
(b) 0.6
(c) 1.0
(d) 1.5
(e) 2.0
(f) 3.0
(g) 4.0

$$\frac{\tilde{T}_2 - \tilde{T}_1}{T_1 - \tilde{T}_1}$$

Fig. 10.2-4 Correction factor for shell and tube heat exchanger. [(*R. W. Bowman, A. C. Moeller, and W. M. Nagle, Trans. ASME,* **62**:283 *(1940).*]

where the correction factor has been calculated and is plotted in Fig. 10.2-4 for the exchanger of Fig. 10.2-3. A similar factor is used for exchangers having other than true countercurrent flow, such as cross flow only, and for multipass exchangers where the number of tube passes exceeds the number of shell passes. The use of Eq. (10.2-7) assumes that the fluid in any one pass through the shell is completely mixed. If the temperature of the shell-side fluid is constant the value for Y is 1.

A multipass heat exchanger with one shell and two tube passes is sketched in Fig. 10.2-5. The correction factor for this exchanger is given in Fig. 10.2-6. Charts are available in standard textbooks for all combinations of shell and tube passes and other types of exchangers.[1]

[1] See, for example, D. Q. Kern, "Process Heat Transfer," McGraw-Hill Book Company, New York, 1950.

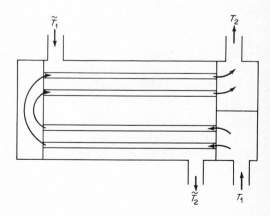

Fig. 10.2-5 Multipass heat exchanger (one shell and two tubes).

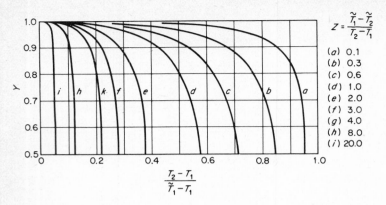

Fig. 10.2-6 Correction factor for shell and tube exchanger with one shell pass and an even number of tube passes. (*Tubular Exchanger Manufacturers Association.*)

Example 10.2-1 Double-pipe heat exchanger The simplest heat exchanger geometry is perhaps that of a double-pipe heat exchanger. A double-pipe heat exchanger is simply a pipe of small diameter contained within a pipe of large diameter. In the outside pipe fluid may flow either cocurrent or countercurrent to the flow of fluid in the inside pipe. These exchangers may be used for condensing and boiling as well as for liquid-liquid heat exchange, but they are probably more commonly applied to liquid-liquid heat exchange. Where large transfer areas and large heat exchange rates are required, double-pipe exchangers are seldom the most economical. However, for so-called trimming applications (that is, applications where the temperature change of one of the streams is relatively slight), the double-pipe exchanger is often a good choice.

Below we consider design of a trimmer exchanger which performs the final temperature control on a stream which enters a process unit.

Double-pipe heat exchange equipment, unlike many other types of equipment, comes in relatively standard sizes and configurations. This means that frequently, upon discontinuing an old process, double-pipe heat exchange equipment is available as salvage which can be utilized in a new process. When existing heat exchange equipment is available the question is often whether or not it can be used to solve a present problem.

Problem statement   In the flowsheet, Fig. 1.3-1, there is need for a trimmer heat exchanger between stream 16 and 22. This exchanger is to raise the temperature of stream 16 from 408 to 427°F by extracting energy from stream 22.

Two hundred feet of standard, 4×3, double-pipe heat exchangers are available within the plant. Is it possible to utilize these exchangers for this task with stream 16 on the tube side? How much of stream 22 will be bypassed under this arrangement? Discuss whether a parallel connection of standard lengths might be more favorable. (See Fig. 10.2-7.)

Stream 16	*Stream 22*
143,400 lb$m$/hr	129,000 lb$m$/hr
502.2 gal/min	461.9 gal/min
sp. gr. = 0.75	sp. gr. = 0.825
ρ = API 50	ρ = API 40
$\hat{C}_p$ = 0.68 Btu/lb$m$ °F	$\hat{C}_p$ = 0.75 Btu/lb$m$ °F
k = 0.075 Btu/ft hr °F	k = 0.075 Btu/ft hr °F

Solution

Stream 16: $Q = (143{,}400 \text{ lb}m/\text{hr})(0.68 \text{ Btu/lb}m \text{ °F})[(425 - 408) \text{ °F}]$

$$= 1.852 \times 10^6 \text{ Btu/hr}$$

Stream 22: $w = \dfrac{1.852 \times 10^6 \text{ Btu/hr}}{(0.75 \text{ Btu/lb}m \text{ °F})[(527 - 520) \text{ °F}]}$

$$= 9.875 \times 10^4 \text{ lb}m/\text{hr}$$

Bypass of stream 22 = $(129{,}000 - 98{,}750)$ lb$m$/hr = 30,250 lb$m$/hr

LMTD: $\Delta T_2 = 551 - 427 = 124$

$\Delta T_1 = 526 - 408 = 118$

$$\text{LMTD} = \frac{124 - 118}{\ln(124/118)} = 121 \text{°F}$$

Stream 16: Re = 5.85×10^4

Pr = 53.5

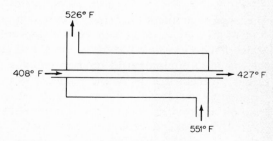

Fig. 10.2-7 Sketch for Example 10.2-1.

From Sieder-Tate equation, Eq. (9.3-57)

$$h_i = 187 \quad \text{Btu/hr ft}^2 \, ^\circ\text{F}$$

$$h_0 = 164 \quad \text{Btu/hr ft}^2 \, ^\circ\text{F}$$

Stream 22: Re $= 3.51 \times 10^4$

$$\text{Pr} = 61.1$$

From Eq. (9.3-59)

$$h_0 = 340 \text{ Btu/hr ft}^2 \, ^\circ\text{F}$$

Assume wall resistance negligible: $U_0 = \dfrac{(164)(340)}{164 + 340}$

$$= 111 \text{ Btu/hr ft}^2 \, ^\circ\text{F}$$

Surface area required: $Q = U_0 A_0 \text{ LMTD}$

$$A_0 = \frac{1.852 \times 10^6 \text{ Btu/hr}}{(111 \text{ Btu/hr ft}^2 \, ^\circ\text{F})(121 \, ^\circ\text{F})} = 138 \text{ ft}^2$$

Surface area per lineal foot $= 0.917 \text{ ft}^2/\text{ft}$

$$\therefore \text{ Length required} = \frac{138}{0.917} = 150 \text{ ft}$$

In series arrangement, in order to meet requirements the velocity must be greater than for the parallel arrangement, and in turn, the heat transfer coefficient will increase, therefore requiring a shorter exchanger with increased pumping power.

□ □ □

We must emphasize that the above example illustrates the fact that double-pipe heat exchange equipment comes in standard sizes; that is, the design calculations that an engineer carries out must be modified to use pieces of equipment that are commercially available. This is an illustration of one of the two general classes of problems that always face the engineer. This example illustrates the question of how one adapts an existing piece of equipment to operate under a new set of conditions. The other problem that faces the engineer is the design of a new piece of equipment to solve an existing problem. Although most academic consideration of problems leans heavily on the treatment of the design case (as does this book), it cannot be emphasized too highly that many of the

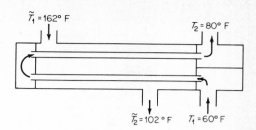

Fig. 10.2-8 Sketch for Example 10.2-2.

problems an engineer faces in practice are those of adapting an existing piece of equipment to run under new conditions.

Example 10.2-2 Shell and tube heat exchanger Many heat exchangers found in process plants are those of the shell and tube type. The number of shell passes and the number of tube passes to be used depend on a relatively fine economic balance between pressure drop and capital costs (the pressure drop, of course, leads to increased pumping costs). At this point we do not wish to immerse ourselves in a great deal of detail; therefore, in the example below we will assume that the number of shell passes and tube passes has already been selected and we will proceed with the determination of the area of such an exchanger.

Problem statement In the natural gasoline plant we must cool a lean oil, shown as stream 22, from 162 to 102°F. The flow rate of the stream is 128,973 lb$m$/hr at these conditions, its specific gravity is 0.7820, and its viscosity is 2.5 cp. We wish to use a shell and tube heat exchanger with one shell-side and two tube-side passes to accomplish this objective. For the cooling fluid we will use cooling water available at 60°F, but to prevent excessive thermal pollution we will assume that the upper limit on exit cooling water is 80°F, since this water will be returned to a stream.[1] We wish to determine the amount of area required for the heat exchanger. (See Fig. 10.2-8.)

Stream 22

129,000 lb$m$/hr

sp. gr. $= 0.782$
$\hat{C}_p = 0.75$ Btu/lb$m$ °F
$\mu = 2.5$ cp

---

[1] In many process applications it may be more economical to use a cooling tower to cool waste water from a variety of operations before discharging or recycling, rather than restricting the design of individual pieces of equipment.

Use 27-in.-ID 1-2 exchanger with 166 $1^1/_4$-in.-OD, 11 BWG tubes laid out on a 1 $^9/_{16}$-in.-square pitch. The overall heat transfer coefficient $U_0 = 136$ Btu/ft^{2}hr °F.[1]

Solution

Energy balance: Oil (22);

$$Q = (129{,}000 \text{ lb}m/\text{hr})(0.75 \text{ Btu/lb}m \text{ °F}) \left[(162 - 102)\text{°F}\right]$$

$$= 5.8 \times 10^6 \text{ Btu/hr}$$

Water: $w = \dfrac{5.8 \times 10^6 \text{ Btu/hr}}{(1 \text{ Btu/lb}m \text{ °F})\left[(80 - 60)\text{°F}\right]} = 2.9 \times 10^5 \text{ lb}m/\text{hr}$

$$\text{LMTD} = \frac{(162 - 80) - (102 - 60)}{\ln\left[(162 - 80)/(102 - 60)\right]} = 60\text{°F}$$

Use Fig. 10.2-6

$$\frac{\tilde{T}_1 - \tilde{T}_2}{T_2 - T_1} = \frac{162 - 102}{80 - 60} = \frac{60}{20} = 3$$

$$\frac{T_2 - T_1}{\tilde{T}_1 - T_1} = \frac{80 - 60}{162 - 60} = \frac{20}{102} = 0.195$$

$$Y = 0.94$$

$$Q = U_0 A_0 Y (\text{LMTD})$$

$$A_0 = \frac{5.8 \times 10^6}{(136 \text{ Btu/ft}^2 \text{hr °F})(0.94)(60 \text{ °F})} = 755 \text{ ft}^2$$

$$L = \frac{755 \text{ ft}^2}{(0.327 \text{ ft}^2/\text{ft})\ 166} = 14 \text{ ft long}$$

Notice that in this example, the Y factor is not far from 1. The Y factor is not an extremely useful device, as can be seen from the charts, because frequently the slope of Y factor versus dimensionless temperature becomes almost infinite, and therefore it is impossible to read the Y factor with any degree of accuracy. The reason for this is that we have forced the form of our mathematical model upon a real system and simply added a correction factor. Therefore the correction factor does not correlate well with the system behavior since we are attempting to describe the

[1] The coefficient was estimated in the previous example from the appropriate design equations for individual heat transfer coefficients. In this case we are using a handbook value for this type of exchanger. In actual practice the value would be estimated from correlations or by measurements similar to those discussed in Sec. 10.1.

system using the wrong model. In cases where more accuracy than can be obtained by the Y factor method is required, and where this is economically desirable, one must go through a detailed integration for the exchanger. This involves more sophisticated mathematical techniques than we have discussed; however, these methods and computer programs to implement them are available.

□ □ □

Example 10.2-3 Economic considerations in heat exchanger design
Heat exchanger design, if one is concerned with only a single exchanger, is frequently not done in an optimal fashion. However, heat exchangers are extremely common process equipment; for example, a refinery uses literally hundreds of exchangers. Under such circumstances, it pays to see that the individual exchangers are designed in optimal fashion. The practicing engineer usually has a computer program available which will optimize heat exchanger design for him; however, it is necessary to understand the principles upon which these programs operate. For this reason, let us reexamine the previous example to see how conclusions change in light of optimal design.

Problem statement Design an optimal-size heat exchanger for Example 10.2-2. Assume that the heat transfer coefficient is constant and that working hours are 24 per day, 300 days per year. The cost of the heat exchanger, including the installation, is $15.00 per square foot of heating area. The yearly depreciation is 10 percent of initial capital. Water cost is 0.0012¢/gal and electricity costs $0.01/kw hr. (For the purposes of this calculation we will not constrain the outlet water temperature to 80°F.) Assume that Δp is constant at 7.0 psi.

Solution We have to solve the problem by trial and error.

$$\text{Power} = V\,\Delta p$$

Assume that the outlet temperature of cooling water is T_{c2}.

$$A = \frac{Q}{U_1 \Delta T_{\text{lm}}} \qquad w = \frac{Q}{\hat{C}_p\,\Delta T_c}$$

$$= \frac{Q}{\hat{C}_p(T_{c2} - T_{c1})}$$

$$V = \frac{1}{62.4}\frac{Q}{\hat{C}_p(T_{c2} - T_{c1})}$$

$$\text{Power} = \frac{1}{62.4} \frac{Q}{\hat{C}_p (T_{c2} - T_{c1})} \Delta p \times 144$$

$$\text{Cost of power} = \frac{1}{62.4} \frac{Q}{\hat{C}_p (T_{c2} - T_{c1})} \Delta p \ 144 \times \frac{0.01}{2.6557 \times 10^6}$$

$$\text{Yearly cost} = \frac{Q}{U \, \Delta T_{1m}} \times 15.00 \times 0.1$$

$$+ \frac{Q \times 24 \times 300}{\hat{C}_p (T_{c2} - T_{c1})} \times 7.48 \, \frac{\text{gal}}{\text{ft}^3} \times 1.2 \times 10^{-5}$$

$$+ \frac{1}{62.4} \frac{Q \times 24 \times 300}{\hat{C}_p (T_{c2} - T_{c1})} \Delta p \ \frac{1.44}{2.6557 \times 10^6}$$

$$= \frac{Q \times 1.5}{U\{(T_{h2} - T_{c2}) - (T_{h1} - T_{c1})/\ln[(T_{h2} - T_{c2})/(T_{h1} - T_{c1})]\}}$$

$$+ \frac{Q}{\hat{C}_p (T_{c2} - T_{c1})} \times 6.46 \times 10^{-1}$$

$$+ \frac{Q}{\hat{C}_p (T_{c2} - T_{c1})} \times \Delta p \times 6.26 \times 10^{-5}$$

$$= \frac{5.8 \times 10^6 \times 1.5}{136 \times \{(120 - T_{c2})/\ln[(162 - T_{c2})/42]\}}$$

$$+ \frac{5.8 \times 10^6}{T_{c2} - 60} \times 6.46 \times 10^{-1}$$

$$+ \frac{5.8 \times 10^6 \times 4.38 \times 10^{-4}}{T_{c2} - 60}$$

$$\text{Yearly cost} = 6.4 \times 10^4 \times \frac{\ln[(162 - T_{c2})/4a]}{120 - T_{c2}} + \frac{3.68 \times 10^6}{T_{c2} - 60}$$

$$\frac{d(\text{yearly cost})}{dT_2} = 0$$

$$= 6.4 \times 10^4 \times \frac{\dfrac{(120 - T_{c2})(-1)}{(162 - T_{c2})} + \ln\left[\dfrac{162 - T_{c2}}{42}\right]}{(120 - T_{c2})^2}$$

$$+ 3.68 \times 10^6 \, \frac{(-1)}{(T_{c2} - 60)^2}$$

$$= 0$$

By trial and error

$$T_{c2} \cong 115$$

$$\Delta T_{ln} = \frac{(162 - 115) - (102 - 60)}{\ln\left[(162 - 115)/(102 - 60)\right]} = 44°F$$

$$A = \frac{5.8 \times 10^6}{136 \times 44} = 1.0 \times 10^3 \text{ ft}^2$$

Length of the exchanger

$$= \frac{1.0 \times 10^3}{0.327 \times 166} = 18.4 \text{ ft}$$

Notice that the exchanger is longer than the original exchanger. The reason a longer exchanger is required is that in our original design we restricted the exit temperature to 80°F. *We notice that for an optimal exchanger the cooling water exit temperature is 115°F.* This is a good example of when optimization is limited by a constraint which cannot be changed at the discretion of the engineer. For this example, if the 80°F were due to legal requirements, we simply could not build the optimal exchanger unless we decided to add a cooling tower to cool the exit water. Note that the pressure drop was an unimportant consideration in the design procedure.

□ □ □

Example 10.2-4 Fin-Fan condenser *Problem statement* Hydrocarbon vapor (Fig. 1.3-1) is condensed in a Fin-Fan still condenser before entering the still reflux accumulator. The condenser is constructed of 1-in.-OD, 18 BWG tubes (fin heights are $5/8$-in., width 0.035 in., with fins spaced at 10 per lineal inch). The tubes are spaced on a $2\,1/4$-in. triangular pitch p. The flow rate of hydrocarbon vapor coming from the distillation tower, Table 1.3-1, is 14,400 lb$m$/hr at 288°F. The vapor condenses at 288°F inside the inner pipe with a coefficient of 500 Btu/hr ft^2°F. Atmospheric air is the cooling medium at 90°F ($\hat{C}_p = 0.25$ Btu/lb$m$°F.) Design a total condenser. (The latent heat of the hydrocarbon vapor at 288°F is 125 Btu/lb$m$.) The value of the overall heat transfer coefficient is estimated from Fig. 10.2-9.[1]

[1] E. M. Cook, *Chem. Eng.,* Aug. 3, **1964**:97.

Solution Heat removed from process stream:

$$Q = (125 \text{ Btu/lb}m)(14,400 \text{ lb}m/\text{hr})$$
$$= 1.8 \times 10^6 \text{ Btu/hr}$$

Assume an air temperature rise of 30°F. If ambient air is at 90°F the outlet air temperature is 120°.

$$\text{LMTD} = \frac{(288 - 90) - (288 - 120)}{\ln[(288 - 90)/(288 - 120)]}$$

$$= \frac{198 - 168}{\ln(198/68)} = \frac{30}{2.3 \log 2.9} = 28°\text{F}$$

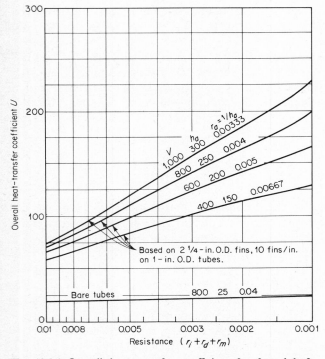

Fig. 10.2-9 Overall heat transfer coefficient for forced-draft units. (*Reprinted by special permission from Chem. Eng., Aug. 3, 1964, copyright (c) 1964 by McGraw-Hill, Inc., New York, N. Y. 10036.*)

To use Fig. 10.2-9 we need the heat transfer resistance. Typical values for condensing light hydrocarbons for $r_i + r_d$ range from 0.005 to 0.008. Since $r_m < 0.001$ use a value for $r_i + r_d + r_m = 0.007$. From Fig. 10.2-9 assume four rows (R) with an air velocity of 600 ft/min. Then $U_0 = 80$ Btu/hr ft^2 °F.

$$A_0 = \frac{1.8 \times 10^6 \text{ Btu/hr}}{(80 \text{ Btu/hr ft}^2 \text{ °F})(28 \text{ °F})} = 800 \text{ ft}^2$$

The bundle face area: $A_f = \dfrac{A_0 \rho}{\pi DR} = \dfrac{(800)(2.25)}{(3.14)(1)(4)} = 144 \text{ ft}^2$

Air flow: $\quad w = \dfrac{1.8 \times 10^6 \text{ Btu/hr}}{(0.211 \text{ Btu/lb}m \text{ °F})(30 \text{ °F})} = 2.5 \times 10^5 \text{ lb}m/\text{hr}$

Air velocity: $V = \dfrac{w}{4.5 A_f} = \dfrac{2.5 \times 10^5 \text{ lb}m/\text{hr}}{(4.5)(144)} = 400 \text{ ft/min}$

Reestimate U on basis of 400 ft/min: $U_0 = 74$ Btu/hr ft^2 °F

Since $\tilde{A}_0 = A_0 \dfrac{74}{80} = 740 \text{ ft}^2$

$$\tilde{A}_f = A_f \frac{74}{80} = 133 \text{ ft}^2$$

$$\tilde{V} = \frac{400}{133} \, 144 = 430 \text{ ft/min}$$

which is close enough.

Since face area is the product of width and length select a standard tube length of 12 ft. Therefore the width is $^{133}/_{12} = 11.2$ ft or 12 ft (standard sizes are 8, 12, 16, etc.). The number of tubes in each of the four rows will be $(12)(12)/2.25 = 64$ tubes or $(4)(64) = 256$ tubes.

The problem may be more complicated since the condensing curve —heat duty versus temperature—may depart from a straight line. If this is the case a "weighted mean temperature difference" is used rather than LMTD.

Notice that certain limitations exist on fan-type coolers which are not present for shell and tube heat exchangers. For example, fan-type coolers must be located in such a position that the resulting air blast is not hazardous to equipment or personnel in the area. Atmospheric air obviously must have free access to the fan.

□ □ □

APPENDIX 10A

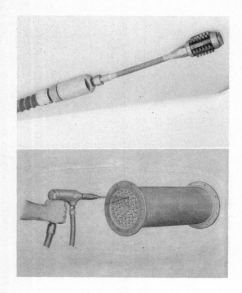

Fig. 10A-1 Tube cleaning equipment. (*Thomas C. Wilson, Inc.*)

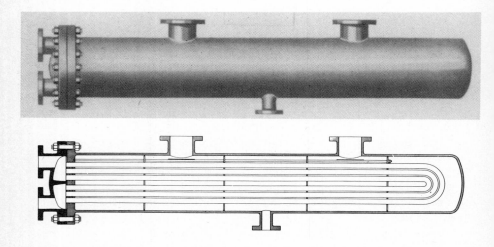

Fig. 10A-2 U-tube removable bundle exchanger. (*Patterson-Kelley Co.*)

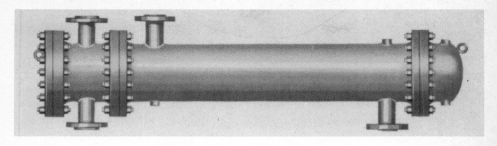

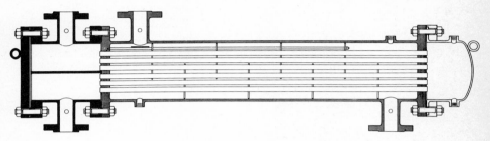

Fig. 10A-3 Straight tube multiple-pass exchanger. (*Patterson-Kelley Co.*)

Fig. 10A-4 Straight tube floating head. Removable pull-through tube bundle, completely assembled with baffles, tie rods, spacers, and floating head. (*Patterson-Kelly Co.*)

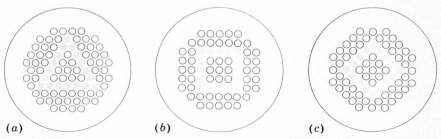

(*a*) (*b*) (*c*)

Fig. 10A-5 Standard tube layout designs. (*a*) Triangular tube pitch, (*b*) square tube pitch, (*c*) oriented square tube pitch. (*Patterson-Kelly Co.*)

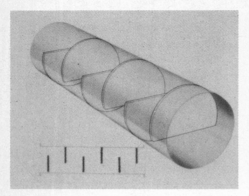

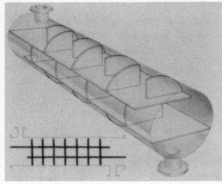

Fig. 10A-6 Segmental baffle designed for up and down flow. (*Patterson-Kelley Co.*)

Fig. 10A-7 Segmental three shell pass baffle design. (*Patterson-Kelley Co.*)

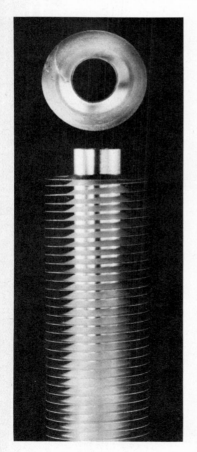

Fig. 10A-8 Finned tube. (*Aerofin.*)

Fig. 10A-9 Finned tube bank. (*Aerofin.*)

Fig. 10A-10 Fin-Fan cooler. (*Young Radiator.*)

Fig. 10A-11 Fin-Fan cooler installation. (*Young Radiator.*)

Fig. 10A-12 Cooling tower. (*Flour Corp.*)

Fig. 10A-13 Cooling tower internals. (*Flour Corp.*)

11
Mass Transfer

Mass transfer is of concern in many areas of engineering and science, but it is probably more important in conjunction with chemical reactions than in any other single area. In many cases, for implementation of chemical reactions on a commercial basis, the capital investment in mass transfer equipment by far exceeds the capital investment associated with the reaction as such.

Why is this so? For several reasons. First, almost no chemical reaction is selective enough to produce only the desired product. Instead, a number of by-products are also produced. Separation of the major product from the by-products (which may also have considerable value) is a problem in mass transfer. Second, to implement a reaction the reacting molecules must somehow get together. Frequently the rate of reaction is exceedingly fast once the reactants "see" each other, and the problem reduces to one of transporting the reactants to the location where the reaction takes place (for example, a catalyst surface). In such cases the reaction is mass transfer limited rather than kinetically limited. Third, many chemicals are handled in solution (for convenience, temperature control, physical and chemical stability, etc.) and the creation of these solutions as well as the recovery of components from the solutions is a mass transfer problem. Fourth, much of the raw material for the chemical and pharmaceutical industry has its genesis in naturally occurring materials (air, natural gas, crude oil, coal, wood, various plant and animal products, etc.) and separation of desired constituents from these highly complex mixtures is again a mass transfer problem. Many other examples could be listed, but hopefully the point has been made that in general an engineer spends as much time moving and sorting molecules as he spends in assembling them into new structures.

Mass transfer is used here in a specialized sense. Mass is certainly transferred when bulk flow occurs, for example, when water flows in a pipe. This is *not*, however, what we mean here by mass transfer. Mass transfer in the sense used here is the migration of a component in a mixture either within the same phase or from phase to phase because of a displacement from physical equilibrium. An example of the difference is in the respiratory system. Air is inhaled and a gas of slightly different composition is exhaled by the lungs, processes which are primarily bulk flow analogous to the transport of air by a blower. This would *not* be called mass transfer by our definition. At the surface of the lung, however, oxygen, one *component* of the air, transfers to the bloodstream because the blood there is not saturated with oxygen—the air and blood are displaced from *physical equilibrium*. This *is* mass transfer by our criterion. Another mass transfer step simultaneously occurs—the transfer of CO_2 from the blood to the air, which is not saturated with CO_2.[1]

Mass transfer can be treated in much the same way as the topic we just discussed, heat transfer, in that we are able to speak of driving forces, resistances, fluxes, boundary layers, etc. The important difference is that a transfer of mass always involves *motion of the medium* (otherwise, no mass would be transferred). This means that the strict analog to heat conduction in a stagnant medium is found by observing events from a coordinate frame *moving* in space rather than *fixed* in space. This complicates the mathematical manipulations somewhat, as we shall soon see.

11.1 MASS TRANSFER BY MOLECULAR DIFFUSION

Because of the second law of thermodynamics, systems not in equilibrium tend toward equilibrium (although perhaps very slowly) as time passes. The difference in the chemical potential of a component between one region in space and another is a measure of the displacement ("distance"

---

[1] This analogy, like all analogies, is somewhat oversimplified as any physiologist well knows. We renew our standing apology to such readers.

in chemical potential corrdinates) from mutual equilibrium, that is, because the chemical potential of a component varies in space.

There are a number of factors that can give rise to a difference in chemical potential: point-to-point concentration differences, pressure differences, temperature differences, and differences in forces caused by external fields (gravity, magnetic, etc.). We can, for example, obtain mass transfer by applying a *temperature* gradient to a system—the Soret effect. (Conversely, we can obtain a flow of heat by imposing a concentration gradient—the Dufour effect.) This effect, important in *thermal diffusion* processes, requires discussion of irreversible thermodynamics for its systematic treatment, and we do not develop the Onsager reciprocal relations which describe this and other coupled effects. We will, therefore, restrict ourselves in this text only to mass transfer which results from differences in concentration, both for the sake of brevity and because this is the effect by far most frequently used in practical problems.

Velocities of components in a mixture

A distinguishing characteristic of problems involving the transport of mass is that the flow velocities of the various molecular species in the mixture can differ. For example, consider the case of some diethyl ether placed in the bottom of a tall cylindrical vessel past the mouth of which is blown ether-free nitrogen (see Fig. 11.1-1). If we assume none of the ether to have evaporated initially, the concentration profile will look like Fig. 11.1-1a. Since the gas phase contains no ether and the liquid phase

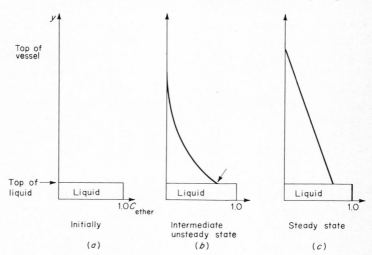

Fig. 11.1-1 Diffusion of vapor from vessel. Concentration in b and c is just slightly less than 1.0 since nitrogen will dissolve in the ether (although *very* little).

no nitrogen, equilibrium is not present. The chemical potential of ether differs in the gas and liquid phases as does that of nitrogen. The nitrogen and ether therefore begin to move in such a way as to bring the system to the equilibrium state. Here the ether moves into the gas phase and the nitrogen into the liquid phase. At some intermediate time the profile may resemble Fig. 11.1-1b. Notice, however, that the movement of nitrogen is limited. As soon as the liquid phase becomes saturated with nitrogen (that is, the chemical potential in liquid and gas phases becomes the same), the nitrogen ceases to move because the driving force has disappeared. The ether, however, is continuously removed by the steady stream of pure nitrogen past the mouth of the flask.

If we wait long enough, the concentration profile will reach a steady state as shown in Fig. 11.1-1c. At this point the ether will evaporate and will be transferred out the end of the vessel at a *steady* rate, and therefore the vaporized ether will have some *constant* velocity with respect to a coordinate system fixed in space. (We assume that the ether liquid level is kept constant.) At steady state, however, the same is *not* true of the nitrogen. After the nitrogen dissolves in the ether to the saturation concentration the nitrogen *no longer moves* in the vessel but is simply a stagnant medium (with respect to axes fixed in space) through which the ether diffuses. Here it is clear that the two components with which we are concerned have *different* velocities.

Until now we have left the concept of velocity of a fluid as something which, for a one-component system, was (hopefully) intuitively obvious. With only *one* component it is not hard to replace the physical reality of lots of molecules banging into one another with the mathematical model describing a *continuum*. Now that we must speak of velocities of components (*not* components of velocity) we must again deal with crude physical reality. This time we will face a more complicated situation, and to describe affairs we will have to introduce several types of velocity.

To attempt to begin at a level where physical intuition has not yet deserted us, let us consider the motion of a single molecule. At any one instant in time, if we forget for the moment what we know about modern physics, the molecule looks like a small lump of matter with a given velocity vector **a** (see Fig. 11.1-2a) referred to a coordinate axis system fixed in space. All is fine because we can clearly define a velocity for the molecule. Unfortunately, the molecule does not maintain this velocity for long, because just as it thinks it knows where it is going it runs into someone (another molecule) who changes its course. Consequently, if we wait a little while we observe something like Fig. 11.1-2b, where the molecule has traveled very *fast*, endured several collisions, and now has velocity **b** but really has not gone very far—only a distance ℓ.

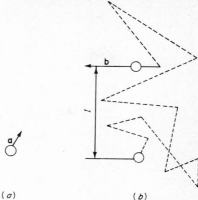

Fig. 11.1-2 Motion of individual molecule.   (*a*)   (*b*)

For mass transfer purposes we are really only interested in how fast the molecule covered distance ℓ, not how fast it moved *between* collisions. Therefore the *apparent* velocity of the molecule is approximately

$$\mathbf{v}_{ij} \cong \frac{\Delta \ell}{\Delta t} \tag{11.1-1}$$

where $\mathbf{v}_{ij}$ refers to the jth molecule (that is, the 1st, 10th, 463d, etc.) of the ith species (e.g., ethanol, oxygen, etc.). Fortunately the velocity at which a molecule hurtles about is so great in gases and liquids under ordinary conditions that it undergoes many collisions in any Δt large enough to be of interest to us. Further, since the many collisions are a stochastic process, $\Delta \ell/\Delta t$ approaches a fairly constant value at small (but not too small) values of Δt. This, then, is the velocity of a molecule for mass transfer purposes.

We are not interested in the velocity of a single molecule, but in the velocity of its chemical species. This we take to be the *number* average velocity:

$$\mathbf{v}_i = \frac{\displaystyle\sum_{j=1}^{n} 1\mathbf{v}_{ij}}{\displaystyle\sum_{j=1}^{n} 1} = \frac{\mathbf{v}_{i1} + \mathbf{v}_{i2} + \cdots + \mathbf{v}_{in}}{(1 + 1 + 1 + 1 + \cdots)} \tag{11.1-2}$$

where n is the number of molecules in some volume ΔV. This average is simply the sum of the velocities of all the molecules of the ith species,

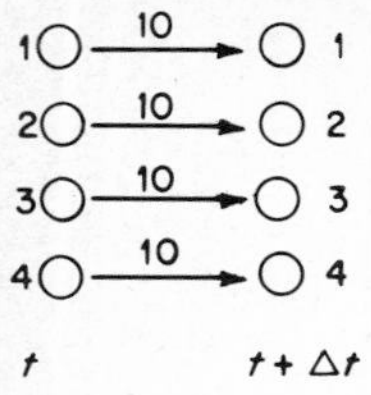

Fig. 11.1-3 Sketch for Example 11.1-1.

divided by the number of molecules of the ith species. Note that it parallels the general definition of average given in Chap. 2 (with weighting factor $= 1$) except that the integrals are replaced by sums. This, then, is what we mean by the velocity of a single species with respect to fixed coordinate axes. To help illustrate what this velocity means physically, we will consider several idealized examples.

Example 11.1-1 Average velocity when individual particles have the same velocity Consider the case of four molecules with identical velocity vectors as shown in Fig. 11.1-3. Since molecules seldom behave this way, and since we wish to extend the analogy below, we might think of these molecules as students walking in the same direction at the same speed, say 10 ft/sec. In this case the application of our definition gives an average velocity of 10 ft/sec as follows:

$$\mathbf{v}_{\text{student}} = \frac{(1)(\mathbf{10}) + (1)(\mathbf{10}) + (1)(\mathbf{10}) + (1)(\mathbf{10})}{1 + 1 + 1 + 1} = \mathbf{10}$$

(The 10s are *vectors* but are easily added without writing them in component form since they all have the same direction.) Notice that the students do not move *relative* to one another since they have the same velocity. Here the calculated velocity is intuitively satisfying because the average velocity multiplied by the time elapsed will give the distance traveled for *each* of the four students.

□ □ □

Example 11.1-2 Average velocity when individual particles have different velocities Now consider a second case, where the velocities of the students *differ* as shown in Fig. 11.1-4. Starting from the same initial configuration, the distance of one student from another *no longer remains the same*. Application of the definition of average velocity gives

$$(a) \qquad \mathbf{v}_{\text{student}} = \frac{(1)(\mathbf{5}) + (1)(\mathbf{15}) + (1)(\mathbf{5}) + (1)(\mathbf{15})}{4} = \mathbf{10}$$

The average velocity is the same, but notice that it does not correspond to the velocity of *any one* of the students. In this sense, it is a *fictitious* velocity. (The mistake of applying average characteristics to individual students is often committed in universities.)

Why are we interested in a fictitious velocity? Because it permits us to predict *average* behavior. Suppose the students each have a mass of 180 lb*m*. The *average* velocity is the velocity at which the *center of mass* of the students moves.

The center of mass for several particles is found by multiplying the mass of each particle times its vector distance from an arbitrary origin, summing all these products (the *first moment* of the masses), and dividing the sum by the sum of all the individual masses. For our case here a convenient origin and coordinate system to choose are the ones shown, which give one of the particles a position vector of zero and run parallel and perpendicular to the motion.

For the original configuration the center of mass is calculated as

$$\mathbf{r}_{cm} = \frac{m_1\mathbf{r}_1 + m_2\mathbf{r}_2 + m_3\mathbf{r}_3 + m_4\mathbf{r}_4}{m_1 + m_2 + m_3 + m_4}$$

$(x_{cm}\mathbf{i} + y_{cm}\mathbf{j})$

$$= \frac{180(0\mathbf{i} + 6\mathbf{j}) + 180(0\mathbf{i} + 4\mathbf{j}) + 180(0\mathbf{i} + 2\mathbf{j}) + 180(0\mathbf{i} + 0\mathbf{j})}{180 + 180 + 180 + 180}$$

Equating components and dividing numerator and denominator by 180

$$x_{cm} = 0$$

$$y_{cm} = \frac{6 + 4 + 2 + 0}{4} = 3$$

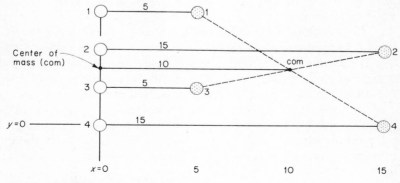

Fig. 11.1-4 Sketch for Example 11.1-2.

Thus the original center of mass is as shown in Fig. 11.1-4.

A similar calculation for the center of mass after time Δt shows that

$$\mathbf{r}_{cm} = (x_{cm}\mathbf{i} + y_{cm}\mathbf{j})$$

$$= \frac{180(5\mathbf{i} + 6\mathbf{j}) + 180(15\mathbf{i} + 4\mathbf{j}) + 180(5\mathbf{i} + 2\mathbf{j}) + 18(15\mathbf{i} + 0\mathbf{j})}{180 + 180 + 180 + 180}$$

Equating components and dividing numerator and denominator by 180

$$(b) \; x_{cm} = \frac{5 + 15 + 5 + 15}{4} = 10$$

$$y_{cm} = \frac{6 + 4 + 2 + 0}{4} = 3$$

The change in position of the center of mass is shown in Fig. 11.1-4. Notice that the distance the center of mass moves (10 units) is equal to the average velocity (10) times elapsed time (1).

The *number* average velocity we defined has been shown above to correspond to the average velocity of the center of mass for a system of particles of *equal mass*. This can be seen by comparing equations (a) and (b)—they are exactly the same. If, however, the masses were *not* equal, equations (a) and (b) would no longer be the same, because we could not have developed equation (b) by multiplying and/or dividing the preceding equation by 180—the coefficients in the numerator would not all be the same, and therefore the weighting factors in the numerator could not all be made to be 1 simultaneously as in equation (a).

From this we can see that instead of mass we could choose *any* property that all the objects possess equally, and our number average velocity will give us the velocity of the center of (first moment of) that property. The important thing is that each object possess the property to the *same extent*, so that the weighting factors in the numerator may all be made equal to 1.

For example, we could define a property called "studentness." Assuming each student possesses "studentness" to the same extent (say 471.3), the number average velocity would give us the velocity of the "center of studentness." (It is obvious that the same procedure will not work with "scholarliness," which is possessed in widely varying degrees.)

In mass transfer we are interested in velocities of *species*, which by definition are collections sharing a common property (for example, chemical composition, molecular weight, etc.). For these cases the number average velocity gives us the rate of motion of the "center of" that property.

☐ ☐ ☐

Example 11.1-3 Number average velocities As a more complicated example of number average velocity, consider oxygen molecules with velocities as shown in Fig. 11.1-5.

To apply Eq. (11.1-2) we remember that to add vectors we add corresponding components; thus, using i to denote oxygen,

$$(v_i)_x = \frac{(v_{i1})_x + (v_{i2})_x + (v_{i3})_x + (v_{i4})_x}{1 + 1 + 1 + 1}$$

$$= \frac{30 \cos 45° + 60 - 40 \cos 30° + 0}{4}$$

$$= \frac{21.2 + 60 - 34.6 + 0}{4} = 18.4$$

$$(v_i)_y = \frac{(v_{i1})_y + (v_{i2})_y + (v_{i3})_y + (v_{i4})_y}{1 + 1 + 1 + 1}$$

$$= \frac{30 \sin 45° + 0 + 40 \sin 30° - 10}{4}$$

$$= \frac{21.2 + 0 + 20 - 10}{4} = \frac{31.2}{4} = 7.8$$

Then $v_i = 18.4\mathbf{i} + 7.8\mathbf{j}$ and $|v_i| = \sqrt{18.4^2 + 7.8^2} = \sqrt{339 + 60.8} = 20$ where $\theta = \arctan (7.8/18.4) = 23°$. This is the velocity of the center of mass of the oxygen molecules.

□ □ □

We now need to define a *single* velocity for a fluid which is made up of *several* components moving at different velocities. Suppose we think of our molecule as being one of a group of students rushing in the same direction from one building to another, and we further suppose the students to be divided into two classes: 150- and 180-lb$m$ students. Assume that the 150-lb$m$ students move at a higher rate of speed than the 180-lb$m$

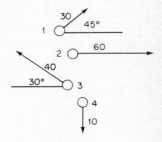

Fig. 11.1-5 Sketch for Example 11.1-3.

students.   This difference in velocity leads to considerable jostling
(collisions), but by applying our definition of species velocity above we
could certainly come up with a "150-lb$m$ student velocity" and a
"180-lb$m$ student velocity" with respect to axes fixed in space. Someone
might just ask us, however, what the "student velocity" is. One reply we
could make to such a question is to ask, "Do you mean how *many*
students went past or what *mass* of students went past?" The reason for
such a reply is that if we are standing still (coordinates fixed in space) we
certainly sense some sort of "student" (fluid) velocity, but this velocity is
generated by two types of students (molecules), 150 lb$m$ (component 1)
and 180 lb$m$ (component 2).

Normally we wish to define the "student" velocity only because we
can use it to *calculate* certain other things, for example, the number rate
of students passing or the mass rate of students passing.

The same question faces us in mass transfer.   Accordingly, we
define *two* velocities with respect to axes fixed in space: (1) a *molar*
average velocity, and (2) a *mass* average velocity.

The molar average velocity is defined as

$$\mathbf{v}^* = \frac{\sum\limits_{i=1}^{n} c_i \mathbf{v}_i}{\sum\limits_{i=1}^{n} c_i} \tag{11.1-3}$$

The mass average velocity is defined as

$$\mathbf{v} = \frac{\sum\limits_{i=1}^{n} \rho_i \mathbf{v}_i}{\sum\limits_{i=1}^{n} \rho_i} \tag{11.1-4}$$

The parallel to the general definition of average is again obvious. In fact,
other average velocities may be defined—the *volume* average velocity, for
example—but we need discuss only the two above for our purposes. The
above velocities are referred to axes *fixed in space*. We sometimes are interested in velocities relative to coordinate axes which themselves have
velocity. We will say more about this later.

From the above definitions we can see the usefulness of the molar
and mass average velocities. The *molar* average velocity when multiplied
by the total molar concentration $\Sigma\ c_i$ yields the total flux in *moles* with respect to axes fixed in space since

$$\sum_{i=1}^{n} c_i \mathbf{v}_i$$

is the sum of the molar fluxes of each individual species, (moles/ft^3) (ft/sec) = (moles/ft^2 sec). Similarly, the *mass* average velocity when multiplied by the total *mass* concentration $\Sigma \, \rho_i$ yields the total *mass* flux $\mathbf{n}_i$ with respect to axes fixed in space.

Our notation in discussing fluxes will be to use *lowercase* letters for *mass* fluxes and *uppercase* letters for *molar* fluxes. For fluxes referred to *stationary* coordinates we use the letter n (or N), while for fluxes referred to *moving* coordinates we use the letter j (or J). For nonstationary coordinate frames, the speed at which the coordinate frame is moving will for us be either the *mass* average velocity or the *molar* average velocity. To distinguish these cases we use an asterisk for fluxes referred to frames moving at the *molar* average velocity. These relations are summarized in Table 11.1-1.

Mechanisms of mass transfer

Mass transfer is carried out by two of the same mechanisms as transport of heat: molecular diffusion, the analog of heat transfer by conduction; and eddy diffusion, the analog of convective heat transfer. The diffusive process is in general much slower than the convective process. We will first study molecular diffusion and then convection.

Fick's law

The basic relation governing mass transfer by molecular diffusion in a binary mixture is called *Fick's law*:

$$\mathbf{J}_A^* = -c\mathscr{D}_{AB} \, \nabla \, x_A \tag{11.1-5}$$

where $\mathbf{J}_A^*$ = molar flux of component A referred to axes moving at the molar average velocity

$\mathscr{D}_{AB}$ = constant of proportionality called the diffusivity of A through B

x_A = mole fraction of A

c = molar density of mixture

Table 11.1-1 Coordinate frame motion

Flux units	Stationary	Mass average velocity	Molar average velocity
Mass	$\mathbf{n}_i$	$\mathbf{j}_i$	$\mathbf{j}^*_i$
Molar	$\mathbf{N}_i$	$\mathbf{J}_i$	$\mathbf{J}^*_i$

In its one-dimensional form the parallel to heat transfer by conduction is clear:

$$J_A^* = - c\mathscr{D}_{AB}\frac{dx_A}{dy}$$

(11.1-6)

since

$$\frac{Q}{A} = -k\frac{dT}{dy}$$

(11.1-7)

Fick's law is basically an empirical law, as is Fourier's law.

Notice that the above flux is with respect to *moving* coordinate axes. Why should the law be written this way? Return for a moment to our analogy between molecules and students. Suppose a nicely ordered group of light students runs blindly through a group of heavy students. Further, suppose the weight difference is so much that the heavy student velocity is relatively unchanged by a collision with a light student.

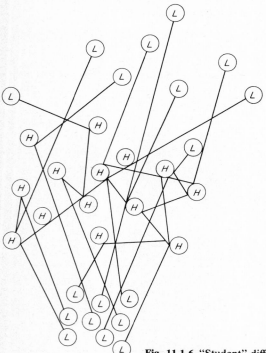

Fig. 11.1-6 "Student" diffusion problem.

The process leads to *diffusion*; that is, the light students get spread out in space compared to their original configuration, as shown in Fig. 11.1-6. The diffusion occurs because of the number of collisions between the individual students (molecules). This is a function of their *relative* velocities rather than of how fast they are running with respect to the ground. For example, if one adds 10 ft/sec to *both* the heavy and the light student velocities, the diffusional process would not change in rate because *relative* velocities are unchanged. This says that we should observe the process by moving at some velocity which is related to *both* the species velocities. Since the diffusion in this simple case is dependent upon the *number* of collisions, which is a function of how *many* people are in the way rather than of how much mass, the appropriate velocity is the *molar* average velocity.

The above analogy is to make the reason for the use of a moving coordinate system *plausible*, not to justify it in general. We have neglected many things in the analogy (such as inelasticity of collisions and intermolecular forces), and so the reader is urged not to apply it blindly.

Use of a moving coordinate system makes the law—as introduced in the form above—*simpler*. If instead the flux is rewritten with respect to coordinate axes *fixed in space* one has an additional apparent flow of magnitude $\mathbf{N}x_A$:

$$\mathbf{N}_A = \mathbf{J}_A^* + \mathbf{N}x_A \tag{11.1-8}$$

or

$$\mathbf{N}_A = -c\mathscr{D}_{AB}\nabla x_A + \mathbf{N}x_A \tag{11.1-9}$$

where $\mathbf{N}$ is the *total* molar flux with respect to stationary coordinates. Note that

$$\mathbf{N} = \mathbf{N}_A + \mathbf{N}_B \tag{11.1-10}$$

by a mass balance. Notice that when the diffusion equation is written with respect to stationary axes [Eq. (11.1-9)] there is an "extra term" which spoils the parallelism to the Fourier equation [Eq. (11.1-7)]. This term also complicates the mathematics considerably.

The fact that the flux of A in Eq. (11.1-8) is made up of a flux from diffusion plus a flux from bulk flow may perhaps be made clearer by considering another analogy. If I constructed a sheet metal tray and put shallow dimples in the bottom at regular intervals, I could fill the tray with a regular pattern of black and white marbles as shown in Fig. 11.1-7. I could then carry the tray horizontally at some velocity, say 1 ft/sec, and

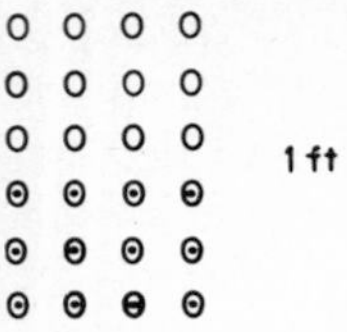

Fig. 11.1-7 Flux of molecules without diffusion.

as I passed an observer standing stationary in space he would observe a flux of white balls with respect to his stationary coordinate system. This flux corresponds to the second term in Eq. (11.1-8). If the end of the tray had a cross section of 1 ft^2, the observer would see a *total flux* **N** of 24 balls/sec ft^2. Multiplying this by the concentration of white balls ($x_A = \frac{1}{2}$) gives the *average* flux of white balls during $\Delta t = 1$ sec as (24) ($\frac{1}{2}$) = 12 white balls/sec ft^2. (Although within Δt the flux of white balls goes from 0 to 24 we are not interested in the *instantaneous* flux—this is the same problem we had before with our *continuum* assumption.)

As yet we have no term corresponding to the *diffusion* term in Eq. (11.1-8). Such a term could be produced if I jiggle the tray as I walk so that the balls *mix*. Consider the two cases sketched in Fig. 11.1-8: the first, without jiggling, the second, with jiggling. In the second case the black balls are farther along on the average—this is the "extra" flux from diffusion that gives the first term in Eq. (11.1-3). Table 11.1-2 gives equivalent forms of Fick's law.

Table 11.1-2 Equivalent form of Fick's law[a]

Flux	Gradient	Form of Fick's first law	
n_A	$\nabla \omega_A$	$n_A - \omega_A(n_A + n_B) = -\rho \mathscr{D}_{AB} \nabla \omega_A$	(a)
N_A	∇x_A	$N_A - x_A(N_A + N_B) = -c \mathscr{D}_{AB} \nabla x_A$	(b)
j_A	$\nabla \omega_A$	$j_A = -\rho \mathscr{D}_{AB} \nabla \omega_A$	(c)
J_A^*	∇x_A	$J_A^* = -c \mathscr{D}_{AB} \nabla x_A$	(d)
j_A	∇x_A	$j_A = -\left(\dfrac{c^2}{\rho}\right) M_A M_B \mathscr{D}_{AB} \nabla x_A$	(e)
J_A^*	$\nabla \omega_A$	$J_A^* = -\left(\dfrac{\rho^2}{c M_A M_B}\right) \mathscr{D}_{AB} \nabla \omega_A$	(f)
$c(v_A - v_B)$	∇x_A	$c(v_A - v_B) = \dfrac{c \mathscr{D}_{AB}}{x_A x_B} \nabla x_A$	(g)

[a]R. B. Bird, W. E. Stewart, and E. N. Lightfoot, "Transport Phenomena," John Wiley & Sons, Inc., New York, 1960.

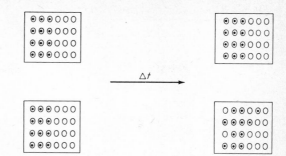

Fig. 11.1-8 Flux of molecules with and without diffusion.

Binary diffusivities

As one might expect from consideration of the mobility of the molecules, diffusivities are generally lower for solids than for liquids, and lower for liquids than for gases. The units of the diffusion coefficient are area per unit time, and most tabulations are in units of cm^2/sec. To convert these units to ft^2/hr one multiplies by 3.87. There are a number of methods for predicting diffusivity a priori.[1]

Diffusion coefficients for solids range from about 10^{-9} to 10^{-1} cm^2/sec, for liquids from about 10^{-6} to 10^{-5} cm^2/sec, and for gases from about 5×10^{-1} to 10^{-1} cm^2/sec. The diffusivity is a *property* and so a function of thermodynamic variables—mainly temperature and concentration, although pressure has a fairly pronounced effect on the diffusivity of gases.

Typical values for diffusion coefficients are listed in Tables 11.1-3,[2] 11.1-4,[3] and 11.1-5.[3]

Solutions of the diffusion equation

The diffusion equation is a partial differential equation which describes conservation of chemical species. Many analytic solutions for simple boundary shapes and various boundary conditions may be found in books.[4,5] Many other solutions are possible by numerical techniques. We will consider only one or two simple situations.

[1] R. W. Bird, W. E. Stewart, and E. N. Lightfoot, "Transport Phenomena," John Wiley & Sons, Inc., New York, 1960.
[2] R. M. Barrer, "Diffusion in and through Solids," The Macmillan Company, New York, 1941.
[3] R. C. Reid and T. K. Sherwood, "The Properties of Gases and Liquids," McGraw-Hill Book Company, New York, 1958.
[4] J. Crank, "The Mathematics of Diffusion," Oxford University Press, London, 1956.
[5] H. S. Carlaw and J. C. Jaeger, "Heat Conduction in Solids," 2d ed., Oxford University Press, London, 1959.

Table 11.1-3   Diffusivities in solids[a]

System	$T, °C$	Diffusivity $\mathscr{D}_{AB}$ $cm^2\ sec^{-1}$
He in SiO_2	20	$2.4 - 5.5 \times 10^{-10}$
He in pyrex	20	$4.5 \ \times 10^{-11}$
	500	$2 \ \times 10^{-8}$
H_2 in SiO_2	500	$0.6 - 2.1 \times 10^{-8}$
H_2 in Ni	85	1.16×10^{-8}
	165	$10.5 \ \times 10^{-8}$
Bi in Pb	20	7.7×10^{-3}
Hg in Pb	20	3.6×10^{-1}
Sb in Ag	20	5.3×10^{-5}
Al in Cu	20	1.2×10^{-2}
Cd in Cu	20	3.5×10^{-9}

[a] R. M. Barrer, "Diffusion in and through Solids," The Macmillan Company, New York, 1941.

One-dimensional equimolar counterdiffusion in rectangular coordinates

If 1 mole of A diffuses in a given direction for each mole of B diffusing in the opposite direction, this is by definition equimolar counterdiffusion. Thus

$$\mathbf{N}_A = -\mathbf{N}_B \qquad\qquad (11.1\text{-}11)$$

Table 11.1-4   Diffusivities in gases[a]
(Pressure about 1 atm)

System	Temperature, °C	$\mathscr{D}_{AB},\ cm^2/sec$
Air-ammonia	0	0.198
Air-benzene	25	0.0962
Air-carbon dioxide	0	0.136
Air-chlorine	0	0.124
Air-ethanol	25	0.132
Air-oxygen	0	0.175
Air-water	25	0.260

[a] R. C. Reid and T. K. Sherwood, "The Properties of Gases and Liquids," McGraw-Hill Book Company, New York, 1958.

Table 11.1-5 Diffusivities in liquids[a]
[*Substance B (solvent)* = *water*]

Solute A	Temperature, °C	Concentration	$\mathscr{D}_{AB} \times 10^5$, cm^2/sec
Acetic acid	12.5	0.01 M	0.91
Carbon dioxide	18.0	0	1.71
Chlorine	18.0	0.1 M	1.40
Glycerol	10.0	0.125 M	0.63
Hydrogen	25.0	0	3.36
Nitrogen	22.0	0	2.02
Oxygen	25.0	0	2.60
Ethanol	25.0	$\bar{x}_A = 0.05$	1.13
		0.10	0.90
		0.275	0.41
		0.50	0.90
		0.70	1.40
		0.95	2.20
n-Butanol	30	$\bar{x}_A = 0.446$	0.267
		0.546	0.437
		0.642	0.560
		0.778	0.920
		0.869	1.24

[a] R. C. Reid and T. K. Sherwood, "The Properties of Gases and Liquids," McGraw-Hill Book Company, New York, 1958.

Therefore from Eq. (11.1-10):

$$N = 0 \tag{11.1-12}$$

Fick's law reduces to, for the one-dimensional case (using the y-direction),

$$N_{Ay} = - c\mathscr{D}_{AB} \frac{dx_A}{dy} \tag{11.1-13}$$

(Remember that x_A is a concentration, *not* a coordinate.)

If we further assume c and $\mathscr{D}_{AB}$ to be independent of concentration and the temperature to be uniform (there is no heat of mixing, etc.), Eq. (11.1-13) may be integrated easily since the variables can be separated:

$$\int_0^L dy = - \frac{c\mathscr{D}_{AB}}{N_{Ay}} \int_{x_{A0}}^{x_{A1}} dx_A \tag{11.1-14}$$

where L = path length for diffusion
x_{A0}, x_{A1} = bounding mole fractions
giving

$$L = -\frac{c\mathcal{D}_{AB}}{N_{Ay}}(x_{A1} - x_{A0}) \tag{11.1-15}$$

(Note that this type of behavior is observed in those binary distillation columns where a mole of B diffuses to the interface and condenses for every mole of A that evaporates and moves away from the interface, that is, those operating with constant molar overflow.)

We can rewrite Eq. (11.1-15) as

$$\underbrace{N_{Ay}}_{\text{flux}} = -\underbrace{\frac{c\mathcal{D}_{AB}}{L}}_{\text{conductance}} \underbrace{(x_{A1} - x_{A0})}_{\text{driving force}} \tag{11.1-16}$$

which is in the form of a flux equaling a conductance times a gradient (driving force).

For gases at moderate pressures we can use partial pressure as driving force instead of mole fraction and

$$N_{Ay} = -\frac{\mathcal{D}_{AB}}{RT}\frac{p_{A1} - p_{A0}}{L} \tag{11.1-17}$$

Example 11.1-4 Equimolar counterdiffusion A mixture of benzene and toluene is supplied as vapor to the bottom of an insulated rectifying column. At one point in the column the vapor contains 80 mole % benzene, and the interfacial liquid contains 70 mole % benzene. The temperature at this point is 89°C. Assuming equilibrium at the interface, and assuming the diffusional resistance to vapor transfer between vapor and liquid to be equivalent to the diffusional resistance of a stagnant vapor layer 0.1 in. thick, calculate the rate of interchange of benzene and toluene between vapor and liquid. The molal latent heats of vaporization of benzene and toluene may be assumed to be equal, and the system is close enough to ideal to use Raoult's law for the equilibrium relationship. The vapor pressure of benzene at 89°C is 988 mm Hg and the diffusivity for toluene-benzene may be assumed to be 0.198 ft²/hr. (Note that one does not really know the equivalent film thickness; this would have to be determined by experimental measurement, but since it depends primarily on the fluid mechanics, it could be determined in a different concentration range or even in a different chemical system.)

Solution In a rectifying column operating without heat loss, the liquid will be at its boiling point. As a result, the toluene condensed liberates sufficient heat to vaporize an equal number of moles of benzene. At any point in the column, therefore, $N_A = -N_B$.

$$p_{A2} = 0.8 \text{ atm}$$

$$p_{A1} = 0.7 \frac{988}{760} = 0.91 \text{ atm}$$

$$N_A = \frac{\mathscr{D}(p_{A1} - p_{A2})}{RTL} = \frac{(0.198)(0.91 - 0.8)}{(0.728)(652.2)(0.1/12)}$$

$$(89°C = 652.2°R)$$

$$\therefore N_A = 5.5 \times 10^{-3} \text{ lb mole benzene/hr ft}^2$$

□ □ □

One-dimensional diffusion of A through stagnant B observed in rectangular coordinates

If B does not move with respect to axes fixed in space we have

$$N_B = 0 \tag{11.1-18}$$

and Fick's law reduces to

$$N_{Ay} = -c\mathscr{D}_{AB} \frac{dx_A}{dy} + N_A x_A \tag{11.1-19}$$

or, after rearranging,

$$\int_0^L dy = -\frac{c\mathscr{D}_{AB}}{N_{Ay}} \int_{x_{A0}}^{x_{A1}} \frac{dx_A}{1 - x_A} \tag{11.1-20}$$

Integrating and substituting limits:

$$N_{Ay} = -\frac{c\mathscr{D}_{AB}}{L} \ln \frac{1 - x_{A0}}{1 - x_{A1}} \tag{11.1-21}$$

In principle this is as far as we need go, since we can calculate the flux from the above relation. Engineers, however, have a compulsion to use equations where they can identify flux, conductance, and gradient terms, and so to conform to convention we multiply and divide by $(x_{A1} - x_{A0})$, substitute x_B for $(1 - x_A)$, and use the definition of log mean to rewrite Eq. (11.1-20) as

$$N_{Ay} = - \frac{c\mathscr{D}_{AB}}{(x_B)_{lm}} \frac{x_{A1} - x_{A0}}{L} \qquad (11.1-22)$$

Note that for very dilute solutions $(x_B)_{lm} \cong 1.0$. This is also an adequate model where B, although not stagnant, moves very slowly with respect to A.

As we mentioned previously partial pressure may be used as the driving force where gases at moderate pressure are diffusing. Thus

$$N_{Ay} = - \frac{p\mathscr{D}_{AB}}{RT(p_B)_{lm}} \frac{p_{A1} - p_{A0}}{L} \qquad (11.1-23)$$

where

$$(p_B)_{lm} = \frac{p_{B1} - p_{B0}}{\ln (p_{B1}/p_{B0})} \qquad (11.1-24)$$

Example 11.1-5   Diffusion of vapor through a stagnant gas   A fan is blowing 32°F air over a container filled to within 1 in. of the top with a solution of ethyl alcohol in water. Assume the concentration of alcohol at the interface is 0.5 and the air flowing across the container does not contain alcohol. Also, assume that the mouth of the container is narrow enough that in effect we have 1 in. of stagnant gas. Will the alcohol content of the solution change significantly in 1 day? ($\mathscr{D}_{AB}$ for this system is about 0.4 ft²/hr.)

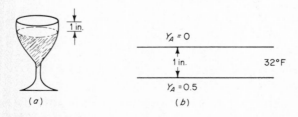

(a)   (b)

Fig. 11.1-9 Sketch for Example 11.1-5.

Solution We replace the real system, Fig. 11.1-9, by the model, Fig. 11.1-9*b*

$$p_{A1} = 0.5 \text{ atm} \qquad p_{A2} = 0 \text{ atm}$$

$$p_{B1} = 0.5 \text{ atm} \qquad p_{B2} = 1 \text{ atm}$$

$$(p_B)_{\text{lm}} = \frac{p_{B2} - p_{B1}}{\ln (p_{B2}/p_{B1})}$$

$$= \frac{1 - 0.5}{\ln(1/0.5)}$$

$$= 0.72 \text{ atm}$$

$$N_A = \frac{\mathscr{D}_{AB} p}{RT} \frac{p_{A1} - p_{A2}}{L}$$

$$= \frac{(0.4)(1)(0.5 - 0)}{(0.728)(492)(^1/_{12})(0.72)}$$

$$= 9.3 \times 10^{-3} \text{ lb mole/hr ft}^2$$

The concentration of alcohol does not change significantly.

□ □ □

11.2 CONVECTIVE MASS TRANSFER

Until this point we have been discussing mass transfer (1) *within a single phase*; (2) by the mechanism of *molecular diffusion*. We now wish to extend our discussion to include *convective* mass transfer, and, in addition, to consider transfer *between phases*.

Most problems in convection are too complicated to permit an analytical solution. You will in more advanced courses investigate some approximate models used to attempt to predict mass transfer—film theory, boundary layer theory, and penetration-renewal theory—but for the moment we will confine ourselves to investigating how we might make use of the predictions made via these theories.

The concentration boundary layer[1]

Consider the case of *mass* transfer to a flat plate in the same way that momentum transfer and heat transfer are considered. Suppose we have a flat plate made of naphthalene (moth crystals, basically), which evapo-

---

[1] For a more complete discussion see W. M. Kays, "Convective Heat and Mass Transfer," McGraw-Hill Book Company, New York, 1966.

rates into a stream of nitrogen. If we plot concentration versus distance from the surface, we get something that looks like Fig. 11.2-1*a*. This looks very little like our profiles in the cases of momentum and energy. We can change this, however, by plotting the *unaccomplished* change in concentration $c_A - c_{AS}$, as in Fig. 11.2-1*b*, and we see that we build up a concentration boundary layer on the plate just as for energy and momentum.

This procedure leads us to believe that by a suitable dedimensionalization of variables we should be able to write equations describing momentum transfer, heat transfer, and mass transfer in such a form that the equations would be identical for cases where the transport takes place by the same mechanism. As a particular example of this procedure, we can consider flow over a flat plate.

Flat plates do not exhibit form drag, so that we do not have to worry about this particular form of momentum transport, which contains no analog in heat transfer and mass transfer. If we restrict ourselves to mass transfer rates sufficiently low that the mass transfer equations written either in stationary coordinates or in coordinates moving at the molar average velocity look the same, we will not have to worry about the bulk flow of the dissolving component affecting flow and complicating mass transfer results. For such a situation, we can write an idealized set of differential equations which involve several assumptions. In particular, for the case presented here, we will consider only two-dimensional flow (that is, we will assume that the flat plate is of infinite width normal to the flow-

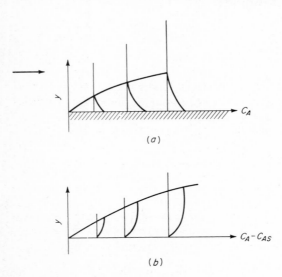

Fig. 11.2-1 Concentration boundary layer.

ing stream), we shall neglect transport by conductive or molecular diffusive mechanisms in all directions except at right angles to the plate surface, and we will assume that the flow is at steady state. Under these circumstances, we can write dedimensionalized differential equations which
describe the flow field, the mass transfer, and the heat transfer. The equation for momentum, energy, or mass transfer has the following dimensionless form (see Sec. 8.2 for a similar discussion)

$$\frac{d^2\Omega}{d\eta^2} + K \frac{f(\eta)}{2} \frac{d\Omega}{d\eta} = 0 \tag{11.2-1}$$

with boundary conditions

At $\eta = 0$, $\Omega = 0$

At $\eta = \infty$, $\Omega = 1$

In this equation Ω is a dimensionless velocity, temperature, or concentration depending on whether a momentum, energy, or mass balance
was used. The coefficient K is dimensionless and η is a dimensionless distance which incorporates y, the distance perpendicular to the surface, and
x, the distance along the surface from the nose. (η is a *similarity* variable.)

$$\eta = y \sqrt{\frac{v_0}{\nu x}} = \frac{y}{x} \sqrt{\text{Re}} \tag{11.2-2}$$

where Re is a Reynolds number based on distance from the nose x and
free-stream velocity v_0. The function $f(\eta)$ is known (but is expressed as a
series).

For mass transfer one obtains [see Eq. (8.2-3) and (8.2-4) for corresponding forms for momentum or heat transfer]

$$\frac{d^2[(\rho_A - \rho_{AS})/(\rho_{A0} - \rho_{AS})]}{d\eta^2} + Sc \frac{f(\eta)}{2} \frac{d[(\rho_A - \rho_{AS})/(\rho_{A0} - \rho_{AS})]}{d\eta}$$

$$= 0 \tag{11.2-3}$$

Notice that Eq. (11.2-3) is identical to the corresponding momentum and
heat transfer equation except for the coefficient of the second term. If we
require the Prandtl number to equal 1 in Eq. (8.2-4) and the Schmidt
number to equal 1 in Eq. (11.2-3) the equations are *identical*, and since
the dedimensionalized boundary conditions are the same, the identical

solution is found for the cases of momentum transfer, heat transfer, and mass transfer.

The solution to these equations can be obtained without great difficulty, and the solution is shown plotted in Fig. 11.2-2. Notice that the ordinate is in each case the unaccomplished change in the dependent variable; for example, v/v_0 is the unaccomplished velocity change (since v equals 0 at the surface of the plate). $(T - T_S)/(T_0 - T_S)$ is the unaccomplished temperature change and $(\rho_A - \rho_{AS})/(\rho_{A0} - \rho_{AS})$ is the unaccomplished concentration change. The abscissa in each case is simply the ratio of the distance from the surface of the plate to the boundary layer thickness at that point. The reason for the peculiar grouping of terms in the abscissa is that from exact solutions to the equations of motion we know that the boundary layer thickness is proportional to $\sqrt{xt}$. (This comes from the solution to the problem of the suddenly accelerated flat plate.) The appropriate time variable for our case is simply the distance from the front of the plate divided by the free-stream velocity. (This is approximately how long it takes a molecule to reach the point in question from the nose of the plate.) This gives

$$\frac{y}{\delta} = \frac{y}{\sqrt{\nu t}} = \frac{y}{\sqrt{\nu x/v_0}} \tag{11.2-4}$$

The meaning of the abscissa can readily be seen by considering what the graph means at a constant value of x. At constant x, for a given fluid flowing at a given velocity,

$$\sqrt{\frac{v_0}{\nu_x}} = \text{constant} \tag{11.2-5}$$

Therefore, the only thing that changes in the abscissa is y, and so the graph gives us a profile of unaccomplished velocity, temperature, or concentration change as a function of y. This profile is not plotted in the way

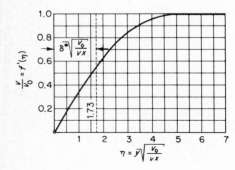

Fig. 11.2-2 Boundary layer. (*H. Schlicting, "Boundary Layer Theory," 4th ed., McGraw-Hill Book Company, New York, 1960.*)

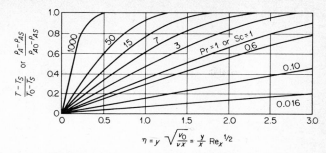

Fig. 11.2-3 Boundary layer solution for flat plate. (*C. O. Bennett and J. E. Myers*, *"Momentum, Heat, and Mass Transfer,"* *McGraw-Hill Book Company, New York, 1962.*)

that you are accustomed to seeing it, because the ordinate and abscissa have been reversed. However, if you take this book and hold it up to the light so that you can view the graph from the back of the page, by rotation of 90° clockwise you will see the profile in the form in which you are accustomed to seeing it plotted.

This is a very specific example, which is included only for illustration. The purpose is to point out that in cases where momentum, heat, and mass are transferred by the *same* mechanism, the *same differential equation* will describe all three processes, with the possible exception that one coefficient may be slightly different in each of the three cases. The coefficient, however, will involve the Prandtl number in one case and the Schmidt number in the other case, and by choosing appropriate values for the Prandtl number and the Schmidt number (which are properties of the fluid only and not of the flow field) one can write the equation in exactly the same form. The physical meaning is that the boundary layers as described in dimensionless variables will coincide for such a case. One can also tabulate solutions for cases in which the Prandtl number and/or Schmidt number are not equal to unity, and such a solution is shown for the specific case of flow past the flat plate in Fig. 11.2-3.

What is the significance of this fact? The utility of these observations to an engineer is that it makes it possible for him to take data in a system other than the one in which he wishes to predict; for example, *in cases where heat transfer and mass transfer occur by the same mechanism*, one can do heat transfer measurements to predict mass transfer, or vice versa. This is frequently desirable. For example, it is often easier to make heat transfer measurements than to make mass transfer measurements. For cases where the Schmidt number or Prandtl number is equal to 1, heat transfer or mass transfer measurements may be used to predict momentum transfer or vice versa (remember that form drag cannot be present).

Again, it is emphasized that this is a general procedure despite the fact that we have chosen a particular example to illustrate this general procedure, namely, the flow over a flat plate.

For most cases of interest, the differential equation describing mass transfer cannot be solved (typically, for cases involving turbulent flow). We still, however can obtain the coefficients in the differential equation by dimensional analysis (assuming that we have sufficient physical insight into the problem to write down the appropriate variables on which to perform this analysis). We therefore can use the analogy between momentum, heat, and mass transfer *despite* the fact that we cannot solve the problem analytically. In effect, by making experimental measurements on our system we are solving the differential equation. We are doing this by an *analog* procedure using the *best possible analog* of our system, that is, the system itself (the homolog). It must be emphasized that the important step in the procedure is the realization that momentum, heat, and mass are being transferred by the same mechanism. Once this has been determined, one solves the problem by either analytical or experimental means depending on which satisfies the requirements of speed, cost, and accuracy for the particular problem.

Example 11.2-1   Comparison of concentration, temperature, and velocity boundary layers   Air with a free-stream velocity of 10 ft/sec and a temperature of 35°F is flowing past a slab of ice whose temperature is 32°F. Calculate the distance from the surface at which the concentration, temperature, and velocity reach a value halfway between the free-stream value and the value at the slab at a point 1 ft from the nose of the slab.

$$\text{Re} = \frac{vx\rho}{\mu} = \frac{10 \times 1 \times 0.0808}{0.017 \times 2.42} = \frac{0.808}{0.04} = 19.7 \qquad \sqrt{19.7} = 4.45$$

$$\text{Pr} = \frac{\hat{C}_\rho \mu}{k} = \frac{0.25 \times 0.017 \times 2.42}{0.014} = 0.74$$

$$\text{Sc} = \frac{\mu}{\rho \mathscr{D}_{AB}} = 0.6$$

From Fig. 11.2-3

$$\eta = \frac{y}{x}\sqrt{\text{Re}} \rightarrow \frac{\eta x}{\sqrt{\text{Re}}} = \frac{\eta}{4.45}$$

For concentration: Sc = 0.6 $\eta = 1.8 \rightarrow y = 0.404$

For temperature: Pr = 0.74 $\eta = 1.7 \rightarrow y = 0.392$

For velocity: $\eta = 1.5 \rightarrow y = 0.337$

□ □ □

Film theory and penetration-renewal theory[1]

Two other approaches to predicting mechanism of mass transfer across an interface are the film theory and penetration-renewal theory. Film theory postulates that there is a stagnant film immediately adjacent to the interface which contains all the resistance to mass transfer found in the system. We now know that most systems do not in fact contain such a film, although the boundary layer can at times act in a very similar fashion. The difficulty in film theory models is that one is always faced with somehow predicting the fictitious film thickness.

Penetration-renewal theory models are among later developments in the theory of mass transfer. Penetration theory models are based on the assumption that there is an identifiable finite mass of fluid which is transported from the free stream to the interface without transferring mass en route, that this finite mass of fluid sits at the interface for a relatively short time with respect to the time required to saturate the mass of fluid with the transferring component, and that the mass of fluid then transfers back to the free stream without transferring mass en route and instantly mixes with the free-stream fluid. This seems a somewhat idealized model, but under some circumstances is the sort of behavior one would expect of eddies that migrate to the interface. The difficulty in penetration-renewal theory models is calculating a rest time for the mass of fluid as it sits at the interface. Penetration theory models have also been combined with film theory models to form the film-penetration theory.

All the models that have been discussed, that is, boundary layer models, film models, and penetration-renewal models, are useful only insofar as they give results which correspond to real mass transfer situations. The accuracy of most mass transfer data at present is perhaps plus or minus 30 percent, and so it is frequently virtually impossible to discriminate as to which of various models gives the best predictions. Most design of the equipment is still done on the basis of highly empirical methods based on the dimensional analysis.

This is not to denigrate the importance of the theoretical approach, since it is only through some theoretical approach that we will achieve ultimate understanding; at the moment, however, most practical results are calculated from what is essentially empiricism.

Diffusion in porous solids

Many chemical systems utilize porous solids (for example, catalyst beds), and so we will consider briefly the effect of porous media on the diffusion coefficient. Diffusion in pores occurs by one or more of three mechanisms: ordinary diffusion, Knudsen diffusion, and surface diffusion. Al-

[1] J. T. Davies, "Mass-Transfer and Interfacial Phenomena," Advances in Chemical Engineering Sciences, vol. IV, p. 1, Academic Press, Inc., New York, 1963.

though we can calculate diffusion in a single pore using the techniques already presented, we normally do not know the details of pore geometry. Therefore, we use the normal diffusion model but use an *effective* diffusion coefficient which accounts for the effect of the porous medium.

Ordinary diffusion takes place when the pores are large compared to the mean free path of the diffusing molecules. However, the effective diffusion coefficient is different from the diffusion coefficient since the porous media have tortuous paths and the coefficient must be corrected by both the areal porosity (free cross-sectional area) and the tortuosity. The tortuosity is a factor that describes the relationship between the actual path length relative to the nominal length of the porous media, taking into account the varying cross section of the pores. The effective diffusion coefficient in a porous medium is given by

$$\mathscr{D}_{\text{eff}} = \frac{\mathscr{D}\theta}{\tau} \tag{11.2-6}$$

where $\mathscr{D}$ is the ordinary diffusion coefficient, θ is the areal porosity, and τ is the tortuosity.

Knudsen diffusion takes place when the size of the pores is of the order of the mean free path of the diffusing molecule. In this situation a molecule collides with the wall as much or more than with other molecules. From kinetic theory for a straight cylindrical pore[1]

$$N = \frac{\mathscr{D}_K}{x_0}(c_1 - c_2) = \frac{\mathscr{D}_K}{RT}\frac{p_1 - p_2}{x_0} = \frac{2r_e v}{3RT}\frac{p_1 - p_2}{x_0} \tag{11.2-7}$$

$$= \frac{2r_e}{3RT}\left(\frac{8RT}{pM}\right)^{1/2}\frac{p_1 - p_2}{x_0}$$

where $\mathscr{D}_K = 9{,}700\, r_e \sqrt{\dfrac{T}{M}}$ $\tag{11.2-8}$

and r_e is the radius of the pore. Usually we correct Eq. (11.2-8) empirically since pores in general are not cylindrical. Defining a mean pore radius

$$r_e = \frac{2V}{S} = \frac{2\theta}{S\rho} \tag{11.2-9}$$

---

[1]C. N. Satterfield and T. K. Sherwood, "The Role of Diffusion in Catalysis," Addison-Wesley Publishing Company, Inc., Reading, Mass., 1963.

Thus

$$\mathscr{D}_{K,\text{eff}} = \frac{\mathscr{D}_K \theta}{\tau} = \frac{8\theta^2}{3\tau S\rho} = \sqrt{\left(\frac{2RT}{\rho M}\right)} = 19,400 \frac{\theta^2}{\tau S\rho} \sqrt{\frac{T}{M}} \quad (11.2\text{-}10)$$

In the transition region between ordinary and Knudsen diffusion an effective diffusion coefficient is defined by

$$N_1 = -\widetilde{\mathscr{D}}_{\text{eff}} \frac{dc}{dx} \qquad (11.2\text{-}11)$$

The expression for $\widetilde{\mathscr{D}}_{\text{eff}}$ is found by integration of Fick's equation for specific boundary conditions.[1,2]

Surface diffusion takes place when molecules adsorbed on solid surfaces are transported over the surface as a result of a two-dimensional concentration gradient on the surface.[3] Surface diffusion normally contributes little to overall transport of mass unless there is a large amount of adsorption.

Dispersion

Dispersion is macroscopic mixing caused by uneven flow. Although dispersion is not primarily due to transport by molecular motion, this type of mixing of two or more components does cause mass transfer. We utilize a diffusion model to describe this macroscopic mixing, replacing the diffusion coefficient with an effective coefficient which is called a dispersion coefficient. (Normally, diffusion effects are included as a small contribution to the dispersion coefficient.[4]) Dispersion results from the bulk motion of two or more components. We can represent macroscopic mixing of two similar but identifiable species in fluid flow using a dispersion coefficient. For example, when fluid A displaces fluid B in a laminar flow in a cylindrical conduit, the fluid at the center of the conduit is moving faster than the fluid near the wall and therefore there is macroscopic mixing of the two species on the average, and this mixing can be described by a dispersion coefficient.

Likewise when one fluid displaces another during flow through a porous medium the two species will mix due to the tortuous nature of the medium. Again the macroscopic mixing of the two species is represented

[1] D. S. Scott and F. A. L. Dullien, *A. I. Ch. E. J.*, **8:**113 (1962).
[2] L. B. Rothfield and C. C. Watson, *A. I. Ch. E. J.*, **9:**19 (1963).
[3] V. G. Levich, "Physicochemical Hydrodynamics," Prentice-Hall, Inc., Englewood Cliffs, N.J., 1962.
[4] S. Whitaker, *A.I.Ch.E. J.*, **13**(3):420 (1967).

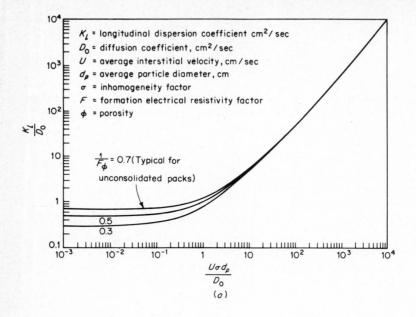

(a)

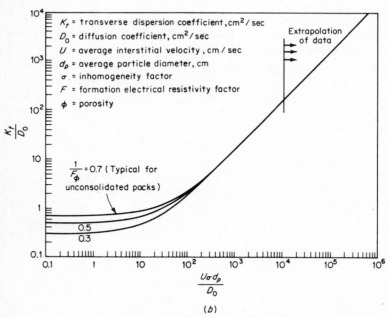

(b)

Fig. 11.2-4 Longitudinal and transverse dispersion coefficients. (*T. K. Perkins and O. C. Johnston, Soc. Petrol. Engrs. J., March 1963.*)

by a dispersion coefficient. As the flow velocity decreases there is less and less mixing by dispersion and relatively more mixing by molecular diffusion. For a porous medium one plots the ratio of the dispersion coefficient to the diffusion coefficient versus a Peclet number as in Fig. 11.2-4[1] The Peclet[2] number vL/D represents the ratio of mass transport by bulk flow to that by dispersion. The Peclet number decreases as velocity decreases, and the ratio of dispersion coefficient to diffusion coefficient becomes constant, since the dispersion coefficient becomes the diffusion coefficient as velocity goes to zero. There are several horizontal lines on Fig. 11.2-4 which result from varying tortuosity of the medium. This means that at these low Peclet numbers mass transfer is dominated by diffusion.

PROBLEMS

11.1 Compare Fick's law of diffusion, Newton's law of viscosity, and Fourier's law of thermal conduction. In each case:

    (a) What is being transported?

    (b) What is the driving force?

    (c) Express the above equations in a form such that the diffusivities all have the same units (ft²/hr).

11.2 Rewrite Fick's law in terms of each of the fluxes in Table 11.1-1.

11.3 Consider the beaker shown below with CCl_4 in the bottom. The CCl_4 evaporates into the pure air. Find the rate of evaporation. Assume equilibrium at the interface. The vapor pressure of CCl_4 is 50 mm Hg at the temperature of the system. The total pressure is 1 atm. The diffusion coefficient of CCl_4 in air is 0.1 cm²/sec.

(Hint: Equilibrium means that Raoult's law can be used at the interface.)

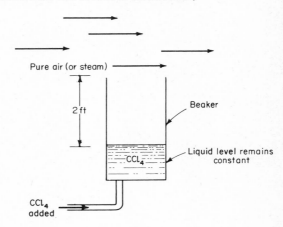

[1]T. K. Perkins and O. C. Johnston, *Soc. Petrol. Engrs. J.*, **1963**:70.
[2]When this dimensionless group incorporates the dispersion coefficient rather than the diffusivity, it is also called the Bodenstein number.

11.4 Consider the same case as in Prob. 11.3 with steam instead of air. The steam condenses and increases the evaporation rate of CCl_4. The latent heat of steam is twice that of CCl_4. Find the rate of evaporation now. Assume that the diffusion coefficient is the same as with air. Evaporation rate should be in moles/hr. Assume ideal gas behavior. Assume steady state.

11.5 For the diffusion of a component A through stagnant B where the cross-sectional area is constant, we showed that

$$(a) \quad N_{Ay} = \frac{-c\mathscr{D}_{AB}}{(x_B)_{lm}} \frac{x_{A0} - x_{A1}}{L}$$

which was derived from

$$(b) \quad N_{Ay} = -c\mathscr{D}_{AB} \frac{dx_A}{dy} + (N_{Ay} + N_{By})x_A$$

Derive an equation which is analogous to (a) in terms of the partial pressure $\bar{p}_A$ and $\bar{p}_B$ ($p = \bar{p}_A + \bar{p}_B$), that is, with a driving force of $(\bar{p}_{A0} - \bar{p}_{A1})$ and p is the total pressure.

11.6 Derive an expression for counterdiffusion at equal *mass* rates.

11.7 Two grams of naphthalene are solidified in the bottom of a 15-in.-long., 1/8-in.-diameter glass tube. The air in the tube is assumed to be saturated at the interface and the room air has no naphthalene in it. How long will it take the solid to disappear at 30°C?

11.8 Calculate the diffusion rate of water vapor from the bottom of a 10-ft pipe, if dry air is blowing over the top of the pipe. The entire system is at 70°F and 1 atm pressure. ($\mathscr{D}_v = 1$ ft^2/hr).

12
Design Equations for Mass Transfer

Most prediction of mass transfer is done in terms of a mass transfer coefficient, just as most prediction of heat transfer is done in terms of a heat transfer coefficient. This is true because most mass transfer situations of interest involve convective mass transfer. In this chapter we discuss mass transfer coefficients, their correlation, and their use. We introduce the concept of an individual-phase mass transfer coefficient and the relationship between individual-phase mass transfer coefficients and overall mass transfer coefficients. We also discuss the relationship of dimensional analysis to equations for the prediction of the mass transfer coefficient, and present design equations for prediction of mass transfer coefficients in various physical situations. We close with a brief discussion of mass transfer with chemical reaction.

12.1 THE MASS TRANSFER COEFFICIENT FOR A SINGLE PHASE

Since we know that in many systems the amount of mass transferred is approximately proportional to the concentration difference, we can write

$$\text{Mass or molar flux} = \text{proportionality factor}$$
$$\times \text{concentration driving force} \qquad (12.1\text{-}1)$$

We call the proportionality factor in such a relation the *mass transfer coefficient*. This is exactly analogous to the way we defined the heat transfer coefficient, and again the usefulness of the coefficient so defined depends on its remaining fairly constant among similar mass transfer problems, and upon our being able to predict (or at least correlate) its performance based on physical parameters.

We must define in a clear fashion precisely what we mean by the *driving force* for each particular case. First let us consider the case of mass transfer within a *single phase*. We usually write for this case

$$N_A = k_z(z_{AS} - z_A) \qquad (12.1\text{-}2)$$

where z_{AS} is the mole fraction of species A at the phase interface, and N_A is the molar flux with respect to fixed coordinates. Note that this coefficient includes *all* the mass transferred (both by diffusion and convection). Since we will be concerned with problems which involve both gas phases and liquid phases, we replace the letter z by the letter x to denote liquid phase, and by the letter y to denote gas phase. The driving force is usually taken so as to give a positive flux. This is *not* the only way to define the mass transfer coefficient. It is, however, the most common way.

In general, the mass transfer coefficient depends on the mass transfer rate. Also, other mass transfer coefficients may be defined with respect to axes *not* fixed in space—for example, moving at the molar average velocity. Since it is our purpose here to introduce the utility of mass transfer coefficients rather than to consider them in detail, we will limit our discussion to low mass transfer rates. In such problems the bulk flow term in Eq. (11.1-9) approaches zero and therefore the mass transfer coefficient is the same for coordinate systems both stationary and moving at the molar average velocity. The reader is cautioned, however, that this assumption is not valid at high mass transfer rates.

For a pure solid such as iodine evaporating into a stream of nitrogen, where Eq. (12.1-2) is applied to the gas phase, y_{AS} is the mole fraction of iodine in the gas phase at the solid surface. This is not a quantity which is easily measured, but experiments have shown that in most cases one can assume that there is *no resistance to mass transfer at the interface*. This is equivalent to the assumption that *the phases immediately on either side of the interface are in equilibrium* (in other words, for no resistance, no displacement from equilibrium is present). Therefore, we would evaluate Eq. (12.1-2) using as y_{AS} the mole fraction of I_2 in a gas mixture at equilibrium with solid I_2 (saturated with N_2) at the pressure and temperature of the interface. Such an assumption is not, of course, valid in the case of a highly contaminated interface.

We must also pick a value to use for z_A. Two cases may be distinguished. First, for *external flows* such as nitrogen flowing past an evaporating sheet of ice, we use for z_a (here y_A) the *free-stream concentration*. In this case, if the nitrogen initially has no water vapor in it, $y_A = 0$. (This is the concentration outside the concentration boundary layer.) Second, for the case of *internal flows*, such as flow of liquid in a pipe where the wall is coated with a dissolving substance, z_A is the *bulk* or *mixing-cup* concentration. The mixing-cup concentration is entirely analogous to the bulk temperature. It is the concentration obtained if the stream is caught in a cup for a short while and then mixed perfectly, and it is defined as

$$\langle z \rangle = \frac{\int_A z\rho(\mathbf{v}\cdot\mathbf{n})\, dA}{\int_A \rho(\mathbf{v}\cdot\mathbf{n})\, dA} \tag{12.1-3}$$

In Chap. 11 we showed that for certain simplified cases of molecular diffusion we could integrate an elementary differential equation and write the resulting expression for mass or molar flux in the following form: conductance times driving force. That is, Eqs. (11.1-16) and (11.1-22) can be written as

$$N_{Ay} = \left[-\frac{c\mathscr{D}_{AB}}{L} \right](x_{A1} - x_{A0}) \tag{12.1-4}$$

and

$$N_{Ay} = \left[-\frac{c\mathscr{D}_{AB}}{L(x_B)_{lm}} \right](x_{A1} - x_{A0}) \tag{12.1-5}$$

The first equation applies to equimolar counterdiffusion and the second to diffusion of A through stagnant B.

The bracketed terms in Eqs. (12.1-4) and (12.1-5) are *exact* expressions for the mass transfer coefficients for these particular situations. In general, however, obtaining such an analytically derived expression is difficult and, in most cases, impossible. The same problems exist in attempting to treat convective heat transfer and momentum transfer (as pointed out in Chaps. 6 and 9); these problems are even more difficult in the case of mass transfer.

Many approaches to the convective mass transfer problem exist: film theory, penetration-renewal theory, boundary layer theory, empirical approaches, etc. These are treated in more advanced courses. Film theory[1,2] postulates the existence next to the interface of an imaginary stagnant film whose resistance to mass transfer is equal to the total mass transfer resistance of the system. The difficulty with this theory is in the calculation of the effective film thickness. Penetration-renewal theory[3-5] assumes that a clump of fluid far from the interface (1) moves to the interface without transferring mass, (2) sits there, stagnant and transferring mass by molecular diffusion for a time short enough that little change in the concentration profile is obtained in the clump, other than very close to the interface, and (3) then moves away from the interface without transferring mass en route and mixes with the bulk fluid instantly. This theory is somewhat more satisfactory in general than film theory, but calculation of the rest time is awkward. Boundary layer theory rests on the solution of a simplified set of differential equations which are approximations to a more nearly correct set of differential equations. Empirical approaches which are merely data correlations serve for specific cases, but give little information about extrapolation.

Example 12.1-1 Calculation of flux from a mass transfer coefficient A thin film of water is flowing down a block of ice with air blowing countercurrent to the water. If the air is dry in the free-stream condition and k_y = 2.0 moles/hr ft², calculate N_A.

Solution

$$N_A = k_y(y_{AS} - y_A)$$

From the fact that the air is dry we know that $y_A = 0$. Since the mass transfer coefficient is given, all that is left to determine is y_{AS}.

Using our assumption of equilibrium at the interface, we know that y_{AS} is the water vapor concentration in equilibrium with liquid water at 32°F saturated with air. The system under these conditions behaves in a

[1] W. Whitman, *Chem. Met. Eng.*, **29**:147 (4) (1923).
[2] W. K. Lewis and W. Whitman, *Ind. Eng. Chem.*, **16**:1215 (1924).
[3] R. Higbie, *Trans. A.I.Ch.E.*, **31**:365 (1935).
[4] P. Danckwerts, *A.I.Ch.E.J.*, **1**:456 (1955).
[5] P. Danckwerts, *Ind. Eng. Chem.*, **43**:1460 (1951).

sufficiently ideal manner that we can probably use Raoult's law to predict equilibrium compositions:

$$y_{AS}p = x_{AS}p'$$

where p is the pressure (here 1 atm) and p' is the vapor pressure of the pure component (here water).

Since the solubility of air in water is slight under those conditions, we will take x_{AS} to be 1.0. The vapor pressure of water at 32°F is 0.0885 psi.

$$y_{AS} = \left(0.0885 \ \frac{lbf}{in.^2}\right)\left(\frac{in.^2}{14.7 \ lbf}\right)\left(\frac{1 \ \text{mole HOH}}{\text{total moles}}\right) = 0.0602$$

Substituting

$$N_A = \left(2.0 \ \frac{\text{moles}}{\text{hr ft}^2}\right)\left[(0.0602 - 0) \ \frac{\text{moles HOH}}{\text{mole total}}\right]$$

$$= 0.120 \ \text{mole/hr ft}^2$$

Note also that the resistance to mass transfer in the *liquid* phase is very nearly zero since the liquid phase is everywhere nearly pure water. We can illustrate this from the equation by writing the flux equation for the *liquid* phase:

$$N_A = k_x(x_A - x_{AS})$$

Since N_A is finite and x_A is almost identical to x_{AS}, k_x is very *large*. Since k_x is a conductance, the resistance is therefore *small*.

□ □ □

As an example of another type of coefficient, a driving force of partial pressure is sometimes used to define a coefficient called k_G.

Example 12.1-2 Mass transfer using partial pressure as a driving force Let k_G denote a mass transfer coefficient based on a driving force of *partial pressure*. Show that $k_y = k_G p$.

Solution By definition

$$N_A = k_G(\bar{p}_A - \bar{p}_{AS}) = k_y(y_A - y_{AS})$$

Using the definition of partial pressure

$$k_G(y_A p - y_{AS} p) = k_y(y_A - y_{AS})$$
$$k_G p = k_y$$

☐ ☐ ☐

Similar manipulation can be done for other k's based on other units. Notice that the units and the numerical value of the mass transfer coefficient change depending on the driving force on which it is based. We encounter this same problem in heat transfer when we change the units in which the driving force is measured—for example, from °F to °C. Another example follows.

Example 12.1-3   Mass transfer using density as a driving force   Let k_ρ denote a mass transfer coefficient based on a driving force of mass density. Develop the relation between this coefficient and k_x.

Solution   By definition

(a) $$n_A = k_\rho(\rho_A - \rho_{AS}) = N_A M_A$$

Notice that the flux is in mass rather than moles. This is the way we usually define this coefficient. If we defined a flux in *mole* units based on a driving force in *mass* units the coefficient would change for each chemical species.

By definition

(b) $$N_A = k_x(x_A - x_{AS})$$

From equations (a) and (b)

(c) $$k_\rho(\rho_A - \rho_{AS}) = k_x(x_A - x_{AS})M_A$$

But

$$\rho_A\left(\frac{\text{mass } A}{\text{unit volume}}\right) = x_A\left(\frac{\text{moles } A}{\text{total moles}}\right) M_A\left(\frac{\text{mass } A}{\text{mole } A}\right) c\left(\frac{\text{total moles}}{\text{unit volume}}\right)$$

Substituting in equation (c)

$$k_\rho(x_A M_A c - x_{AS} M_A c) = k_x(x_A - x_{AS})M_A$$

or

$$k_\rho = \frac{k_x}{c}$$

Note again that the units of k_ρ and k_x are *not the same.* The units are

$$[k_x] = [(\text{moles})/(\text{time})(\text{area})(\text{mole fraction})] = \left[\frac{\text{moles}}{L^2 t}\right]$$

$$[k_\rho] = [(\text{mass})/(\text{time})(\text{area})(\text{mass/volume})] = \left[\frac{L}{t}\right]$$

□ □ □

12.2　THE OVERALL MASS TRANSFER COEFFICIENT

Most of the time we are interested in mass transfer between two *fluid* phases (see Fig. 12.2-1). In heat transfer and momentum transfer there is almost always a solid boundary involved in the transfer path (e.g., the pipe wall). In mass transfer, however, we deal with cases in which we are transferring a component from a gas and/or a liquid to another liquid without any solid barrier between the two phases. Not only does this mean that the boundary becomes severely distorted, but also the precise location of the boundary is not always known.

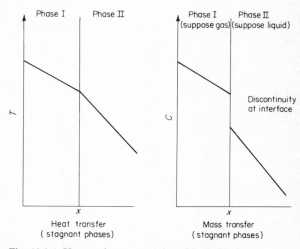

Fig. 12.2-1 Heat and mass transfer with static phase.

In considering the mass transfer problem we are also confronted with a profound difference from the heat transfer situation in that the *variable of interest is no longer usually continuous across the interface.* For heat transfer, if equilibrium exists immediately on either side of phase interfaces, the *temperature* distribution is continuous across phase boundaries. We know from thermodynamics, however, that two phases may be in equilibrium and possess *different concentrations*—all that is necessary is that the partial molar free energies (chemical potentials) of the given component in each phase be the same. In fact, a good part of our study of physical equilibrium in thermodynamics is aimed at predicting the function m in

$$y_A = mx_A \tag{12.2-1}$$

where y_A, x_A = equilibrium concentrations

m = function of thermodynamic state variables

Raoult's and Henry's laws are examples of efforts in this direction where m is a constant. In general m *is not* constant.

Raoult's law is

$$\bar{p}_A = y_A p = x_A p' \tag{12.2-2}$$

or

$$y_A = \frac{p'}{p} x_A \tag{12.2-3}$$

which implies

$$m = \frac{p'}{p} \tag{12.2-4}$$

while Henry's law

$$\bar{p}_A = y_A p = H x_A \tag{12.2-5}$$

yields

$$m = \frac{H}{p} \tag{12.2-6}$$

We would like now to define an overall coefficient of *mass* transfer in much the same way that one can define an overall coefficient of *heat transfer*, that is, in terms of the *overall driving force*. The obvious overall driving force is $y_A - x_A$, by analogy to the heat transfer case, but this is a function of m. We can see this by rewriting as

$$y_A - x_A = (y_A - y_{AS}) + (y_{AS} - x_{AS}) + (x_{AS} - x_A) \qquad (12.2\text{-}7)$$

by adding and subtracting y_{AS}, x_{AS}. However, using Eq. (12.2-1),

$$y_{AS} - x_{AS} = m x_{AS} - x_{AS} = x_{AS}(m - 1) \qquad (12.2\text{-}8)$$

and substituting in Eq. (12.2-7),

$$y_A - x_A = (y_A - y_{AS}) + x_{AS}(m - 1) + (x_{AS} - x_A) \qquad (12.2\text{-}9)$$

Note that as we approach equilibrium, y_A and x_A approach y_{AS} and x_{AS} respectively, and

$$(y_A - x_A)_{eq} = x_{AS}(m - 1) \qquad (12.2\text{-}10)$$

This is not at all the behavior we wish for a driving force, for we want a driving force that goes to *zero* at equilibrium, not some finite value. This leads us to define for our driving force one of two quantities

$$\text{Driving force} = x_A - x_A^* \quad \text{or} \quad y_A - y_A^* \qquad (12.2\text{-}11)$$

where $x_A^* = y_A/m$ $\qquad (12.2\text{-}12)$

$$y_A^* = m x_A \qquad (12.2\text{-}13)$$

depending on whether we are using x or y. In other words, *the starred quantities are those values of x or y that would be in equilibrium with the bulk fluid of the other phase*. Note that this driving force goes to zero at equilibrium.

The concentrations x_A^* and y_A^* are not normally present anywhere in the real system for at such a point the driving force would be zero; they are fictitious concentrations introduced for convenience.

This driving force leads us to define an *overall* mass transfer coefficient as

$$N_A = K_z(z_A - z_A^*) \qquad (12.2\text{-}14)$$

where we substitute x for z to treat liquids and y for z to treat gases.

We can visualize the preceding by means of Fig. 12.2-2, where we use y to denote the concentration in one phase and x the concentration in the other. We plot on these coordinates the equilibrium line [Eq. (12.2-1)] which we have shown as curved, since *in general m is not constant.*

We now ask ourselves how we find x_A^*, y_A^*, x_{AS}, and y_{AS} if we are given two bulk concentrations x_A and y_A. It is not difficult to locate x_A^* and y_A^* immediately—we simply apply the definition—for example, that x_A^* is the composition that would be in equilibrium with y_A. Since the equilibrium curve is a locus of equilibrium compositions we find x_A^* as the abscissa of the equilibrium curve for the ordinate y_A. The location of y_A^* is done similarly.

To locate the interfacial composition x_{AS}, y_{AS}, we use the fact that we must obtain the same value for N_A regardless of whether we calculate using k_x or k_y

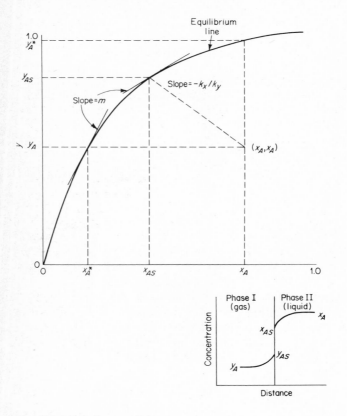

Fig. 12.2-2 Interface conditions.

$$N_A = k_x(x_{AS} - x_A) = k_y(y_A - y_{AS}) \tag{12.2-15}$$

Rearranging

$$\frac{-k_x}{k_y} = \frac{y_A - y_{AS}}{x_A - x_{AS}} \tag{12.2-16}$$

This is the equation of a straight line through (x_A, y_A) with slope $-k_x/k_y$. The line also passes through (x_{AS}, y_{AS}), and this point, assuming equilibrium at the interface, also lies on the equilibrium curve. We may therefore locate (x_{AS}, y_{AS}) by drawing a line through (x_A, y_A) with slope $-k_x/k_y$ and taking its intersection with the equilibrium curve as shown.

Example 12.2-1 Calculation of interface composition In absorption of NH_3 by water in a wetted-wall column, the mass transfer coefficient in the gas phase was determined to be 0.12 lb mole/hr ft² atm and that in the liquid phase to be 8.1 lb mole/hr ft² atm. At one point in the column, laboratory analysis of a gas sample gave the amount of NH_3 in the gas phase as 20% (mole basis) and a liquid sample analyzed as 5% (mole basis). Estimate the composition of NH_3 at the interface. Column instruments give the pressure as essentially atmospheric and the temperature as 68 °F.

Solution The equilibrium data for the system are shown plotted in Fig. 12.2-3. We know from the samples that $y = 0.20$, $x = 0.05$ is a point on

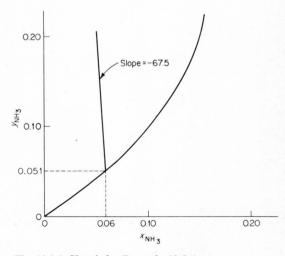

Fig. 12.2-3 Sketch for Example 12.2-1.

the line represented by Eq. (12.2-16). The slope of this line is

$$\frac{-k_x}{k_y} = \frac{-8.1}{0.12} = -67.5$$

Drawing a line through the point with the calculated slope as shown, we find that it intersects the equilibrium line at a point which gives $x_{AS} = 0.06$, $y_{AS} = 0.051$.

□ □ □

12.3 RELATION BETWEEN OVERALL AND SINGLE-PHASE MASS TRANSFER COEFFICIENTS

By using exactly the same procedure one uses to develop overall heat transfer coefficients from single-phase coefficients one can *sometimes* develop similar relations between the overall and single-phase mass transfer coefficients. *The procedure is first to write the overall driving force as the sum of individual driving forces.* Let us use the liquid phase for our basis as an example:

$$x_A - x_A^* = x_A - \frac{y_A}{m} \qquad (12.3\text{-}1)$$

Adding and subtracting x_{AS} twice from each side and substituting y_{AS}/m for x_{AS},

$$x_A - x_A^* = (x_A - x_{AS}) + \left(x_{AS} - \frac{y_{AS}}{m}\right) + \left(\frac{y_{AS}}{m} - \frac{y_A}{m}\right) \qquad (12.3\text{-}2)$$

Notice that the above procedure will hold only for cases where m evaluated at y_A and m evaluated at y_{AS} are the same, i.e., basically for *straight* equilibrium lines as shown in Fig. 12.3-1. But our assumption of equilibrium at the interface says that the middle term is zero if m *at these two points is the same* (again, considering only *straight* equilibrium lines):

$$x_A - x_A^* = (x_A - x_{AS}) + \frac{1}{m}(y_{AS} - y_A) \qquad (12.3\text{-}3)$$

Substituting from the defining expressions (12.1-2) and (12.2-14),

$$\frac{N_A}{K_x} = \frac{N_A}{k_x} + \frac{N_A}{mk_y} \qquad (12.3\text{-}4)$$

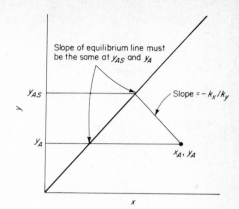

Slope of equilibrium line must
be the same at y_{AS} and y_A

y_{AS}

Slope $= -k_x/k_y$

y_A

x_A, y_A

Fig. 12.3-1 Assumption in the overall coefficient.

$$\frac{1}{K_x} = \frac{1}{k_x} + \frac{1}{mk_y} \tag{12.3-5}$$

These relations are true *only* for a straight equilibrium line because we
have assumed in Eqs. (12.3-3) and (12.3-4) that m is the same at y_{AS} and
y_A. If the equilibrium line is *not* straight this procedure is invalid. In
other words, one cannot develop a simple overall coefficient in terms of
the individual coefficients. Similarly, starting with $y_A^* - y_A$ we can show
that

$$\frac{1}{K_y} = \frac{1}{k_y} + \frac{m}{k_x} \tag{12.3-6}$$

Note that if either $1/k_y \ll m/k_x$ or vice versa, one phase controls
the process just as in heat transfer. For example, if $1/k_y \to 0$ ($k_y \to \infty$) the
liquid phase controls.

Nowhere in any of this material have we shown how to calculate k
or K from the physical structure (fluid mechanics, etc.). This will be our
topic in the next section.

**Example 12.3-1   Overall mass transfer coefficient for series resistan-
ces**   In Example 12.2-1 calculate the overall mass transfer coefficient
based on the liquid phase.

Solution

$$\frac{1}{K_x} = \frac{1}{k_x} + \frac{1}{mk_y}$$

From Fig. 12.2-3 m is about 0.85

$$\frac{1}{K_x} = \frac{1}{8.1} + \frac{1}{(0.85)(0.12)}$$

$$\frac{1}{k_x} = 0.123 + 9.80 = 9.93$$

$$K_x = 0.101 \frac{\text{moles}}{\text{hr ft}^2}$$

This is an approximation because the equilibrium line is not straight.

□ □ □

Example 12.3-2   Controlling resistance for mass transfer   For the system treated in Examples 12.2-1 and 12.3-1, determine the controlling resistance for mass transfer.

Solution   In Example 12.3-1 we showed that

$$\frac{1}{k_x} = \frac{1}{8.1} = 0.123$$

and

$$\frac{1}{mk_y} = \frac{1}{(0.85)(0.12)} = 9.80$$

Further, $1/K_x$ represents the resistance to mass transfer (since K_x is a *conductance*). Since

$$\frac{1}{K_x} = \frac{1}{k_x} + \frac{1}{mk_y}$$

by far the largest contribution to the resistance comes from the *gas* phase, and so the gas phase controls.

The same answer is obtained by looking at Eq. (12.3-6):

$$\frac{1}{K_y} = \frac{1}{k_y} + \frac{m}{k_x}$$

$$= \frac{1}{0.12} + \frac{0.85}{8.1}$$

$$= 8.33 + 0.105$$

Again, the gas phase controls.

The pertinence of this to the engineer is that to increase mass transfer rates in the equipment it is probably more profitable to examine modification of the gas process than the liquid. It also shows that fairly large changes can probably be made on the *liquid* side without upsetting (that is, significantly changing) the present operation of the column.

□ □ □

12.4 DIMENSIONAL ANALYSIS FOR MASS TRANSFER

We now must show how to calculate the mass transfer coefficients we have discussed in the above section. We do this, just as for heat transfer coefficients, in terms of design equations which are usually written in dimensionless terms. To introduce these design equations, we will first examine the significant dimensionless groups involved with mass transfer.

For mass transfer by molecular diffusion and *forced* convection the pertinent variables are a characteristic length D, the diffusivity $\mathscr{D}_{AB}$, the velocity of the fluid v, the mass density of the fluid ρ, the viscosity of the fluid μ, and the mass transfer coefficient k_ρ. For the case of *natural* convection one also must include the acceleration due to gravity g, and the mass density difference $\Delta\rho_A$. If we carry out a dimensional analysis in a precisely parallel manner to Chap. 3, we can find the significant dimensionless groups to be

$$\text{Sh} = \frac{k_\rho D}{\mathscr{D}_{AB}} \qquad \text{Sherwood number} \qquad\qquad (12.4\text{-}1)$$

$$\text{Sc} = \frac{\mu}{\rho \mathscr{D}_{AB}} \qquad \text{Schmidt number} \qquad\qquad (12.4\text{-}2)$$

$$\text{Re} = \frac{D v \rho}{\mu} \qquad \text{Reynolds number} \qquad\qquad (12.4\text{-}3)$$

$$\text{Fr} = \frac{v^2}{gD} \qquad \text{Froude number} \qquad\qquad (12.4\text{-}4)$$

$$\text{Gr} = \frac{gD^3 \rho \,\Delta\rho_A}{\mu^2} \qquad \text{Grashof number for mass transfer} \qquad (12.4\text{-}5)$$

Example 12.4-1 Dimensional analysis for convective mass transfer For the following variables which appear in convective mass transfer—k_ρ, D, $\mathscr{D}_{AB}$, μ, ρ, v, g, $\Delta\rho_A$—show that one obtains the dimensionless groups given above.

The fundamental dimensions in the MLt system are

$$k_\rho \qquad \frac{M}{L^2 t} \frac{L^3}{M} = \frac{L}{t}$$

$$D \qquad L = \tilde{Q}_1$$

$$\mathscr{D}_{AB} \qquad L^2/t$$

$$\mu \qquad M/Lt = \tilde{Q}_2$$

$$\rho \qquad M/L^3 = \tilde{Q}_3$$

$$v \qquad L/t$$

$$g \qquad L/t^2$$

$$\Delta \rho_A \qquad M/L^3$$

Solution Following the procedure in Chap. 3:

$$D^a \mu^b \rho^c k_\rho = \pi_1$$

$$L^a \left(\frac{M}{Lt}\right)^b \left(\frac{M}{L^3}\right)^c \frac{L}{t} = \pi_1$$

For length

$$a - b - 3c + 1 = 0$$

For mass

$$b + c = 0$$

For time

$$-b - 1 = 0$$

Therefore

$$a = 1$$
$$b = -1$$
$$c = 1$$

and

$$\pi_1 = \frac{D\rho}{\mu} k_\rho$$

Next:

$$D^a \mu^b \rho^c \mathscr{D}_{AB} = \pi_2$$

$$L^a \left(\frac{M}{Lt}\right)^b \left(\frac{M}{L^3}\right)^c \frac{L^2}{t} = \pi_2$$

and for length

$$a - b - 3c + 2 = 0$$

For mass

$$b + c = 0$$

For time

$$-b - 1 = 0$$

and so

$$a = 0$$
$$b = -1$$
$$c = 1$$

and

$$\pi_2 = \frac{\rho}{\mu} \mathscr{D}_{AB} = \frac{1}{Sc}$$

Next:

$$D^a \mu^b \rho^c v = \pi_3$$

$$L^a \left(\frac{M}{Lt}\right)^b \left(\frac{M}{L^3}\right)^c \frac{L}{t} = \pi_3$$

For length

$$a - b - 3c + 1 = 0$$

For mass

$$b + c = 0$$

For time

$$-b - 1 = 0$$

and so

$$a = +1$$
$$b = -1$$
$$c = 1$$

and

$$\pi_3 = \frac{Dv\,\rho}{\mu} = \text{Re}$$

Next:

$$D^a \mu^b \rho^c g = \pi_4$$

$$L^a \left(\frac{M}{Lt}\right)^b \left(\frac{M}{L^3}\right)^c \frac{L}{t^2} = \pi_4$$

For length

$$a - b - 3c + 1 = 0$$

For mass

$$b + c = 0$$

For time

$$-b - 2 = 0$$

and

$$a = 3$$
$$b = -2$$
$$c = 2$$

so

$$\pi_4 = \frac{D^3 \rho^2}{\mu^2} g$$

Next:

$$D^a \mu^b \rho^c \Delta\rho_A = \pi_5$$

$$L^a \left(\frac{M}{Lt}\right)^b \left(\frac{M}{L^3}\right)^c \frac{M}{L^3} = \pi_5$$

For length

$$a - b - 3c - 3 = 0$$

For mass

$$b + c + 1 = 0$$

For time

$$-b = 0$$

and so

$$a = 0$$
$$b = 0$$
$$c = -1$$

and

$$\pi_5 = \frac{\Delta \rho_A}{\rho}$$

This is a perfectly acceptable set of dimensionless groups. However, it is customary to work with the groups listed earlier (which are more readily identified with coefficients in the differential equations normally used). We obtain these groups as follows:

$$\frac{\pi_1}{\pi_2} = \frac{D\rho\, k_\rho}{\mu} \frac{\mu}{\rho \mathscr{D}_{AB}} = \frac{k_\rho D}{\mathscr{D}_{AB}} = \text{Sh}$$

$$\frac{1}{\pi_2} = \frac{\mu}{\mathscr{D}_{AB}\rho} = \frac{\nu}{\mathscr{D}_{AB}} = \text{Sc}$$

$$\pi_3 = \frac{Dv\,\rho}{\mu} = \text{Re}$$

$$\frac{\pi_3{}^2}{\pi_4} = \frac{\mu^2}{D^3 \rho^2 g} \frac{D^2 v^2 \rho^2}{\mu^2} = \frac{v^2}{Dg} = \text{Fr}$$

$$\pi_4 \pi_5 = \frac{D^2 \rho^2 g}{\mu^2} \frac{\Delta \rho_A}{\rho} = \frac{D^3 \rho g}{\mu^2} \Delta \rho_A = \text{Gr}$$

We may write our Sherwood number in terms of k_x using the result from Example 12.1-3 that

$$k_\rho = \frac{k_x}{c}$$

and so

$$\mathrm{Sh} = \frac{k_x D}{c \mathscr{D}_{AB}}$$

□ □ □

The usual functional relationships used to correlate data are

$$\mathrm{Sh} = f(\mathrm{Re},\mathrm{Sc}) \tag{12.4-6}$$

for forced convection [paralleling (Eq. 9.3-3)], and

$$\mathrm{Sh} = f(\mathrm{Gr},\mathrm{Sc}) \tag{12.4-7}$$

for natural convection [paralleling (Eq. 9.3-7)]. The Froude number is normally unimportant in mass transfer correlations.

If we assume that mass and heat are transferred by the *same mechanism* (by the same mechanism we mean that the dimensionless profiles of velocity, temperature, and concentration are similar—in other words, they come from the same differential equation), we see that the above amounts to a replacement of the Nusselt number by the Sherwood number and of the Prandtl number by the Schmidt number. This indicates that the Sherwood number, by analogy to the Nusselt number, is the ratio of total *mass* transferred to mass transferred by molecular diffusion, and that the Schmidt number, by analogy to the Prandtl number, represents the molecular diffusion of momentum to that of *mass*.

This all suggests that *if the mechanism of mass and heat transfer is the same*, we can use our correlations for heat transfer to predict mass transfer by the simple expedient of substituting Sh and Sc for Nu and Pr respectively. We can, in fact, do this (for both laminar and turbulent flow) but only under fairly restrictive assumptions. We can even extend the technique to using *momentum* transfer data to predict mass or heat transfer, but, as was mentioned earlier, only under much more stringent restrictions because form drag has no counterpart in mass or heat transfer (see Chap. 11).

We have thus far avoided a very distasteful topic, namely, that our mass transfer coefficient as defined is a function of some things not in our list of variables from which we performed our dimensional analysis. For example, we showed in our treatment of molecular diffusion that the mass transfer coefficient depends on whether we have equimolar counterdiffusion or diffusion of A through stagnant B. In addition, the general coefficient we defined includes both molecular diffusion and bulk flow, and the bulk flow term is very sensitive to the flow field, i.e., the momentum transfer.

In fact, for a rigorous treatment we have to define *several* mass transfer coefficients, each appropriate to a different situation. The mass transfer coefficient as we have defined it holds for a general mass transfer situation. It suffers, however, from the difficulty that it depends on mass transfer rate because increasing rates of mass transfer distort velocity and concentration profiles. These effects are described by nongradient terms in the differential equations; that is, terms that do not contain concentration driving forces. These terms cannot be described well by gradient models such as we postulate in Eqs. (12.1-2) and (12.2-14). For this reason mass transfer coefficients sometimes are defined which describe only the gradient contribution to transfer (as opposed to bulk flow) or perhaps total transfer only for specified bulk flow situations such as equimolar counterdiffusion. These coefficients usually are designated by some type of superscript.[1] Our objective here, however, is to introduce the reader to the utility and prediction of mass transfer coefficients, and so we will limit our discussion to situations which satisfy the following criteria:

1. Constant physical properties
2. Very small net bulk flow at the interface
3. No chemical reactions in the fluid
4. No viscous dissipation (that is, negligible temperature rise produced by viscous dissipation)
5. No radiant energy interchange
6. No pressure, thermal, or forced diffusion

Under these restrictions the several mass transfer coefficients we must define for a deeper discussion all reduce to the same number, and so we will be *consistent* and *correct* in what is to come later but not *general*. The above set of restrictions, although severe, also frequently gives useful approximations for real problems since the uncertainty of mass transfer calculations is great. In exchange for the restrictions we gain vastly simplified notation and discussion.

[1] R. B. Bird, W. E. Stewart, and E. N. Lightfoot, "Transport Phenomena," John Wiley & Sons, Inc., New York, 1960.

12.5   DESIGN EQUATIONS

Using the restrictions above, we have for the mass transfer version of the
Reynolds analogy, paralleling Eqs. (9.3-22) and (9.3-23),

$$\frac{Sh}{ScRe} = \frac{f}{2} \tag{12.5-1}$$

As in the case of heat transfer this model has severe restrictions.

For *laminar* flow past a flat plate, an approximate expression for
fluids with Sc > 0.6 is

$$Sh_x = 0.33 Re_x^{1/2} Sc^{1/3} \tag{12.5-2}$$

where the characteristic length in Sh and Re is the distance from the nose
of the plate, and the Sherwood number is the *local* one in that it contains
the local mass transfer coefficient and the distance from the nose of the
plate.

Integrating over the surface of the plate yields an expression for
mean Sherwood number as

$$Sh_L = 0.66 Re_L^{1/2} Sc^{1/3} \tag{12.5-3}$$

where the characteristic length is now the length of the plate, and the mass
transfer coefficient is a *mean* coefficient.

**Example 12.5-1   Average mass transfer coefficient from local coeffi-
cient**   Given the equation for local mass transfer coefficient for flow over
a flat plate

$$Sh_x = 0.33 Re_x^{1/2} Sc^{1/3}$$

calculate an expression for the average coefficient.

Solution   By an average coefficient we mean a number that, multiplied by
an appropriate driving force and the area of the plate, gives the same mass
transfer rate as integrating the product of the local coefficient, local driving force, and a differential area over the surface of the plate. In equation
form

$$k_{av}(y_{AS} - y_A)_{av} A = \int_A k_{loc}(y_{AS} - y_A) \, dA$$

(If we have a *constant* driving force as along a homogeneous plate, this is the obvious driving force to associate with the average coefficient. The choice of average driving force is, however, arbitrary, as can be seen by considering the case of mass transfer in a pipe where the driving force changes along the pipe. In this case, in fact, average coefficients based on both log mean and arithmetic mean driving forces are used.)

Substituting for our local coefficient

$$k_{av}(y_{AS} - y_A)A = \int_A 0.33 \frac{\mathscr{D}_{AB}}{x} \sqrt{\frac{xv_0\rho}{\mu}} \left(\frac{\nu}{\mathscr{D}_{AB}}\right)^{1/3} (y_{AS} - y_A)\, dA$$

The only variable under the integral by our assumption of constant conditions is x, the distance to the front of the plate.

Letting W be the width of the plate and L the length:

$$k_{av}(y_{AS} - y_A)WL = 0.33\, \mathscr{D}_{AB} \sqrt{\frac{v_0\rho}{\mu}} \left(\frac{\nu}{\mathscr{D}_{AB}}\right)^{1/3} (y_{AS} - y_A)W \int_0^L \frac{dx}{x^{1/2}}$$

Integrating

$$k_{av}L = 0.33\, \mathscr{D}_{AB} \sqrt{\frac{v_0\rho}{\mu}} \left(\frac{\nu}{\mathscr{D}_{AB}}\right)^{1/3} 2L^{1/2}$$

or

$$k_{av}L = 0.66\, \mathscr{D}_{AB} \sqrt{\frac{Lv_0\rho}{\mu}} \left(\frac{\nu}{\mathscr{D}_{AB}}\right)^{1/3}$$

which may be rearranged as

$$Sh_L = 0.66 Re_L^{1/2} Sc^{1/3}$$

where the subscript on the Sherwood number indicates that it contains the *average* coefficient and is based on the total plate length. The subscript on the Reynolds number means that it is based on the total plate length.

□ □ □

Mass transfer for *laminar* flow with developed velocity distribution in a *pipe* may be treated using the corresponding heat transfer curves, Fig. 9.3-2, by substituting Sherwood number for Nusselt number and Schmidt number for Prandtl number.

One local expression for *turbulent* flow past a *flat plate* is

$$\text{Sh}_x = 0.0202\text{Re}_x{}^{4/5} \tag{12.5-4}$$

where the characteristic length is again the distance from the nose of the plate, and the expression is valid only for fluids with $\text{Sc} = 1.0$. For turbulent flow in pipes we will use the Reynolds analogy here, although several extensions of this treatment are available.

Packed beds are frequently used as devices for contacting gas and liquid or liquid and liquid. Note that the mass transfer we will discuss with respect to packed beds is between the flowing phases and not between the packing in the bed and either of the phases. Because of the fact that the interface is so ill-defined in a packed bed, a slightly different approach to the mass transfer coefficient is taken. For flows in packed beds it is seldom possible to determine the interfacial area. Since it is difficult to separate the interfacial area effect as a separate entity, it is usual to determine only the *product* of the interfacial area per unit volume and the mass transfer coefficient.

The interfacial area per unit volume of *empty* tower we denote as a:

$$a = \frac{A_i}{V_t} \tag{12.5-5}$$

where A_i = interfacial area

V_t = volume of empty tower

$V_t = A_{xs} \times \text{height}$

A_{xs} = cross-sectional area of empty tower

and seldom know its value. If, however, we rewrite our defining equation for the mass transfer coefficient (where $\mathcal{W}_A$ is the molar flow rate of species A)

$$N_A = \frac{\mathcal{W}_A}{A_i} = k_z(z_{AS} - z_A) \tag{12.5-6}$$

Rearranging and multiplying and dividing by A_{xs} (xs is the cross section of the empty tower)

$$\mathcal{W}_A = k_z \frac{A_i}{A_{xs} \times \text{height}} (A_{xs} \times \text{height})(z_{AS} - z_A) \tag{12.5-7}$$

or

$$\mathcal{W}_A = k_z a V_t \, (z_{AS} - z_A) \tag{12.5-8}$$

This gives us an equation with which we can determine $k_z a$ (either $k_x a$ or $k_y a$). We can take a packed bed, measure $\mathcal{W}_A$ by analyzing both the inlet and outlet of either of the two fluids, get z_A from a bulk fluid sample, obtain $V_{empty\ tower}$ from the physical dimensions of the tower, and from this information calculate an average $k_z a$. As the height of tower over which we perform the measurements is decreased to a differential size, we approach calculation of the local $k_z a$. Note: We cannot separate k_z and a unless in some manner we can determine interfacial area. This means that $k_z a$, in addition to varying with all the factors that influence the mass transfer coefficient, will also vary with factors which change the interfacial area per unit volume, such as packing method (for example, stacked, dry dumped, wet dumped, vibrated, etc.), packing configuration (rings, saddles, etc.), packing size, and so on.

The quantity used to correlate mass transfer in packed beds is usually not $k_z a$ per se, but another quantity known as the height of a transfer unit H. This quantity combines superficial fluid rate with the mass transfer coefficient and interfacial area per unit volume. The reason for using this quantity, which has the units of length, is that this combination of variables arises very naturally in an integration when one designs packed columns. We consider this integration later; for now, we simply state the definitions for H and present design equations. The definitions are

$$H_G = \frac{G}{k_y a} \tag{12.5-9a}$$

$$H_L = \frac{L}{k_x a} \tag{12.5-9b}$$

where G and L are superficial molar velocities (moles per unit time per unit cross section of empty tower), and a is the interfacial area per unit tower volume. (See Sec. 13.2 for calculation of height of a transfer unit.) Flow in packed beds can be treated using Tables 12.5-1 and 12.5-2. The equations are (where G' and L' have units of lb$m$/hr ft^2 and μ has units lb$m$/ft hr)

$$H_G = \alpha G'^\beta L'^\gamma \, \mathrm{Sc}^{0.5} \tag{12.5-10a}$$

$$H_L = \phi \left(\frac{L'}{\mu}\right)^\eta \mathrm{Sc}^{0.5} \tag{12.5-10b}$$

Note: These are dimensional equations. The constants for these equations can be determined from Tables 12.5-1 and 12.5-2.

H can also be defined to incorporate the *overall* mass transfer coefficient, as follows:

$$H_{OG} = \frac{G}{K_y a} \tag{12.5-11a}$$

$$H_{OL} = \frac{L}{K_x a} \tag{12.5-11b}$$

Mass transfer from spheres can be treated using Eq. (9.3-64) rewritten as

$$\text{Sh} = 2.0 + 0.6\,\text{Sc}^{1/3}\,\text{Re}^{1/2} \tag{12.5-12}$$

Mass transfer in drops and bubbles is a function of internal circulation and solutions are available primarily for limiting cases at present. We will not treat this topic here because of its complexity, although it is one of the most important to engineers from a practical standpoint.

Example 12.5-2 Calculation of mass transfer coefficients Compare the empirical equation developed by Sherwood and Gilliland[1]

$$\text{Sh} = 0.023\,\text{Re}^{0.83}\,\text{Sc}^{1/3}$$

with Eq. (9.3-56) for the calculation of the mass transfer coefficient for air flowing at 50°F and at a bulk velocity of 50 ft/sec in a 1-in.-ID tube coated with naphthalene.

Solution Equation (9.3-56) reads, substituting Sh for Nu and Sc for Pr,

$$\text{Sh} = 0.023\,\text{Re}^{0.8}\,\text{Sc}^{1/3}$$

for air at 50°F

$$\rho = 0.078 \text{ lb}m/\text{ft}^3$$

$$\mu = 1.2 \times 10^{-5} \text{ lb}m/\text{ft sec}$$

$$\mathscr{D}_{AB} = 0.2 \text{ ft}^2/\text{hr}$$

[1]T. K. Sherwood and E. R. Gilliland, *Ind. Eng. Chem.*, **26:**516 (1934).

Table 12.5-1 Constants^a for determining H_G in Eq. (12.5-10a)

Type of packing	α	β	γ	Range of values, lb/hr ft^2	
				G'	L'
Raschig rings:					
$^3/_8$-in.	2.32	0.45	−0.47	200–500	500–1500
1-in.	7.00	0.39	−0.58	200–800	400–500
	6.41	0.32	−0.51	200–600	500–4500
2-in.	3.82	0.41	−0.45	200–800	500–4500
Berl saddles:					
$^1/_2$-in	32.4	0.30	−0.74	200–700	500–1500
	0.811	0.30	−0.24	200–700	1500–4500
1-in.	1.97	0.36	−0.40	200–800	400–4500
1$^1/_2$-in.	5.05	0.32	−0.45	200–1000	400–4500

^{a} From R. E. Treybal, "Mass-transfer Operations," p. 239, McGraw-Hill Book Company, New York, 1955.

We neglect the presence of the naphthalene and use the density and viscosity of air above in the Reynolds and Schmidt numbers.

$$\text{Re} = \left(\frac{1}{12}\ \text{ft}\right)\left(50\ \frac{\text{ft}}{\text{sec}}\right)\left(\frac{0.078\ \text{lb}m}{\text{ft}^3}\right)\left(\frac{\text{ft sec}}{1.2 \times 10^{-5}\ \text{lb}m}\right) = 27{,}000$$

$$\text{Sc} = \left(1.2 \times 10^{-5}\ \frac{\text{lb}m}{\text{ft sec}}\right)\left(\frac{\text{ft}^3}{0.078\ \text{lb}m}\right)\frac{\text{hr}}{0.2\ \text{ft}^2}\left(3{,}600\ \frac{\text{sec}}{\text{hr}}\right) = 2.77$$

The Sherwood and Gilliland equation gives

$$\text{Sh} = (0.023)(27{,}000)^{0.83}(2.77)^{1/3}$$

$$\text{Sh} = 154$$

Table 12.5-2 Constants^a for determining H_L in Eq. (12.5-10b)

Type of packing	ϕ	η	Range of L'
Raschig rings:			
$^3/_8$-in.	0.0018	0.46	400–15,000
1-in.	0.010	0.22	400–15,000
2-in.	0.012	0.22	400–15,000
Berl saddles:			
$^1/_2$-in.	0.0067	0.28	400–15,000
1-in.	0.0059	0.28	400–15,000
1$^1/_2$-in.	0.0062	0.28	400–15,000

^{a} From T. K. Sherwood and F. A. L. Holloway, *Trans. A.I.Ch.E.*, **36**:39 (1940).

and Eq. (9.3-56) gives

$$Sh = (0.023)(27,000)^{0.8}(2.77)^{1/3}$$

$$Sh = 113$$

This is a fairly large discrepancy but it is not bad in light of the low accuracy obtainable with many mass transfer calculations.

□ □ □

Example 12.5-3   Dissolution of a sphere   Estimate the mass transfer coefficient for a $^1/_8$-in. sphere of glucose dissolving in a water stream flowing at 0.5 ft/sec. The temperature is 25°C. Diffusivity of glucose in water at 25°C is 0.69×10^{-5} cm²/sec.

Solution   Properties of water at 25°C:

$$\mu = 0.9 \text{ cp}$$

$$\rho = 62.2 \text{ lb}m/\text{ft}^3$$

We neglect the effect of the glucose on the physical properties of the water

$$Re = \left[\frac{1}{(8)(12)}\text{ ft}\right]\left(0.5\,\frac{\text{ft}}{\text{sec}}\right)\left(62.2\,\frac{\text{lb}m}{\text{ft}^3}\right) \times$$

$$\left[\frac{\text{ft sec}}{(0.9)(6.72\times10^{-4})\,\text{lb}m}\right] = 536$$

$$Sc = \left[(0.9)(6.72\times10^{-4})\,\frac{\text{lb}m}{\text{ft sec}}\right]\left(\frac{\text{ft}^3}{62.2\,\text{lb}m}\right)\left(\frac{\text{sec}}{0.69\times10^{-5}\,\text{cm}^2}\right) \times$$

$$\left(\frac{\text{cm}^2\,\text{hr}}{3.87\,\text{sec ft}^2}\right)\left(3,600\,\frac{\text{sec}}{\text{hr}}\right)$$

$$= 2.18$$

Modifying the Froessling correlation, Eq. (9.3-64), for mass transfer:

$$Sh = 2.0 + 0.6\,Sc^{1/3}\,Re^{1/2}$$

$$Sh = 2.0 + 0.6\,(2.18)^{1/3}\,(536)^{1/2}$$

$$\text{Sh} = 20.0 = \frac{kD}{\mathscr{D}_{AB}}$$

$$k = \left[(20.0)\,(0.69 \times 10^{-5})\,\frac{\text{cm}^2}{\text{sec}}\right]\left(3.87\,\frac{\text{ft}^2}{\text{hr}}\,\frac{\text{sec}}{\text{cm}^2}\right)\left[\frac{(8)\,(12)}{\text{ft}}\right]$$

$$k = 0.051 \text{ ft/hr}$$

□ □ □

12.6   MASS TRANSFER WITH CHEMICAL REACTION

As we have mentioned previously, many applications of mass transfer are found in conjunction with chemical reactions. In such situations either the mass transfer or the chemical reaction rate can be the controlling factor, but more frequently neither is completely dominant. In this section we consider briefly the effects of chemical reaction on the apparent mass transfer rate. In general, treatment of simultaneous mass transfer and chemical reaction is extremely complicated, and so we will attempt only to give a qualitative insight into such effects.

As an example of how chemical reaction can affect mass transfer rates, consider the situation shown in Fig. 12.6-1a. On the left-hand side

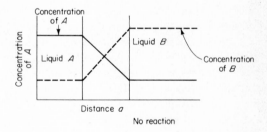

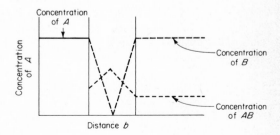

Fig. 12.6-1 Instantaneous irreversible reaction mass transfer in a stagnant film.

of a porous barrier, we maintain a nearly pure liquid, component A. On the right-hand side of the barrier, we maintain nearly pure liquid B. We assume both these solutions to be perfectly stirred so that no concentration gradients can appear in the liquid reservoirs. Concentration profiles in the porous barrier will look the same as if the barrier were a stagnant film of liquid. We will (for one-dimensional diffusion) get concentration profiles which will look as shown. Liquid A will continuously diffuse through the barrier in one direction and liquid B will diffuse in the opposite direction. Since the cross-sectional area normal to the diffusion path remains constant, and we are assuming that densities, diffusion coefficients, etc., do not vary, the profiles will be linear. Now let us consider the case where liquids A and B follow the reaction

$$A + B \rightarrow AB \tag{12.6-1}$$

and let us assume that this reaction is instantaneous and irreversible. This means that as soon as a molecule of A "sees" a molecule of B they immediately react. The concentration profiles will look something like those shown in Fig. 12.6-1b. A will still diffuse into the porous barrier and B will diffuse from the opposite side, but as soon as A and B meet each other product AB is formed. This product must now diffuse out of the porous barrier into the bulk liquids. A diffuses from left to right, B diffuses from right to left, and the product formed by their reaction diffuses out *both* to the left and to the right to the bulk solutions. The important thing to notice is that the net effect of the chemical reaction has been to steepen the concentration gradient for both component A and component B. Since everything else remains constant, this means that diffusion rates (mass transfer rates) will be increased in the presence of the chemical reaction.

It is possible to solve analytically the case where component A is in the form of a solid flat plate and component B is in the fluid flowing past this plate. We can compare the Sherwood number for this situation with the Sherwood number obtained from our correlation for mass transfer with flow over a flat plate, Eq. (12.5-3). If we write the mean Sherwood number for the case where the Schmidt number of component A and the Schmidt number of component B are identical we can write the mean Sherwood number as[1]

$$\text{Sh} = \left(1 + \frac{C_B}{C_{AS}}\right) 0.664 \, \text{Sc}_A^{1/3} \, \text{Re}_L^{1/2} \tag{12.6-2}$$

---

[1] We omit the development. See S. K. Friedlander and M. Litt, Diffusion Controlled Reaction in a Laminar Boundary Layer, *Chem. Eng. Sci.*, **7**:229 (1958).

where C_{AS} is the concentration of A in the fluid at the interface in moles per unit volume, and C_B is the free-stream concentration of B in the same units.

Notice that this is simply Eq. (12.5-3) multiplied by $(1 + C_B/C_{AS})$. Since the multiplying factor is always greater than 1, for this case chemical reaction *always* speeds up mass transfer. The treatment can be extended to differing Schmidt numbers, different stoichiometry, etc.

These are only very elementary examples selected from a vast class of problems which involve mass transfer with chemical reaction. These examples have assumed instantaneous chemical reaction, which meant that A and B were never present in solution at the same point. In general, with slower reaction, one does not have a sharp zone in which chemical reaction takes place; rather, this zone can be spread across some distance in space. Several cases are sketched in Fig. 12.6-2. The stoichiometric coefficients of A and B in the chemical reaction are also involved in the calculation. Finally, most reactions either absorb or give off energy, and

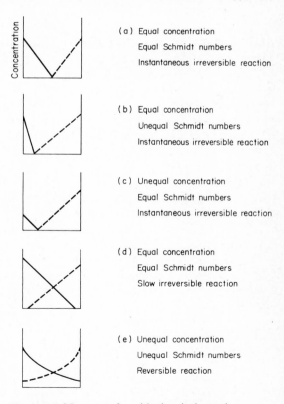

(a) Equal concentration
    Equal Schmidt numbers
    Instantaneous irreversible reaction

(b) Equal concentration
    Unequal Schmidt numbers
    Instantaneous irreversible reaction

(c) Unequal concentration
    Equal Schmidt numbers
    Instantaneous irreversible reaction

(d) Equal concentration
    Equal Schmidt numbers
    Slow irreversible reaction

(e) Unequal concentration
    Unequal Schmidt numbers
    Reversible reaction

Fig. 12.6-2 Mass transfer with chemical reaction.

the mass transfer problem with chemical reaction is frequently coupled to the heat transfer problem.

Example 12.6-1 Acidization of an oil well To increase the porosity and thus the production of oil wells several stimulation techniques are used. In limestone reservoirs, which usually contain natural fractures, acid is pumped into the formation to react with the limestone and increase porosity. Assume 0.2 mole fraction aqueous hydrochloric acid is pumped into a fractured limestone formation and assume the fractures are short and wide so that the model for flow developing over a flat plate is fairly applicable. The velocity in the fractures is said to be about 1 ft/sec. Using the flat-plate model, discuss the mass transfer situation.

Solution To include the effect of stoichiometry of the chemical reaction Eq. (12.6-2) can be modified as follows:[1]

$$\text{Sh} = \left(1 + \frac{aC_B}{bC_{AS}}\right) 0.664 \, \text{Sc}_A^{1/3} \, \text{Re}_L^{1/2}$$

where a and b are the stoichiometric coefficients in the reaction

$$aA + bB \rightarrow \text{products}$$

Here our reaction is

$$CaCO_3 + 2HCl \rightarrow CaCl_2 + H_2O + CO_2$$

and so $a = 1$, $b = 2$.

We next evaluate the factor $(1 + aC_B/bC_{AS})$. The solubility of $CaCO_3$ in water is about 4×10^{-6} mole fraction. Since we can relate mole fraction and molar density by

$$C_i = x_i C$$

where x is mole fraction and C is total molar density, if we assume the total molar density to be the same at the interface and under free-stream conditions we can write

$$1 + \frac{aC_B}{bC_{AS}} = 1 + \frac{ax_B C}{bx_{AS} C} = 1 + \frac{ax_B}{bx_{AS}}$$

[1] Friedlander and Litt, *ibid.* This solution assumes equal Schmidt numbers for A and B, which is not correct, but for this illustration the difference is unimportant.

Evaluating:

$$1 + \frac{ax_B}{bx_{AS}} = 1 + \frac{(1)(0.2)}{(2)(4 \times 10^{-6})} = 2.5 \times 10^4$$

The tremendous enhancement of the mass transfer rate is logical because in the absence of the reaction we would simply be trying to dissolve $CaCO_3$ in water. Since the solubility is so low, there would be an extremely small concentration gradient (between surface and free-stream conditions) to drive the mass transfer. With chemical reaction, the driving force is still low, but it is applied over a much shorter distance, since the reaction takes place very close to the surface. The *gradient*, therefore, is much larger.

This example is obviously extremely oversimplified, but it was chosen to give a qualitative feeling for mass transfer with chemical reaction. The interested reader is referred to the original journal article for details on heat effects, unequal Schmidt numbers, etc.

□ □ □

In general, the reader should remember that mass transfer will usually be enhanced by the presence of chemical reaction. Chemical reaction also furnishes a further benefit in operations such as absorption of a component of a gas phase into a liquid phase, since the chemical reaction permits the liquid phase to absorb more of the transferring component than would be possible in the absence of chemical reaction.

PROBLEMS

12.1 If k_G denotes a mass transfer coefficient based on a driving force of *partial pressure* $[N_A = k_G(\bar{p}_A - \bar{p}_{AS})]$, show that

$$k_x = k_G p$$

where p is the total pressure and x the mole fraction.

12.2 Develop the flux equation for mass transfer using a driving force based on *mass* concentration and a coefficient based on molar concentration.

12.3 The equilibrium equation for acetone distributed between air and water is

$$y_A = 1.13 \, x_A$$

The bulk concentration of acetone in the liquid is 0.015 and in the air 0.0.

$$k_x = 30 \text{ lb moles/hr ft}^2$$

$$k_y = 15 \text{ lb moles/hr ft}^2$$

Find $y_{AS}, x_{AS}, N_A, K_x, K_y$ graphically.

12.4 Solve Prob. 12.3 algebraically.

12.5 In a dilute mass transfer system the equilibrium data can be represented by

$$y = 15.3x$$

If $k_x = 5.0$ and $k_y = 1.0$, which resistance controls?

12.6 Mass is transferred from a cylindrical naphthalene bar with a diameter of $1/_2$ in. and length of 6 in. suspended in a 50°F airstream moving with a velocity of 40 ft/sec. Estimate the initial mass transfer coefficient.

12.7 Air is blowing through a pipe whose walls are wetted with a film of water. The temperature is uniform at 70°F. At a point where $y_A = 0.001$, calculate the flux of water evaporated from the pipe wall.

y_A = bulk concentration of HOH

$D = 4$ in.

$v = 50$ ft/sec

$\mathscr{D}_{AB} = 0.26 \text{ cm}^2/\text{sec}$

13
Mass Transfer Applications

This chapter parallels Chaps. 7 and 10. As in those two chapters, we have just finished investigating a process—here mass transfer—which is usually described for design purposes in terms of coefficients. In this chapter, as before, we show how one can obtain experimental data for determining coefficients.

It is almost always possible to make experimental measurements which will yield (although perhaps only very approximately) the overall mass transfer coefficient, just as experimental determination of the heat transfer coefficient and friction factor can be performed. Obviously, the expense in equipment, supplies, salaries, and time makes it preferable to *calculate* rather than experimentally determine those coefficients. In general, this is more difficult for mass transfer than for heat transfer or momentum transfer.

We present here only a sample of how mass transfer coefficients may be determined experimentally. The purpose is to show how such measurements, which underlie most design equations, are made—not to tabulate the many experimental techniques. We do not illustrate at all the attempts at a priori calculation of mass transfer coefficients because these techniques are more properly treated in more advanced courses in transport phenomena.

We also present selected examples of design of mass transfer equipment (specifically packed columns). An engineer must remember always that the specific techniques he learns must all ultimately merge in the design of a complete system, and further, that the elements of such a system, even though frequently assumed to be independent, are seldom so.

13.1 EXPERIMENTAL DETERMINATION OF MASS TRANSFER COEFFICIENTS[1]

Mass transfer coefficients (including interfacial area) can be determined by measurements in a packed column. Individual mass transfer coefficients may be determined in a wetted-wall column. In this section (as in the parallel chapters 7 and 10) we present an experimental determination of height of a liquid transfer unit and associated overall mass transfer coefficients (including interfacial area). The example following the experimental discussion presents the experimental determination of an individual gas phase mass transfer coefficient.

Carbon dioxide was absorbed from 20% mixture with nitrogen by continuous countercurrent contact with water at 28°C in a packed column. The experimental setup is shown in Fig. 13.1-1. The column was 3 $^5/_8$-in.-ID glass randomly packed to a height of 34.75 in. with $^3/_8$-in. glass Raschig rings. The gas inlet was inserted $^1/_2$ in. above the grid support and the liquid level was maintained at the grid support to minimize end effects. Water entered the column at a constant flow rate through 16 downcomers with a short liquid head to seal out air. Precision pressure regulation controlled the gas flow rate through a thermostat and into the column with flow rates determined with manometers. Continuous gas analyses were made by a thermal conductivity measurement. As shown in Fig. 13.1-1 gas samples were removed at various locations in the tower.

The experimental procedure follows. Water at the desired flow rate and temperature flowed through the tower, and gas at the desired flow rate and temperature flowed into the bottom of the tower. After a sufficient time passed to allow for steady-state operation, samples of the gas were obtained and their analysis determined. Each run was normally repeated three times to ensure reproducibility. The operating conditions for the particular run given here we listed in Table 13.1-1. The experimental gas composition at the five sampling points for run 43 are shown in

[1]M. I. Brittan and E. T. Woodburn, *A.I.Ch.E. J.*, **12**:541 (1966).

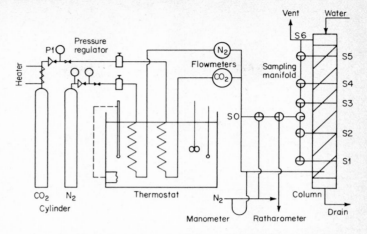

Fig. 13.1-1 Experimental setup for determination of mass transfer coefficients. [*M. I. Brittan and E. T. Woodburn, A.I.Ch.E. J.,* **12**:*541 (1966).*]

Fig. 13.1-2. Similar data were collected for all runs. Material balances of the solute were run on all streams including a determination of the free carbon dioxide content of the feed water.

The overall mass transfer data as represented by H_{OL} and $K_L a$ were determined from the analysis with

$$\int_{x_0}^{x_1} \frac{dx}{x^* - x} = \frac{K_L a \rho l}{L} = \frac{l}{H_{OL}} \tag{13.1-1}$$

This equation can be evaluated graphically or analytically if the Henry's law constant for the particular condition is known. The results of these

Table 13.1-1 Operating conditions for experiment of Fig. 13.1-1[a]

Run no.	L', lbm/hr ft²	G', lbm/hr ft²	p_r, mm Hg
22	9,195	9.58	623.5
30	7,356	5.75	623.5
43	5,517	5.76	624.1
73	6,436.5	9.56	622.0

[a] M. I. Brittan and E. T. Woodburn, *A.I.Ch. E. J.,* **12**:541 (1966).

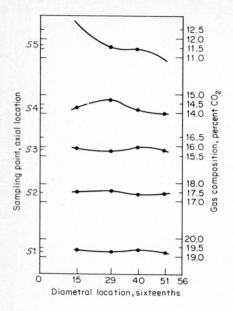

Fig. 13.1-2 Transverse concentration profiles, run 43. [*M. I. Brittan and E. T. Woodburn, A.I.Ch.E. J.*, **12**:*541 (1966)*.]

measurements and calculations are summarized in Table 13.1-2 for the four runs of Table 13.1-1. The mass transfer coefficient data of Table 13.1-2 can be represented by the usual power-function correlation, such that

$$K_L a = 3.12 \ L'^{0.31} \qquad\qquad (13.1\text{-}2)$$

We should note that at the top of the column there appears to be a profile in concentration with respect to radius but the profile flattens out halfway through the column. When we use Eq. (13.1-1) we assume a flat

Table 13.1-2   Mass transfer data from experiment of Fig. 13.1-1[a]

L', lbm/hr ft²	9,195		8,275.5		7,356		5,517	
	$H_{OL.}$ ft	$K_L a$, hr⁻¹	$H_{OL.}$ ft	$K_L a$, hr⁻¹	$H_{OL.}$ ft	$K_L a$, hr⁻¹	$H_{OL.}$ ft	$K_L a$, hr⁻¹
S0 to S6	2.48	59.7	2.46	54.2	2.24	52.8	1.88	47.4
S0 to S6	2.75	53.9	2.64	50.5	2.39	49.6	1.95	45.5
		10.6		7.3		6.5		4.1
S1 to S5	2.64	56.0	2.48	53.7	2.45	48.3	1.97	45.0

[a] M. I. Brittan and E. J. Woodburn, *A.I.Ch. J.*, **12**:541 (1966).

concentration profile; thus the values from the upper sampling points (54 and 55) would have to be corrected, probably using an axial diffusion model. The need for correction of such profiles is an active research area.[1]

Example 13.1-1 Experimental determination of a gas phase mass transfer coefficient in a wetted-wall column A manufacturing plant carries out a liquid phase reaction using dilute gaseous ammonia in air. It is suspected that the rate of formation of product is being controlled not by the rate of the liquid phase reaction, but by the slow rate of mass transfer of ammonia from the gas phase to the liquid. Preliminary calculations support this hypothesis. Before modifying the full-scale equipment to attempt to improve the gas-to-liquid transfer step, it has been decided to run a small-scale simulation and determine the gas phase mass transfer coefficient to verify if the hypothesis is well founded.

The plant-scale reaction is carried out in a wetted-wall column. The column to be run in the laboratory was designed to give dynamically similar performace using the dimensional analysis techniques discussed earlier with the result that a 1-in.-ID column 50 in. long was used in which ammonia in air was passed countercurrent to a film of acid flowing down the walls. The acid was chosen to react rapidly with the ammonia and therefore to make the mass transfer resistance in the liquid phase negligible, as we discussed in the section on mass transfer with chemical reaction.

Experimental results follow. Calculate the mean mass transfer coefficient for the gas phase based on a driving force in mole fraction.

Inlet gas phase ammonia mole fraction	0.01
Outlet gas phase ammonia mole fraction	0.0032
Inlet ammonia/air flow rates lb moles/hr	0.75

Solution We can write

$$N_A = k_y(y_A - y_{AS})_{av}$$

where

$$(y_A - y_{AS})_{av} = \frac{(y_A - y_{AS})_0 + (y_A - y_{AS})_1}{2}$$

or

$$\mathscr{W}_A = k_y A(y_A - y_{AS})_{av}$$

[1] M. I. Brittan and E. T. Woodburn, *A.I.Ch.E. J.*, **12**:541 (1966).

If the rate of chemical reaction with the acid is extremely high and liquid phase mixing is adequate, x_{AS} will be very nearly zero and therefore y_{AS} will be nearly zero. The interfacial area will be approximately that of the inside surface of the pipe if the film is thin. We calculate k_y as

$$k_y = \frac{\mathscr{W}_A}{A(y_A - y_{AS})_{av}} = \frac{\mathscr{W}_A}{\pi DL \ (y_{AO} + y_{A1})/2}$$

where by a mass balance on the air/NH_3 phase in the column

$$\mathscr{W}_A = \left(0.75 \ \frac{\text{total moles}}{\text{hr}}\right)\left(\frac{0.01 \text{ moles } NH_3}{\text{total moles}}\right)$$

$$- \left(0.75 \ \frac{\text{total moles}}{\text{hr}}\right)\left(\frac{0.99 \text{ moles air}}{\text{total moles}}\right)\left(\frac{0.0032 \text{ moles } NH_3}{1 - 0.0032 \text{ moles air}}\right)$$

$$= 0.0075 \ \frac{\text{moles } NH_3}{\text{hr}} - \frac{0.0024 \text{ moles } NH_3}{\text{hr}}$$

$$= 0.0051 \text{ moles } NH_3/\text{hr}$$

Therefore

$$k_y = \left(\frac{0.0051 \text{ moles } NH_3/\text{hr}}{\pi \ (^1/_{12} \text{ ft})(^{50}/_{12} \text{ ft})}\right) \Big/ \left(\frac{0.01 + 0.0032}{2} \ \frac{\text{moles } NH_3}{\text{total moles}}\right)$$

$$= 0.71 \text{ total mole/hr ft}^2$$

□ □ □

13.2 SELECTED APPLICATIONS

In this section we will be concerned with mass transfer application, in particular absorption calculation associated with packed towers. The objective as in similar chapters (Chaps. 4, 7, 10) is to present some practical examples of problems in this case of mass transfer. In order to illustrate these principles we must consider design of mass transfer columns. These designs are not detailed but will illustrate the principles discussed in Chaps. 11 and 12.

Although we will discuss the associated equipment in the natural gasoline plant, Fig. 1.3-1, we do not work any specific examples since the plant absorber operates at high pressure, equilibrium data are not readily available, and the calculations are very complicated. The simpler associated examples illustrate the principles.

Design of mass transfer columns

Despite the heading, we will not do detailed design calculations so much as determine the parameters necessary to proceed with these calculations. We consider specific types of mass transfer equipment which are intended as examples of more general classes of equipment.

Many large-scale mass transfer operations in industry are concerned with the transfer of a component between two essentially immiscible phases. A classic illustration is gas absorption and desorption in dilute systems. For example, small amounts of SO_2 may be removed from air by absorption at nearly atmospheric pressure in an amine of low vapor pressure. The low vapor pressure of the amine means that little amine is transferred from liquid to gas; the low operating pressure ensures that little air dissolves in the amine; therefore, we transfer essentially only SO_2.

This sort of operation is frequently carried out in a column or tower, with the gas entering at the bottom and flowing countercurrent to the liquid, which enters at the top. If mass transfer rate is the only consideration (that is, assuming that there is not some sort of chemical reaction going on which liberates large amounts of heat which must be removed from the system), the tower is filled with some sort of device to increase the interfacial area between the gas and liquid and, therefore, to promote better mass transfer. These devices may take the form of trays of various types: bubble cap, sieve, jet, grid, etc. Alternatively, the tower may simply be packed with small solids of varying shapes. Examples of various types of plates and of various types of packing are shown in references.[1]

The choice of type of device to use for tower internals may be governed by a number of considerations. For example, if the pressure drop through the tower is large enough that compression cost becomes significant, a packed column may be preferable to a plate-type column because of the decreased compression cost. Packed columns, however, have a tendency to channel, and also suffer from the disadvantage that (if they are used with liquids that form deposits not readily removable by chemical cleaning) manual cleaning is extremely difficult. The weight of the packing for a packed column can sometimes become a limiting consideration—sometimes it becomes large enough to crush the packing at the bottom unless intermediate supports are placed in the column.

If reasonably rapid response for control purposes is required, packed towers have the advantage of less liquid holdup than plate-type towers. Obviously, the cost of the tower internals is also a consideration. Many of the decisions regarding which type of tower internals to use are based on qualitative or rule-of-thumb criteria. Here we wish to stress the

[1] See also R. H. Perry, C. H. Chilton, and S. D. Kirkpatrick (eds.), "Chemical Engineers' Handbook," 4th ed., pp. 14–15, 18–26, McGraw-Hill Book Company, New York, 1963.

important steps in tower design rather than confusing the reader with the myriad of details in the complete process, and so in what follows we will regard the choice of type of tower internals as having already been determined.

Design problems for mass transfer equipment come in many different forms; however, they all embody the same essential features. We will now introduce the method of calculation by considering a typical example. The simplified form of the usual gas absorption problem might be as follows. Given:

1. A gas stream entering at a given rate $\mathcal{W}_G$ containing some given fraction of component A, y_{A0}. (We will use the subscript 0 to refer to the concentrated end of the tower and subscript 1 for the dilute end.)
2. A specified exit gas concentration of component A, y_{A1} (alternatively, the percent or fraction to be removed may be given).
3. A specified inlet fraction of A in the scrubbing liquid, x_{A1}.
4. A specified size and type of packing.

Calculate the height and diameter of tower required to achieve the separation. (In practice, of course, it is usually also the job of the engineer to select the best type of packing, to decide on the allowable exit concentration, to optimize the economics, etc. We are not interested in this much detail for our objectives here.)

A typical continuous contacting device is sketched in Fig. 13.2-1. We will discuss countercurrent gas absorption as our example; treatment for other cases is directly analogous; that is, the equations developed here hold for continuous contacting of immiscible phases, whether this is done

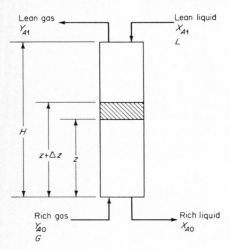

Fig. 13.2-1 Countercurrent gas absorber.

in a spray column, packed column, plate column, wetted-wall column, or any other type of device which contacts immiscible phases.

The gas enters at a molar velocity of G (moles per hour per foot-squared empty tower) and contains y_{A0} mole fraction component A. This mole fraction is assumed to be small so that our assumptions with regard to the mass transfer coefficient will be satisfied. The carrier gas B is assumed insoluble in the liquid and the liquid is assumed to have a negligible vapor pressure. The *cross-sectional* area of the *empty* tower is denoted by A_{xs}.

We have two separate tasks before us: first, to determine the height of the tower, and second, to determine its diameter. We will consider these tasks in the reverse of the order listed.

Determination of liquid-to-gas ratio

The problem of sizing the tower diameter is purely one of fluid mechanics; crudely, it is simply necessary to have sufficient room for the liquid and gas. To solve this fluid mechanics problem we need to know both the gas and the liquid rates. In our problem the gas rate is given; we must determine the liquid rate.

We can visualize the situation somewhat more easily by using a graph. In Fig. 13.2-2, we show the *equilibrium* curve for our system. (This must be given or obtained through literature values, experiment, or calculation, a problem not in transfer operations, but in thermodynamics.[1]) Points on this curve, as the name implies, are phrases in *equilibrium*. We are concerned here with the dilute case and so we show only the lower portion of the curve.

Note that we can plot from the given information a point (x_{A1}, y_{A1}) representing the conditions at the *top* of the tower. Since we are transferring *from the gas to the liquid* the concentration in the gas phase must be *larger* than the concentration that would be in equilibrium with the liquid for the driving force to be in the correct direction. (If it were *lower*, we would transfer material from the liquid to the gas, and our problem would be one in *desorption* or *stripping of a liquid*.) Note that this is true in Fig. 13.2-2. This is true not only for the *top* end of the tower, but for the gas and liquid phases at *any* point in the tower.

Now let us ask ourselves *how much liquid* is required to absorb the A which is transferred. Obviously this is not a unique amount since we can absorb the A in a little or a lot of liquid. There is, however, a *lower limit* to the amount of liquid we can use. We can see this by looking at our driving force for mass transfer. At any point in the tower, we find the in-

[1]See K. Denbigh, "The Principles of Chemical Equilibrium," 2d ed., Cambridge University Press, New York, 1966.

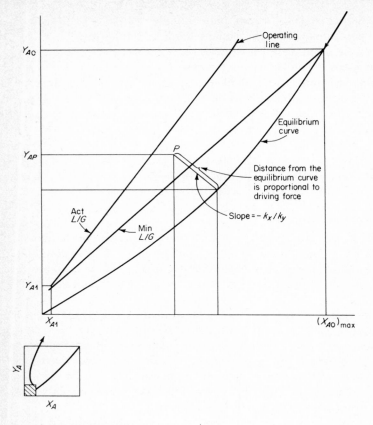

Fig. 13.2-2 Gas absorption.

terfacial composition by drawing a straight line through (x_A, y_A) with slope $-k_x/k_y$ and taking its intersection with the equilibrium curve, as shown for an arbitrary point P. The length of this diagonal line in Fig. 13.2-2 is proportional to the driving force, since for any point in the tower we choose, the length of a corresponding line is proportional to $y_A - y_{AS}$ and $x_A - x_{AS}$, the driving forces in the system. We cannot ever *cross* the equilibrium line by simply mixing two phases, as in a flask, however, because when we reach the equilibrium line our driving force goes to zero and mass transfer stops. (Our process has no appreciable "inertia" to carry it past equilibrium.) Of course, as the driving force becomes smaller and smaller, the mass flux also becomes smaller and smaller, and so more and more surface area is required to transfer mass at the same rate.

What does this say about our minimum liquid rate? It says that *even if we had a tower containing infinite interfacial area*, i.e., infinite height, *if we maintain the required outlet gas concentration* we can at *best* reduce our driving force to something which approaches zero at one or more points in the tower, but we *cannot* cross the equilibrium line. If we *attempt* to cross the equilibrium curve, that is, if we *reach* it, mass transfer stops at that point.

How does this affect the rate of the liquid we use? Assuming constant L and G, an overall mass balance on component A for the tower shows that

$$LA_{xs}x_{A1} + GA_{xs}y_{A0} = LA_{xs}x_{A0} + GA_{xs}y_{A1} \qquad (13.2\text{-}1)$$

For constant cross-sectional area (the usual case):

$$G(y_{A0} - y_{A1}) = L(x_{A0} - x_{A1}) \qquad (13.2\text{-}2)$$

Since $GA_{xs} = \mathscr{W}_G$, and y_{A0}, y_{A1}, and x_{A1} are specified, to make L as small as possible means that we want $x_{A0} - x_{A1}$ as large as possible, that is, x_{A0} as large as possible. This corresponds to approaching equilibrium. In other words, the minimum liquid rate will be found where we *first* reach equilibrium at *some point* in a tower of infinite height as we decrease liquid rate, *assuming we continue to transfer the same amount of mass.*

This means that we need the locus of composition of the liquid phase for any given liquid rate. We can find this locus by writing a material balance for a system boundary around the top of the tower and through an arbitrary intermediate point at height z. Neglecting radial composition variation:

$$\begin{array}{cc} \text{Input} & \text{Output} \\ Lx_{A1} + Gy_A = & Lx_A + Gy_{A1} \end{array} \qquad (13.2\text{-}3)$$

or

$$y_A = \frac{L}{G} x_A + \left(y_{A1} - \frac{L}{G} x_{A1}\right) \qquad (13.2\text{-}4)$$

This is the equation of a line (called the *operating* line) which gives the locus of liquid phase compositions as a function of gas phase composition or vice versa. In other words, it relates the compositions of passing phases. The line passes through (x_{A1}, y_{A1}). It also passes through (x_{A0}, y_{A0}) as is easily demonstrated. For L/G constant, the line is straight. For our assumptions of immiscibility of carrier phases and dilute systems, L and G will be very nearly constant and we will take the line to be straight.

For the moment we have assumed that $\mathscr{W}_G$, x_{A1}, and y_{A1} are given.

For any value of $\mathscr{W}_L$, therefore, we can plot the operating line by drawing it through (x_{A1}, y_{A1}) with slope

$$\frac{\mathscr{W}_L}{\mathscr{W}_G} = \frac{LA_{xs}}{GA_{xs}} = \frac{L}{G} \tag{13.2-5}$$

Here we consider only problems with L and G essentially constant. If L and G vary through the tower, we must write Eq. (13.2-1) as

$$L_1 A_{xs} x_{A1} + G_0 A_{xs} y_{A0} = L_0 A_{xs} x_{A0} + G_1 A_{xs} y_{A1} \tag{13.2-6}$$

The criterion for minimum liquid rate is the same, but it is not so easily seen as for the case of constant L and G. The operating line, Eq. (13.2-4), would become

$$y_A = \frac{L}{G} x_A + \left(\frac{G_1}{G} y_{A1} - \frac{L_1}{G} x_{A1} \right) \tag{13.2-7}$$

and its slope would vary throughout the tower.

We now ask ourselves: "If we start with a straight line of large slope through (x_{A1}, y_{A1}) and decrease the slope (that is, decrease L), when will we first reach an equilibrium condition within the tower?" Since the tower is bounded by concentrations y_{A0} and y_{A1}, obviously the line labeled $(L/G)_{\min}$ is the first to satisfy this requirement. We can measure the slope of this line on the graph and calculate W_L as

$$(LA_{xs})_{\min} = (\mathscr{W}_L)_{\min} = GA_{xs} \left(\frac{L}{G} \right)_{\min} = \mathscr{W}_G \left(\frac{L}{G} \right)_{\min} \tag{13.2-8}$$

Note that it is not necessary for equilibrium lines of different curvature that the first intersection be at the *end* of the tower.

Since to operate the tower with this liquid rate and satisfy y_{A1} we would require an infinite area for mass transfer, we must actually operate at a somewhat higher liquid rate. The actual liquid rate is obtained by balancing the capital costs and expenses involved; here, we will assume that we will use $\mathscr{W}_L = 1.5 (\mathscr{W}_L)_{\min}$, a not untypical figure. We then would obtain an actual operating line as shown. Since the operating line equation is valid at the ends of the column as well as within the column, x_{A0} is located as shown. (The value of x_{A0} could also be obtained analytically via a mass balance around the entire tower on component A.)

We now know both liquid and gas rates and can proceed to size the tower cross-sectional area A_{xs}. If our problem were one of desorption rather than absorption, all would proceed as above except the operating line would fall *below* the equilibrium curve and we would look for $(L/G)_{\max}$, not $(L/G)_{\min}$.

Calculation of tower diameter

As we mentioned earlier, the tower must be of sufficient diameter to accommodate the gas and liquid. Let us now be more precise in our statements.

    If we have a packed column operating at a given liquid rate and we gradually increase the gas rate, we find that after a certain point the gas rate is so high that the drag on the liquid is sufficient to keep the liquid from flowing freely down the column. Liquid begins to accumulate and tends to block the entire cross section for flow (so-called loading). This, of course, both increases the pressure drop and prevents the packing from mixing the gas and liquid effectively, and ultimately some liquid is even carried back up the column. This undesirable condition, known as *flooding*, occurs fairly abruptly and the *superficial gas* velocity at which it occurs is called the *flooding velocity*. We wish to design towers to avoid reaching the flooding velocity, and so we usually try to operate at perhaps half the flooding velocity by making the tower diameter sufficiently large.

    Flooding velocities for countercurrent gas absorption towers can be calculated using Fig. 13.2-3 in conjunction with Table 13.2-1. Note that

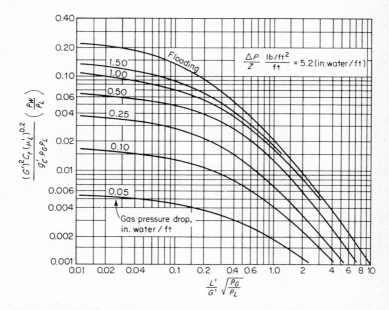

Fig. 13.2-3 Flooding and pressure drop in packed towers. (*Norton Company, formerly U.S. Stoneware.*)

Table 13.2-1 Characteristics of random packings[a]

Packing	Nominal size, in.										
	$1/4$	$3/8$	$1/2$	$5/8$	$3/4$	1	$1 1/4$	$1 1/2$	2	3	$3 1/2$
Raschig rings: ceramic[b]											
C_f	1,000	750	640	380	255	160	125	95	65	37	
ϵ	0.73	0.68	0.63	0.68	0.73	0.73	0.74	0.71	0.74	0.78	
a_p	240	155	111	100	80	58	45	38	28	19	
Metal $1/32$**-in. wall**											
C_f	700		300	258	185	115					
ϵ	0.69		0.84		0.88	0.92					
a_p	236		128		83.5	62.7					
$1/16$**-in. wall**											
C_f			340	290	230	145	110	82	57	37	
ϵ			0.73		0.78	0.85	0.87	0.90	0.92	0.95	
a_p			118		71.8	56.7	49.3	41.2	31.4	20.6	
Pall rings: plastic											
C_f				97		52		32	25		16
ϵ				0.88		0.90		0.905	0.91		
a_p				110		63.0		39	31		23.4
metal											
C_f				71		48		28	20		
ϵ				0.902		0.938		0.953	0.964		
a_p				131.2		66.3		48.1	36.6		
Intalox saddles, ceramic:											
C_f	600		265			130	98		52	40	
ϵ	0.75		0.78			0.77	0.775		0.81	0.79	
a_p	300		190			102	78		59.5	36	
Berl saddles, ceramic:											
C_f	900		380			170	110		65	45	
ϵ	0.60		0.63			0.66	0.69		0.75	0.72	
a_p	274		142			82	76		44	32	

[a]Table from the Norton Co., formerly the United States Stoneware Company. Data are for wet-dumped packing in 16- and 30-in.-ID towers.
[b]Nominal size and wall thickness, in inches, for ceramic Raschig rings are, respectively, $1/4$, $1/32$; $3/8$, $1/16$; $1/2$ to $3/4$, $3/32$; 1, $1/8$; $1 1/4$ and $1 1/2$, $3/16$; 2, $1/4$; 3, $3/8$.

the proper units must be used as the plot is not dimensionless.
    The procedure is as follows:
1. Calculate the abscissa of Fig. 13.2-3. (Note: L' and G' are lb$m$/hr ft^2.)
2. Read the ordinate. (μ' in cp.)
3. From Table 13.2-1 determine C_f for the appropriate packing.
4. Calculate G'.
5. Calculate A_{xs}.

Calculation of tower height

Calculation of tower height may be done via either H_G, H_L, H_{OG}, or H_{OL}; the procedures are completely parallel. We will use H_G below.

We must furnish sufficient interfacial area in the tower that the required amount of mass may be transferred with the existing driving force. For a given interfacial area per unit volume and a given tower cross section, this is equivalent to specifying the height.

Consider a material balance for A on the *gas phase* for a disk of some thickness Δz as shown in Fig. 13.2-1:

$$A_{xs} G y_A|_z = A_{xs} G y_A|_{z+\Delta z} + k_y a A_{xs} \, \Delta z \, (y_A - y_{AS}) \qquad (13.2\text{-}9)$$

Rearranging

$$\frac{G y_A|_z - G y_A|_{z+\Delta z}}{\Delta z} = k_y a (y_A - y_{AS}) \qquad (13.2\text{-}10)$$

Taking the limit as $\Delta z \to 0$

$$-\frac{dG y_A}{dz} = k_y a (y_A - y_{AS}) \qquad (13.2\text{-}11)$$

but since G is constant under our assumptions

$$\frac{-dy_A}{dz} = \frac{k_y a}{G} \, (y_A - y_{AS}) = \frac{y_A - y_{AS}}{H_G} \qquad (13.2\text{-}12)$$

Now we can observe from the last chapter that H_G is proportional to G, L, and Sc, all of which are constant for cases we consider here. We can, therefore, write, since H_G is a constant,

$$\int_0^Z dz = H_G \int_{y_{A1}}^{y_{A0}} \frac{dy_A}{y_A - y_{AS}} \qquad (13.2\text{-}13)$$

$$Z = H_G \int_{y_{A1}}^{y_{A0}} \frac{dy_A}{y_A - y_{AS}} = H_G n_G \qquad (13.2\text{-}14)$$

where Z is the tower height. The integral on the right may be integrated graphically by calculating H_G and H_L and observing that

$$\frac{H_G}{H_L} = \frac{G}{k_y a} \frac{k_x a}{L} = \frac{k_x}{k_y} \frac{G}{L} \qquad (13.2\text{-}15)$$

Since we know L and G

$$\frac{-k_x}{k_y} = -\frac{L}{G}\frac{H_G}{H_L} \qquad (13.2\text{-}16)$$

This permits us to get y_{AS} as outlined above, given some values of y_A, and we can plot $1/y_A - y_{AS}$ versus y_A and take the area under the curve between y_{A1} and y_{A0}. The value of this integral is called the *number of gas phase transfer units* n_G. We then multiply by H_G to obtain the height of the tower. (Note that n_G is dimensionless.) If we had used H_L, we would write n_L for the integral; for H_{0G}, n_{0G}; for H_{0L}, n_{0L}. The latter two are called the number of overall transfer units. (The driving force used with overall transfer units is $z_A - z_A^*$.)

Example 13.2-1   Calculation of interfacial concentration   To evaluate the integral which determines the number of gas phase or liquid phase transfer units, n_G or n_L, we must know the interfacial composition of the liquid and gas. Let us now consider an example of how one finds these compositions when the height of a gas phase transfer unit and the height of a liquid phase transfer unit are known. (These must, of course, be determined either from experimental data or from design equations such as we have already considered.)

Problem statement   A packed column is used as shown to desorb component A from low concentration in a liquid into a stream of carrier gas. If $H_G = 2$ ft and $H_L = 1$ ft, at the point in the tower where $y_A = 0.01$, what is x_{AS}? *Equilibrium data:* $y = 10x$. (See Fig. 13.2-4.)

Solution   The most easily visualized way of solving this problem is a graphical one. We know that we can write

(a) $$\frac{-k_x}{k_y} = \frac{y_A - y_{AS}}{x_A - x_{AS}}$$

and obtain k_x/k_y from

$$\frac{k_x}{k_y} = \frac{G}{k_y a}\frac{k_x a}{L}\frac{L}{G} = \frac{H_G}{H_L}\frac{L}{G} = \frac{2}{1}\,5 = 10$$

where we have been given H_G, H_L, and L/G. The straight line (a) goes through the point (x_{AS}, y_{AS}) with slope of -10. We know that x_{AS}, y_{AS} also lies on the equilibrium line, which we are given as

(b) $$y_{AS} = 10x_{AS}$$

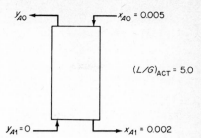

Fig. 13.2-4 Sketch for Example 13.2-1.

Therefore, we can find the composition desired, x_{As}, as the intersection of lines (a) and (b). We can easily plot equation (b) because it must go through the origin and the slope is 10. The question is how to plot equation (a). We know that the slope of this line is $-k_x/k_y = -10$ but we need a point on the line to plot it.

Examination of equation (a) shows that the line also goes through (x_A, y_A). We know, however, that the operating line also passes through this point, and in addition we know that the ordinate of one point on the line (a) is $y_A = 0.01$. We can, therefore, find the interface composition by

1. Plotting the equilibrium line
2. Plotting the operating line
3. Plotting equation (a) through the point on the operating line where $y_A = 0.01$
4. Determining x_{As} from the intersection of equation (a) with the equilibrium line

See Fig. 13.2-5. Plot the equation of the equilibrium line, which

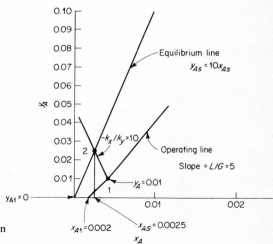

Fig. 13.2-5 Graphical calculation for Example 13.2-1.

passes through (0,0) with a slope of 10. Next the operating line is plotted by noting that it passes through (0.002, 0) (conditions at bottom of tower) with a slope L/G of 5.0. We next locate the point on the operating line for which $y_A = 0.01$, as shown. Through this point, which is also on the line represented by equation (a), we plot equation (a) by drawing a line with slope $-k_x/k_y = -10$. The abscissa of the point where line (a) strikes the equilibrium line is the desired answer $x_{AS} = 0.0025$.

The same problem may, of course, be solved algebraically since our graphical construction corresponds to the algebraic operations which solve the following equations simultaneously:

1. The equilibrium line
   $$y_{AS} = 10x_{AS}$$
2. Equation (a)
   $$-10 = \frac{y_A - y_{As}}{x_A - x_{As}}$$
3. The operating line, Eq. (13.2-4)
   $$y_A = 5x_A + [0 - (5)(0.002)]$$
   or
   $$y_A = 5x_A - 0.01$$
   and
4. The equation
   $$y_A = 0.01$$

The above are four equations in four unknowns. We may solve them in the same sequence we did graphically.

First, substitute the fourth equation into the operating line equation to give

$$0.01 = 5x_A - 0.01$$

or

$$x_A = 0.004$$

This corresponds to locating point 1. We substitute this result into the second equation, equation (a), to give

$$-10 = \frac{0.01 - y_{As}}{0.004 - x_{As}}$$

and solve simultaneously with the equation of the equilibrium line by sub-

MASS TRANSFER APPLICATIONS

stituting for y_{As}, yielding

$$-10 = \frac{0.01 - 10x_{As}}{0.004 - x_{As}}$$

Solving:

$$x_{As} = 0.0025$$

As before—this corresponds to locating point 2.

This illustrates the fact that finding interfacial compositions when heights of transfer units are known rather than mass transfer coefficients is essentially the same procedure.

□ □ □

Example 13.2-2 Analytical determination of number of transfer units— straight operating and equilibrium lines Until this point we have suggested evaluating the integral which determines the number of transfer units by graphical means. Obviously, if we can obtain analytical equations which describe the operating and equilibrium lines, and if these equations are sufficiently well behaved, we can perform the integration analytically. (If they are not sufficiently well behaved, we could, of course, integrate the equations using a computer, which is equivalent to doing them graphically.) We will illustrate in the example below the development of an explicit equation for calculation of the number of transfer units for a particularly simple case, that is, the case of both a straight operating line and a straight equilibrium line. The operating line will frequently approach straight-line behavior because in very dilute systems which satisfy the conditions we have set forth in this text, little enough mass is transferred that for all practical purposes L and G will remain essentially constant throughout the column, and therefore the operating line will remain straight. If in addition the solutions with which we deal are sufficiently ideal we know that an equilibrium relationship of the form of Henry's law or Raoult's law will hold; these are straight-line equilibrium relationships. This may or may not be the case even in dilute solutions depending on the interaction of the solute and solvent molecules.

Problem statement For the specific case of straight operating and equilibrium lines, develop an analytical equation for the number of transfer units explicit in the terminal conditions.

Solution We wish to develop an analytical expression for the integral.

Now $y_A - y_A^*$ is the difference between two straight lines and therefore $y_A - y_A^*$ is a linear function of x_A. Since x_A is a linear function of y_A, $y_A - y_A^*$ is also a linear function of y_A. Thus

$$\frac{d(y_A - y_A^*)}{dy_A} = \text{constant} \qquad \text{(slope of the line } y_A - y_A^* \text{ versus } y_A \text{)}$$

Since this is constant the differential may be replaced by the difference

$$\frac{d(y_A - y_A^*)}{dy_A} = \frac{(y_{A0} - y_{A0}^*) - (y_{A1} - y_{A1}^*)}{y_{A0} - y_{A1}}$$

Now substitute and integrate

$$n_{OG} = \frac{y_{A0} - y_{A1}}{(y_{A0} - y_{A0}^*) - (y_{A1} - y_{A1}^*)} \int_{y_{A1}}^{y_{A0}} \frac{d(y_A - y_A^*)}{y_A - y_A^*}$$

$$n_{OG} = \frac{y_{A0} - y_{A1}}{(y_A - y_A^*)_{\text{lm}}}$$

where

$$(y_A - y_A^*)_{\text{lm}} = \frac{(y_A - y_A^*)_0 - (y_A - y_A^*)_1}{\ln\left[(y_A - y_A^*)_0 / (y_A - y_A^*)_1\right]}$$

There is another approach to the above derivation; that is, rather than substituting for the independent variable in the integration, substitute for y_{As} in terms of the equilibrium relationship and the operating line equation. This procedure, however, leads to the same results and is somewhat more cumbersome from an algebraic point of view.[1]

□ □ □

This is only one example of a number of cases which can be integrated analytically. For example, there are many equations which describe curved equilibrium lines which will lead to expressions that can be evaluated analytically. We also can have instances in which the operating line is curved and can also be integrated analytically. Our purpose here is not to provide an exhaustive tabulation of all the solutions but rather to point out the fact that one should not go to great lengths to obtain

[1] A. P. Colburn, *Trans. A.I.Ch.E.* **35**:211 (1934).

numerical integrations without first checking to see if analytical integration is possible.

Example 13.2-3 Effect of change of L/G on outlet composition We have thus far considered the problem only of column *design*, that is, how one constructs a column to perform a given separation. Unfortunately, perhaps the more frequent problem which concerns the engineer in production or development in industry is that of how an existing column will perform under a changed set of operating conditions. This problem arises, for example, in cases where an existing column is to be converted for use in production of a different product. In this example, we attempt to give some insight into the way this changes the overall calculations in the problem.

Problem statement If the L/G in Example 13.2-1 is changed to 0.1, assuming the same input rate of gas, and the same inlet gas concentration, what is the outlet gas concentration?

Solution From Eq. (13.2-10)

$$\frac{H_{G3}}{H_{G2}} = \left(\frac{G_3}{G_2}\right)^\beta \left(\frac{L_3}{L_2}\right)^\gamma$$

where the subscript 3 refers to Example 13.2-1 and the subscript 2 refers to Example 13.2-3 but $(G_3/G_2)^\beta = 1$ since the gas rate is unchanged.
    From Eq. (13.2-11)

$$\frac{H_{L3}}{H_{L2}} = \left(\frac{L_3}{L_2}\right)^n$$

Therefore

$$H_{G3} = 1.17 \underbrace{\left(\frac{9.1}{3.95}\right)^{-0.47}}_{0.675} = 0.79 \text{ ft}$$

$$H_{L3} = 0.92 \underbrace{\left(\frac{9.1}{3.95}\right)^{0.46}}_{1.47} = 1.36 \text{ ft}$$

$$H_{OG} = H_G + \frac{mG}{L} H_L = 0.79 + \underbrace{\frac{1.5}{9.1}}_{0.21} (1.36) = 1.0 \text{ ft}$$

$$n_{OG} = \frac{Z}{H_{OG}} = \frac{3.16}{1} = 3.16 = \frac{y_{A0} - y_{A1}}{(y_A - y_A^*)_{lm}}$$

$$3.16 = \frac{0.05 - y_{A1}}{\dfrac{(0.05 - 0.003) - y_{A1}}{\ln\left[(0.05 - 0.003)/y_{A1}\right]}}$$

$$3.16 = \frac{0.05 - y_{A1}}{\dfrac{0.047 - y_{A1}}{\ln\left(0.047/y_{A1}\right)}}$$

By trial and error $y_{A1} = 0.002$.

□ □ □

As is obvious from the procedure above, the calculation of the performance of an existing column under a new set of operating conditions is somewhat more complicated than that of designing a new column because of the trial and error procedure involved.

In general, if we consider the problem of changing L/G on an existing tower we first must account for the change in height of a transfer unit. Since the product of H with n gives the tower height Z, which is fixed, n is therefore fixed. This fixes the value of an integral, and so the problem is one of solving a definite integral with an unknown limit, given its value. Note that the presence of the driving force term in the denominator of the integral prevents the operating line from reaching the equilibrium line for any finite column because this would imply an infinite n since the integral would go to infinity.

In the following example we use data from Table 1.3-1; however, the tower design is not necessarily the design that would be used if the plant were to be modified. The purpose of the example is to apply the mass transfer principles discussed earlier. In this particular instance because of the high pressure requirements for the plant we have chosen to consider a separate application. Following the example, some comments are made as to how a plant modification would proceed. Calculation of such a modification is outside the scope of this text.

Example 13.2-4 Absorber calculation Use the data from the flowsheet of Table 1.3-1 for stream 5 (but at a pressure of 100 psia) and do a rough design calculation to size a packed absorption column to remove C_3 hydrocarbons from a portion of stream 5, the overhead vapor from the existing absorber. (Note that the existing absorber essentially absorbs *all* the hydrocarbons in the absorber oil followed by flash separations.)

The absorber oil is approximately a kerosene in composition. The demand for the new product stream is to be about 10,000 MSCF/day. Design the absorber to perform this separation assuming that 90% of the C_3 fraction is to be recovered and that $L/G = 1.5\,(L/G)_{min}$. Use 1-in. Raschig rings for packing.

Solution Since the C_3 hydrocarbons are a relatively minor constituent of stream 5, we will apply our methods for dilute systems. The equilibrium data for this system are shown in Fig. 13.2-6.

The entering mole fraction of C_3 can be computed from the flowsheet as

$$y_{A0} = \frac{47.33 \text{ moles/hr}}{10,919 \text{ moles/hr}} = 0.00433$$

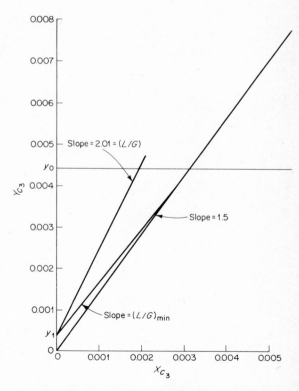

Fig. 13.2-6 Graphical calculation for Example 13.2-4.

Since the system is so dilute, recovery of 90% of the C_3 implies that

$$y_{A1} = \frac{(0.1)(0.00433)(10,919)}{(10,919) - (0.9)(0.00433)(10,919)}$$

$$\cong (0.1)(0.00433) = 0.000433$$

Assume that $x_{A1} = 0$. We plot this point on Fig. 13.2-6 and find $(L/G)_{min} = 1.34$. The operating L/G is then

$$\frac{L}{G} = (1.5)(1.34) = 2.01$$

and the operating line may be drawn as shown.

The number of transfer units is obtained from

$$n_{OG} = \frac{y_{A0} - y_{A1}}{(y_{A0} - y_A^*)_{lm}}$$

$$n_{OG} = \frac{0.00433 - 0.000433}{\frac{(0.00433 - 0.00291) - (0.000433 - 0)}{\ln[(0.00433 - 0.00291)/(0.000433 - 0)]}}$$

$$n_{OG} = 4.69$$

We now must calculate H_{OG}. Since

$$H_{OG} = \frac{G}{K_y a}$$

$$H_{OG} = \frac{G}{a[1/(1/k_y + m/k_x)]} = \frac{G}{1/(1/k_y a + m/k_x a)}$$

$$= \frac{G}{1/[(1/G)H_G + (m/L)H_L]}$$

$$= H_G + \frac{mG}{L} H_L$$

and we can get H_G and H_L from our design equations in Chap. 12. In order to use these equations

$$H_G = \alpha G'^\beta L'^\gamma Sc^{0.5} \qquad H_L = \phi\left(\frac{L'}{\mu}\right)^\eta Sc^{0.5}$$

we must know L and G to select the proper values of the parameters. We therefore do a flooding calculation to size tower diameter first.

To get the flooding velocity we use Fig. 13.2-3. The densities are about 52 lb$m$/ft^3 for the absorption oil and about 0.67 lb$m$/ft^3 for the gas phase at column conditions.

$$G' = \left(10{,}000{,}000 \frac{\text{ft}^3}{\text{day}}\right)\left(\frac{\text{mole}}{359 \text{ ft}^3}\right)\left(\frac{16.8 \text{ lb}m}{\text{mole}}\right)\left(\frac{\text{day}}{24 \text{ hr}}\right) = 19{,}500 \text{ lb}m/\text{hr}$$

$$L' = \left(10{,}000{,}000 \frac{\text{ft}^3}{\text{day}}\right)\left(\frac{\text{mole gas}}{359 \text{ ft}^3 \text{ gas}}\right)\left(\frac{2.01 \text{ mole liquid}}{\text{mole gas}}\right)\left(\frac{135 \text{ lb}m \text{ oil}}{\text{mole}}\right)$$

$$\frac{L'}{G'}\sqrt{\frac{\rho_G}{\rho_L}} = \frac{315{,}000}{19{,}000}\sqrt{\frac{0.67}{52}} = 1.83$$

From Fig. 13.2-3

$$\frac{G'^2 C_f \mu'_L{}^{0.2}}{g_c \rho_G \rho_L}\frac{\rho_w}{\rho_L} = 0.012$$

From Table 13.2-1, assuming wet-packed stoneware Raschig rings,

$$C_f = 160$$

The liquid viscosity is about 20 cp.

$$G'^2 = \frac{0.012 g_c \rho_G \rho_L}{C_f \mu'_L{}^{0.2}}\frac{\rho_L}{\rho_w} = \left[(0.012)(32.2)\frac{\text{lb}m \text{ ft}}{\text{lb}f \text{ sec}^2}\right]\left[\frac{(3{,}600)^2 \text{ sec}^2}{\text{hr}^2}\right]$$

$$\frac{52}{62.4}\left(\frac{0.67 \text{ lb}m}{\text{ft}^3}\right)\left(\frac{\text{lb}m}{\text{ft}^3}\right)\frac{1.0}{(160)(20 \text{ cp})^{0.2}}$$

$$G'^2 = 500{,}000$$

$$G' = 707 \text{ lb}m/\text{hr ft}^2 \qquad \text{at flooding}$$

Let us arbitrarily choose to run at gas velocity one-third of flooding. Therefore

$$G' = \frac{w_G}{A} = 236 \text{ lb}m/\text{hr ft}^2$$

and

$$L' = 236\frac{315{,}000}{19{,}500} = 3{,}800 \text{ lb}m/\text{hr ft}^2$$

and column diameter is

$$A = \frac{\pi D^2}{4} = \frac{w_G}{G'} = \frac{19{,}500}{236}$$

$$D^2 = 105 \text{ ft}^2$$

$$D = 10.25 \text{ ft}$$

We can now calculate H_L and H_G

$$H_G = \alpha \, G'^{\beta} L'^{\gamma} \, Sc^{0.5}$$

where

$$\alpha = 6.41$$

$$\beta = 0.32$$

$$\gamma = -0.51$$

and μ is about 0.012 cp

$\mathscr{D}$ is about 1.5×10^{-5} ft²/sec

$$H_G = (6.41)(236)^{0.32}(3{,}800)^{-0.51} \left[\frac{(0.012)(6.72 \times 10^{-4})}{0.67} \right.$$

$$\left. \frac{\text{lb}m \; \text{ft}^3}{\text{ft sec lb}m} \; \frac{\text{sec}}{1.5 \times 10^{-5} \; \text{ft}^2} \right]$$

$$H_G = 0.445 \text{ ft}$$

$$H_L = \phi \left(\frac{L'}{\mu} \right)^{\eta} Sc^{0.5}$$

where $\phi = 0.01$

$$\eta = 0.22$$

$\mathscr{D}$ is about 1×10^{-5} cm²/sec

$$H_L = 0.01 \left[\left(\frac{3{,}800 \; \text{lb}m}{\text{hr ft}^2} \right) \left(\frac{\text{ft hr}}{(20)(2.42) \; \text{lb}m} \right) \right]^{0.22}$$

$$\left[\frac{(20)(2.42)}{52} \frac{\text{lb}m}{\text{ft hr}} \right] \left(\frac{\text{ft}^3}{\text{lb}m} \right) \left(\frac{\text{hr}}{(2 \times 10^{-5}) \, 3.87 \; \text{ft}^2} \right)$$

$$H_L = 2.88$$

Then $H_{OG} = H_G + \dfrac{mG}{L} H_L$

$$= 0.445 + \frac{1.5}{2.01} 2.88 = 2.60 \text{ ft}$$

Therefore the column height is $Z = n_{OG} H_{OG} = (5.23)(2.60) = 13.6$ ft and, as calculated, the diameter is 10.25 ft. This height, of course, is the depth of the packing and does not include space for a liquid distribution at the top or room for gas and liquid to disengage.

This column would be a possible column to use in a stage absorption for the natural gasoline plant where several absorbers are used in series each at lower pressures. The absorber in the flowsheet of Table 1.3-1 is a tray tower and operates at a pressure of 1,050 psia. If we were to produce a stream from streams 5 and 7 like the one in the example it would be possible to accomplish this by one-stage contact between all the oil and 15 percent of stream 5 in a chilled presaturator with two theoretical trays.

□ □ □

Example 13.2-5 Economic optimization of an absorber The total problem of optimizing absorber design is a matter of minimizing some sort of cost function which is the sum of costs associated with the unit, such as depreciation, power, maintenance, raw materials, makeup absorbent, operator salaries, etc. Normally maintenance, raw material, and salaries are constant regardless of absorber size. The interaction of all the factors listed, plus many more we have not listed, makes true optimal design a matter for a large computer. In economics, however, as with other types of problems, we pick out limiting steps and concentrate our attention on them so as to reduce the number of variables to be handled.

For purposes of illustration, we will only consider optimizing the cost of removing propane from the stream in Example 13.2-4 versus the cost in lost propane if it is not removed.

Let us assume that the loss of propane in Example 13.2-4 will entail a net cost of $0.2¢/$lb while the cost of the tower is about \$2,000 per foot. Assuming that H_{OG} will not change drastically, that the life of the equipment is 10 years, and that straight-line depreciation is used, what recovery of propane should we specify, as opposed to the 90 percent recovery arbitrarily chosen in Example 13.2-5? Use 340 operating days per year and $L/G = 2.01$.

Solution The pertinent annual cost will be

$$AC = \text{depreciation} + \text{lost propane cost}$$

We wish to minimize this quantity

$$\text{Depreciation} = \frac{\text{tower cost}}{10}$$

$$= \frac{(\text{cost/ft}) \, Z}{10}$$

$$= \frac{(\text{cost/ft}) \, n_{OG} H_{OG}}{10}$$

$$= \frac{2,000}{10} \, 2.60 \, n_{OG} = 532 n_{OG}$$

The cost of lost propane is

$$\text{Lost propane cost} = \left(10,000,000 \, \frac{\text{ft}^3}{\text{day}}\right) \left(340 \, \frac{\text{days}}{\text{yr}}\right) \left(\frac{\text{mole}}{359 \, \text{ft}^3}\right) 0.00433$$

$$\frac{16.8 \, \text{lb}m}{\text{mole}} \, (1 - \text{fraction recovered}) \, (0.0025)$$

$$= 1,720 \, (1 - \text{fraction recovered})$$

This makes the annual cost

(a) $\text{AC} = 532 n_{OG} + 1,720 \, (1 - \text{fraction recovered})$

but n_{OG} is a function of fraction recovered:

$$n_{OG} \frac{y_{A0} - y_{A1}}{(y_A - y_A^*)_{\text{lm}}}$$

and

$$y_{A1} \cong y_{A0} \, (1 - \text{fraction recovered})$$

and so letting $F = \text{fraction recovered}$

(b) $$n_{OG} = \frac{y_{A0} - y_{A0}(1 - F)}{\dfrac{(y_{A0} - y_{A0}) - [y_{A0}(1 - F) - y_{A1}^*]}{\ln\{(y_{A0} - y_{A0}^*)/[y_{A0}(1 - F) - y_A^*]\}}}$$

but

(c) $x_{A1} = 0$

and so

$$(d) \qquad y_{A1}^* = (1.5)(0) = 0$$

and y_{A0}^* may be found by using the operating line (assumed straight) to find x_{A0} and then the equilibrium line (assumed straight) to find y_{A0}^* :

$$y_{A0} = \frac{L}{G} x_{A0} + \left(y_{A1} - \frac{L}{G} x_{A1} \right)$$

or

$$x_{A0} = \frac{G}{L} (y_{A0} - y_{A1}) + x_{A1}$$

and

$$y_{A0}^* = m x_{A0} = 1.5 x_{A0}$$

so that

$$y_{A0}^* = 1.5 \left[\frac{G}{L} (y_{A0} - y_{A1}) + x_{A1} \right]$$

or

$$(e) \qquad y_{A0}^* = 1.5 \left[\frac{G}{L} (y_{A0} - y_{A1})(1 - F) + x_{A1} \right]$$

The set of equations (a) to (e) may be solved analytically to yield annual cost as a function of F alone, and then the optimum F may be obtained by differentiating and setting the derivative equal to zero. A simpler method is to use a computer, assume a variety of values of F, and solve as follows:

1. Calculate y_{A0}^* from equation (e) using the assumed F, equation (c), and the known values of $G/L = 1/2.01$ and $y_{A0} = 0.00433$.
2. Calculate n_{oG} from equation (b) using the value found in step 1 for y_{A0}^*, assumed F, equation (d), and $y_{A0} = 0.00433$.
3. Calculate annual cost from equation (a).

Table 13.2-2   Fortran computer program for absorber optimization

```
PRINT 10
GTOL = 1./2.01
F = 0.
XA1 = 0.
YA1A = 0.
YAO = .00433
DO 1 I = 1, 19
F = F + .05
YA1 = YAO*(1.−F)
YAOA = 1.5*(GTOL*(YAO−YA1) + XA1)
TLM = ((YAO−YAOA)−(YA1−YA1A))/(ALOG(YAO − YAOA)/
1 (YA1−YA1A)))
TNOG = (YAO−YA1)/TLM
AC1 = 532*TNOG + 1720.*(1.−F)
PRINT 12,F,TNOG,AC1
1 CONTINUE
10 FORMAT ('             F           NOG          ANNUAL COST')
12 FORMAT(4E15.3)
END
```

**Table 13.2-3   Results of absorber optimization using
computer program in Table 13.2-2**

F	NOG	ANNUAL COST[a]
5.000E−02	5.228E−02	1.662E+03
1.000E−01	1.096E−01	1.606E+03
1.500E−01	1.726E−01	1.554E+03
2.000E−01	2.424E−01	1.505E+03
2.500E−01	3.200E−01	1.460E+03
3.000E−01	4.068E−01	1.420E+03
3.500E−01	5.047E−01	1.387E+03
4.000E−01	6.159E−01	1.360E+03
4.500E−01	7.434E−01	1.342E+03
5.000E−01	8.912E−01	1.334E+03
5.500E−01	1.065E+00	1.340E+03
6.000E−01	1.27E+00	1.364E+03
6.500E−01	1.522E+00	1.412E+03
7.000E−01	1.833E+00	1.491E+03
7.500E−01	2.231E+00	1.617E+03
8.000E−01	2.761E+00	1.813E+03
8.500E−01	3.512E+00	2.126E+03
9.000E−01	4.686E+00	2.665E+03
9.500E−01	6.942E+00	3.779E+03

[a] Annual cost is in dollars.

Tables 13.2-2 and 13.2-3 show this calculation for 19 equally spaced values of F. The lowest annual cost is at about $F = 0.5$, as opposed to $F = 0.9$ as originally used. The values for n_{OG} are also shown.

If we optimize the absorber in Table 1.3-1 we would also have to consider how the recovery from stream 7 is changed.

□ □ □

PROBLEMS

13.1 An aqueous solution containing 2 mole percent of a volatile organic is to be stripped with steam. This aqueous solution is fed to the top of a packed column and steam is fed to the bottom. Ninety-five percent of the entering organic is to be stripped from the water solution by the steam. The equilibrium relation for this situation is given by

$$y = 0.1\sqrt{x}$$

where y and x are mole fractions. In actual operation L/G remains approximately constant because of compensating heat effects.

(a) Plot the equilibrium line. Determine the *maximum* value of L/G which is possible for column operation.

(b) Using the actual operating value of L/G equal to 0.5, plot the operating line. Find the outlet concentration y_1 for this case. Write the equation for the operating line also.

(c) Find the number of overall liquid phase transfer units based on the mole fraction. Can you find this analytically?

(d) Given that K_x has been found experimentally for the above conditions to be 20 lb moles/hr ft³, find n_{OL} for a liquid rate of 500 lb moles/hr ft² and determine the height of the packed section.

13.2 Graphically calculate the interface composition where $y_A = 0.002$ in the tower shown below.

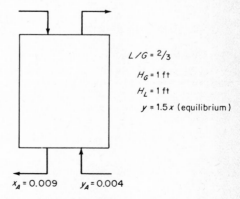

$L/G = 2/3$
$H_G = 1$ ft
$H_L = 1$ ft
$y = 1.5x$ (equilibrium)

$x_A = 0.009$ $y_A = 0.004$

Calculate y_A^* graphically at the point where $x_A = 0.007$. What is the minimum concentration that the entering liquid can have for the tower operating as shown?

13.3 A packed column is to be used to desorb component A from low concentration in a liquid into a stream of carrier gas. The column is operated *cocurrently*.

(a) Develop the *general* equation of the operating line for cocurrent flow *in the form* $y = mx + b$, assuming constant L and G.

(b) What is the limiting L/G in molar units?

(c) Is the above L/G_{max} or L/G_{min}?

Equilibrium data: $y = 0.3x$. Use the nomenclature shown; justify your answers.

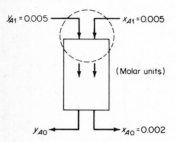

$y_{A1} = 0.005$ $x_{A1} = 0.005$

(Molar units)

y_{A0} $x_{A0} = 0.002$

13.4 The column problem 13.3 is now operated *countercurrently* as shown below. At the point in the tower where $y_A = 0.01$, what is x_{AS}? For this mode of operation, $H_G = 2$ ft and $H_L = 1$ ft.

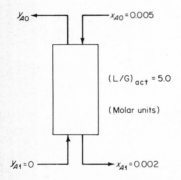

y_{A0} $x_{A0} = 0.005$

$(L/G)_{act} = 5.0$

(Molar units)

$y_{A1} = 0$ $x_{A1} = 0.002$

13.5 A tower is packed with 1-in. ceramic Raschig rings and treats 50,000 ft^3 of entering gas per hour. The gas contains 2 mole percent of SO_2 and is scrubbed with initially SO_2 free water so that 90 percent of the SO_2 is removed. The tower pressure is 1 atm and the temperature is 70°F. If the actual L/G rate is to be 25 percent greater than the minimum and the gas velocity is to be 50 percent of the flooding velocity, what should the tower diameter be? The equilibrium curve is

$$y_A = 1.1x_A$$

You can use the flooding correlation in Fig. 18-51 on p. 18–30 of Perry's handbook (4th ed.). For 1-in. Raschig rings $a/\epsilon^3 = 159$. (Also $\phi = 1.0$ for this situation.)

13.6 It is desired to reduce the mole fraction of NH_3 in waste gas stream from $y = 0.01$ to $y = 0.001$ as shown in the sketch. The equilibrium relation is $y = 1.66x$.

(a) *Sketch* the equilibrium line.

(b) Sketch in the values $y = 0.01$ and $y = 0.001$.

(*c*) Indicate the operating line point on your sketch that corresponds to the top of the tower.

(*d*) Indicate the operating line point on your sketch that corresponds to the bottom of the tower if minimum L/G is used.

(*e*) Sketch the operating line if $L/G = 2\ L/G_{min}$.

(*f*) Sketch and calculate the concentration of the outlet water for (*e*) above.

(*g*) If H_{OG} is 1.6 ft, what will the height be?

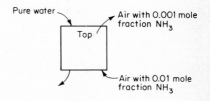

Index

Index